T0211144

Introduction to Linear Algebra

Introduction to Linear Algebra

Rita Fioresi
University of Bologna

Marta Morigi
University of Bologna

CRC Press
Taylor & Francis Group
Boca Raton London New York

CRC Press is an imprint of the
Taylor & Francis Group, an **informa** business

A CHAPMAN & HALL BOOK

First edition published 2022
by CRC Press
6000 Broken Sound Parkway NW, Suite 300, Boca Raton, FL 33487-2742

and by CRC Press
2 Park Square, Milton Park, Abingdon, Oxon, OX14 4RN

© 2022 Casa Editrice Ambrosiana

Authorized translation from Italian language edition published by CEA – Casa Editrice Ambrosiana, A Division of Zanichelli editore S.p.A.

CRC Press is an imprint of Taylor & Francis Group, LLC

Library of Congress Cataloging-in-Publication Data

Names: Fioresi, Rita, 1966- author. | Morigi, Marta, author.
Title: Introduction to linear algebra / Rita Fioresi, University of
Bologna, Marta Morigi, University of Bologna.
Description: First edition. | Boca Raton : Chapman & Hall/CRC Press, 2021.
| Includes bibliographical references and index.
Identifiers: LCCN 2021019430 (print) | LCCN 2021019431 (ebook) | ISBN
9780367626549 (hardback) | ISBN 9780367635503 (paperback) | ISBN
9781003119609 (ebook)
Subjects: LCSH: Algebras, Linear.
Classification: LCC QA184.2 .F56 2021 (print) | LCC QA184.2 (ebook) | DDC
512/.5--dc23
LC record available at https://lccn.loc.gov/2021019430
LC ebook record available at https://lccn.loc.gov/2021019431

ISBN: 978-0-367-62654-9 (hbk)
ISBN: 978-0-367-63550-3 (pbk)
ISBN: 978-1-003-11960-9 (ebk)

Typeset in LM Roman
by KnowledgeWorks Global Ltd.

Contents

Preface

This textbook comes from the need to cater the essential notions of linear algebra to Physics, Engineering and Computer Science students. We strived to keep the abstraction and rigor of this beautiful subject and yet give as much as possible the intuition behind all of the mathematical concepts we introduce. Though we provide the full proofs of all of our statements, we introduce each topic with a lot of examples and intuitive explanations to guide the students to a mature understanding.

This is not meant to be a comprehensive treatment on linear algebra but an essential guide to its foundation and heart for those who want to understand the basic concepts and the abstract mathematics behind the powerful tools it provides.

A short tour of our presentation goes as follows.

Chapters 1, 13 and 14 are independent from each other and from the rest of the book. Chapter 1 and/or Chapter 13 can be effectively used as a motivational introduction to linear algebra and vector spaces. Chapter 14 contains some further topics like the principle of induction and Euclid's algorithm, which are essential for computer science students, but they can easily be omitted and the chapter is independent from the rest of the book.

Chapters 2, 3, 4 and 5 introduce the basic notions concerning vector spaces and linear maps, while Chapters 6, 7, 8 and 9 further develop the theory to reach the question of eigenvalues and eigenvectors. A minimal course in linear algebra can end after Chapter 6, or even better after Chapter 9. In the remaining Chapters 10, 11 and 12 we study scalar products, the Spectral Theorem and quadratic forms, very important for physical and engineering applications.

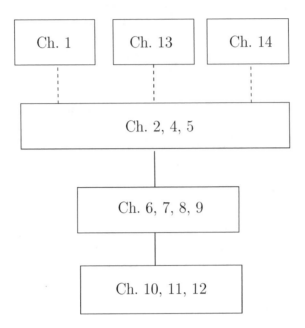

Acknowledgements. First of all, we want to thank our students in Computer Science and Physics, who have taught us how to explain in the most accessible way this beautiful subject, keeping its necessary abstraction. We invite our readers to be passionate, like some of their predecessors, and grasp the perfection of this theory.

We also thank the Department of Mathematics, which has supported us through these years of teaching and also our many students who alerted us about the typos of the previous version: if this book is improved it is also and because of their contribution. We last thank Prof. Faglioni for valued technical support and also the Taylor and Francis staff for helping us to finalize our work.

Finally, our special thanks is offered to our families, whose encouragement and support have made this book possible.

CHAPTER 1

Introduction to Linear Systems

In this chapter, we discuss how to solve linear systems with real coefficients using a method known as *Gaussian algorithm*. Later on, we will also use this method to solve other questions; at the same time, we will interpret linear systems as special cases of a much deeper theory.

1.1 LINEAR SYSTEMS: FIRST EXAMPLES

A *linear equation* is an equation where the unknowns appear with degree 1, that is an equation of the form:

$$a_1 x_1 + a_2 x_2 + \ldots + a_n x_n = b, \tag{1.1}$$

where $a_1, a_2, \ldots, a_n$ and b are assigned numbers and $x_1, x_2, \ldots, x_n$ are the unknowns. The numbers $a_1, \ldots, a_n$ are called *coefficients* of the linear equation, b is called *known term*. If $b = 0$ the equation is said to be *homogeneous*. A *solution* of the equation (1.1) is a n-tuple of numbers $(s_1, s_2, \ldots, s_n)$ that gives an equality when put in place of the unknowns. For example $(3, -1, 4)$ is a solution of the equation $2x_1 + 7x_2 - x_3 = -5$ because $2 \cdot 3 + 7 \cdot (-1) - 4 = -5$.

A *linear system of m equations in n unknowns* $x_1, x_2, \ldots, x_n$ is a set of m linear equations in n unknowns $x_1, x_2, \ldots, x_n$ that must be simultaneously satisfied:

$$\begin{cases} a_{11}x_1 + a_{12}x_2 + \cdots + a_{1n}x_n = b_1 \\ a_{21}x_1 + a_{22}x_2 + \cdots + a_{2n}x_n = b_2 \\ \vdots \\ a_{m1}x_1 + a_{m2}x_2 + \cdots + a_{mn}x_n = b_m \end{cases} \tag{1.2}$$

The numbers $a_{11}, \ldots, a_{1n}, \ldots, a_{m1}, \ldots, a_{mn}$ are called the system coefficients, while $b_1, \ldots, b_m$ are called the known terms. If $b_i = 0$ for every $i = 1, \ldots, m$, the system is said to be *homogenous*. A *solution* of the linear system (1.2) is a n-tuple $(s_1, s_2, \ldots, s_n)$ of numbers that satisfies all the system equations. For example $(1, 2)$

is the solution of the linear system

$$\begin{cases} x_1 + x_2 = 3 \\ x_1 - x_2 = -1 \end{cases}$$

In this book, we will deal exclusively with linear systems with *real coefficients* that is, systems of the form (1.2) in which all the coefficients a_{ij} of the unknowns and all known terms b_i are real numbers. The solutions that we will find, therefore, will always be ordered n-tuples of real numbers.

Given a linear system, we aim at answering the following questions:

1. Does the system admit solutions?

2. If so, how many solutions does it admit and what are they?

In certain cases, it is particularly easy to answer these questions. Let us see some examples.

Example 1.1.1 Consider the following linear system in the unknowns x_1, x_2:

$$\begin{cases} x_1 + x_2 = 3 \\ x_1 + x_2 = 1 \end{cases}$$

It is immediate to observe that the sum of two real numbers cannot be simultaneously equal to 3 and 1. Thus, the system does not admit solutions. In other words, when the conditions assigned by the two equations of the system are incompatible, then the system does not have solutions.

The example above justifies the following definition.

Definition 1.1.2 A system is said to be *compatible* if it admits solutions.

Example 1.1.3 Consider the following linear system in the unknowns x_1, x_2:

$$\begin{cases} x_1 + x_2 = 3 \\ x_2 = -1 \end{cases}$$

Substituting in the first equation the value of x_2 obtained from the second one, we get: $x_1 = 3 - x_2 = 3 + 1 = 4$. The system is therefore compatible and admits a unique solution: $(4, -1)$. In this example, two variables are assigned (the unknowns x_1 and x_2), and two conditions are given (the two equations of the system). These conditions are compatible, that is they are not contradictory, and are "independent" meaning that they cannot be obtained one from the other. In summary:

Two real variables along with two compatible conditions give one and only one solution.

Example **1.1.4** Now consider the linear system in the unknowns x_1, x_2:

$$\begin{cases} x_1 + x_2 = 3 \\ 2x_1 + 2x_2 = 6. \end{cases}$$

Unlike what happened in the previous example, here the conditions given by the two equations are not "independent", in the sense that the second equation is obtained by multiplying the first by 2. The two equations give the same relation between the variables x_1 and x_2. Then, solving the linear system means simply solving the equation $x_1 + x_2 = 3$. This equation certainly has solutions: for example, we saw in the previous example that $(4, -1)$ is a solution, but also $(1, 2)$ or $(0, 3)$ are solutions. Exactly how many solutions are there? And how can we find them out? In this case, we have two variables and one condition on them. This means that a variable is free to vary in the set of real numbers, which are infinitely many. The equation allows us to express a variable, say x_2, as a function of the other variable x_1. The solutions are all expressible in the form: $(x_1, 3 - x_1)$. With this, we mean that the variable x_1 can take all the infinite real values, and that in order for the equation $x_1 + x_2 = 3$ to be satisfied, it must be $x_2 = 3 - x_1$. A more explicit way, but obviously equivalent, to describe the solutions, is $\{(t, 3 - t)|t \in \mathbb{R}\}$. Of course, we could decide to vary the variable x_2 and express x_1 as a function of x_2. In that case, we would give the solutions in the form $(3 - x_2, x_2)$, or equivalently we say that the set of solutions is: $\{(3 - s, s)|s \in \mathbb{R}\}$. In summary:

Two real variables along with one condition give infinite solutions.

Definition 1.1.5 Two linear systems are called *equivalent* if they have the same solutions.

In Example 1.1.4 we observed that the linear system

$$\begin{cases} x_1 + x_2 = 3 \\ 2x_1 + 2x_2 = 6 \end{cases}$$

is equivalent to the equation $x_1 + x_2 = 3$. Of course, being able to understand if two systems are equivalent can be very useful; for example, we can try to solve a linear system by reducing it to an equivalent one, but easier to solve.

In the next section, we introduce some useful concepts to simplify the way we write a linear system.

1.2 MATRICES

Given two natural numbers m, n, a $m \times n$ *matrix* with real coefficients is a table of mn real numbers placed on m rows and n columns. For example:

$$\begin{pmatrix} 5 & -6 & 0 \\ 4 & 3 & -1 \end{pmatrix}$$

is a 2×3 matrix.

If $m = n$ the matrix is said to be *square* of order n. For example

$$\begin{pmatrix} 1 & 0 \\ \frac{2}{3} & 3 \end{pmatrix}$$

is a square matrix of order 2.

Denote by $M_{m,n}(\mathbb{R})$ the set of $m \times n$ matrices with real coefficients and simply by $M_n(\mathbb{R})$ the set of square matrices of order n with real coefficients.

Given a matrix A, the number that appears in the i-th row and j-th column of A is called the (i,j) entry of A.

For example, in the matrix

$$A = \begin{pmatrix} 5 & -6 & 0 \\ 4 & 3 & -1 \end{pmatrix}$$

the $(1,3)$ entry is 0, while the $(2,2)$ entry is 3. Of course, two $m \times n$ matrices A and B are equal if their entries coincide, that is, if the (i,j) entry of A coincides with the (i,j) entry of B, for every $i = 1, \ldots, m$ and for every $j = 1, \ldots, n$.

Given a generic $m \times n$ matrix, we can write it synthetically as follows:

$$A = \begin{pmatrix} a_{11} & a_{12} & \cdots & a_{1n} \\ a_{21} & a_{22} & \cdots & a_{2n} \\ \vdots & \vdots & & \vdots \\ a_{m1} & a_{m2} & \cdots & a_{mn} \end{pmatrix},$$

where a_{ij} is the (i,j) entry, $i = 1, \ldots, m$, $j = 1, \ldots, n$.

We now want to define the product rows by columns between two matrices A and B, in the case where the rows of A have the same length as the columns of B.

If A is a $m \times s$ matrix and B is a $s \times n$ matrix, we define the product c_{ij} of the i-th row of A and j-th column of B in the following way:

$$c_{ij} = \begin{pmatrix} a_{i1} & a_{i2} & \cdots & a_{is} \end{pmatrix} \begin{pmatrix} b_{1j} \\ b_{2j} \\ \vdots \\ b_{sj} \end{pmatrix} = a_{i1}b_{1j} + a_{i2}b_{2j} + \ldots + a_{is}b_{sj},$$

which we also write as:

$$c_{ij} = \sum_{h=1}^{s} a_{ih}b_{hj}.$$

In practice, we have multiplied, in order, the coefficients of the i-th row of A by the coefficients of the j-th column of B, then we have added the numbers obtained.

For example, if we have

$$A = \begin{pmatrix} 1 & 0 & 3 & -1 \\ 0 & -2 & 2 & 1 \\ 1 & 0 & -1 & 0 \end{pmatrix} \qquad B = \begin{pmatrix} 0 & 1 \\ -3 & 5 \\ 1 & 0 \\ 2 & -1 \end{pmatrix}$$

then

$$c_{12} = 1 \cdot 1 + 0 \cdot 5 + 3 \cdot 0 + (-1) \cdot (-1) = 2$$

$$c_{31} = 1 \cdot 0 + 0 \cdot (-3) + (-1) \cdot 1 + 0 \cdot 2 = -1.$$

At this point we define the product of A and B as

$$C = AB = \left(c_{ij} \right)_{\substack{i=1,\dots,m \\ j=1,\dots,n}}.$$

The matrix C is the product of A and B, and it is a $m \times n$ matrix.

In the previous example, we have that

$$C = \begin{pmatrix} 1 & 0 & 3 & -1 \\ 0 & -2 & 2 & 1 \\ 1 & 0 & -1 & 0 \end{pmatrix} \begin{pmatrix} 0 & 1 \\ -3 & 5 \\ 1 & 0 \\ 2 & -1 \end{pmatrix} = \begin{pmatrix} 1 & 2 \\ 10 & -11 \\ -1 & 1 \end{pmatrix}.$$

We note that, in general, the number of rows of AB is equal to the number of rows of A, and the number of columns of AB is equal to the number of columns of B.

We also observe that the product of a $m \times n$ matrix and a $n \times 1$ matrix (i.e. a vector in $\mathbb{R}^n$) results in a $m \times 1$ matrix, that is, a vector in $\mathbb{R}^m$.

Proposition 1.2.1 *The product operation between matrices enjoys the following properties:*

1. *Associative, that is, $(AB)C = A(BC)$ where A, B, C are matrices such that the products that appear in the formula are defined.*

2. *Distributive, that is, $A(B+C) = AB + AC$, provided that the sum and product operations that appear in the formula are defined.*

Proof. The proof is a calculation and amounts to applying the definition. We show only the associativity of the product. Consider $A \in M_{m,s}(\mathbb{R})$, $B \in M_{s,r}(\mathbb{R})$, $C \in M_{r,n}(\mathbb{R})$. We observe that:

$$(AB)_{iu} = \sum_{h=1}^{s} a_{ih} b_{hu}, \quad (BC)_{hj} = \sum_{u=1}^{r} b_{hu} c_{uj};$$

then

$$((AB)C)_{ij} = \sum_{u=1}^{r} (AB)_{iu} c_{uj} = \sum_{u=1}^{r} \left(\sum_{h=1}^{s} a_{ih} b_{hu} \right) c_{uj} =$$

$$\sum_{u=1}^{r} \sum_{h=1}^{s} a_{ih} b_{hu} c_{uj} = \sum_{h=1}^{s} \sum_{u=1}^{r} a_{ih} b_{hu} c_{uj} =$$

$$\sum_{h=1}^{s} a_{ih} \left(\sum_{u=1}^{r} b_{hu} c_{uj} \right) = \sum_{h=1}^{s} a_{ih} (BC)_{hj} = (A(BC))_{ij}.$$

The proof of distributivity is similar. ■

Note that the product operation between matrices is not commutative. Even if the product AB between two matrices A and B is defined, the product BA may not be defined. For example if

$$A = \begin{pmatrix} 1 & 0 \\ 2 & 1 \\ -1 & 0 \end{pmatrix}, \quad B = \begin{pmatrix} 1 & -1 \\ 0 & 1 \end{pmatrix}$$

we have that

$$AB = \begin{pmatrix} 1 & -1 \\ 2 & -1 \\ -1 & 1 \end{pmatrix},$$

while BA is not defined. Similarly if

$$A = \begin{pmatrix} 1 & 2 \\ 0 & -3 \end{pmatrix}, \quad B = \begin{pmatrix} -1 & 1 \\ 0 & 2 \end{pmatrix}$$

we have that

$$AB = \begin{pmatrix} -1 & 5 \\ 0 & -6 \end{pmatrix}, \quad BA = \begin{pmatrix} -1 & -5 \\ 0 & -6 \end{pmatrix}.$$

1.3 MATRICES AND LINEAR SYSTEMS

Let us now see how it is possible to use matrices and the product rows by columns to describe a linear system.

Consider a linear system of the form:

$$\begin{cases} a_{11}x_1 + a_{12}x_2 + \cdots + a_{1n}x_n &= b_1 \\ a_{21}x_1 + a_{22}x_2 + \cdots + a_{2n}x_n &= b_2 \\ \quad\quad\quad \vdots & \\ a_{m1}x_1 + a_{m2}x_2 + \cdots + a_{mn}x_n &= b_m \end{cases}$$

We can write this system in matrix form as follows:

$$\begin{pmatrix} a_{11}x_1 + a_{12}x_2 + \cdots + a_{1n}x_n \\ a_{21}x_1 + a_{22}x_2 + \cdots + a_{2n}x_n \\ \vdots \\ a_{m1}x_1 + a_{m2}x_2 + \cdots + a_{mn}x_n \end{pmatrix} = \begin{pmatrix} b_1 \\ b_2 \\ \vdots \\ b_m \end{pmatrix}$$

and then using the product rows by columns in the following way:

$$\begin{pmatrix} a_{11} & a_{12} & \cdots & a_{1n} \\ a_{21} & a_{22} & \cdots & a_{2n} \\ \vdots & \vdots & & \vdots \\ a_{m1} & a_{m2} & \cdots & a_{mn} \end{pmatrix} \begin{pmatrix} x_1 \\ x_2 \\ \vdots \\ x_n \end{pmatrix} = \begin{pmatrix} b_1 \\ b_2 \\ \vdots \\ b_m \end{pmatrix}$$

or, more synthetically, as

$$A\underline{x} = \underline{b},$$

where $A = (a_{ij})$ is the $m \times n$ matrix which has as entries the coefficients of the unknowns,

$$\underline{x} = \begin{pmatrix} x_1 \\ x_2 \\ \vdots \\ x_n \end{pmatrix}$$

is the column of the n unknowns, and

$$\underline{b} = \begin{pmatrix} b_1 \\ b_2 \\ \vdots \\ b_m \end{pmatrix}$$

is the column of m known terms. The matrix $A = (a_{ij})$ is called the *incomplete matrix* associated with the system and the matrix

$$(A|\underline{b}) = \begin{pmatrix} a_{11} & a_{12} & \cdots & a_{1n} & b_1 \\ a_{21} & a_{22} & \cdots & a_{2n} & b_2 \\ \vdots & \vdots & & \vdots & \vdots \\ a_{m1} & a_{m2} & \cdots & a_{mn} & b_m \end{pmatrix}$$

is called the *complete* matrix associated with the system.

Example 1.3.1 Consider the linear system

$$\begin{cases} 2x_1 + \sqrt{2}x_2 - x_3 = 2 \\ x_1 - x_3 = 1 \end{cases}$$

in the unknowns x_1, x_2, x_3.

Then the incomplete matrix and the complete matrix associated with the system are, respectively:

$$A = \begin{pmatrix} 2 & \sqrt{2} & -1 \\ 1 & 0 & -1 \end{pmatrix} \text{ and } (A|\underline{b}) = \begin{pmatrix} 2 & \sqrt{2} & -1 & 2 \\ 1 & 0 & -1 & 1 \end{pmatrix}.$$

Using matrices is simply a more convenient way to write and deal with linear systems. Each row of the complete matrix associated with a linear system is equivalent to an equation in which the unknowns are *implied*.

Definition 1.3.2 A matrix is said to be in *row echelon form* or *staircase form* if the following conditions are met:

(a) rows consisting of zeros, if any, are found at the bottom of the matrix;

(b) the first nonzero element of each (nonzero) row is located to the right of the first nonzero element of the previous row.

Example 1.3.3 The matrix

$$A = \begin{pmatrix} 1 & -1 & -1 & 2 & -4 \\ 0 & 0 & -1 & 3 & 5 \\ 0 & 0 & 0 & \frac{1}{3} & 1 \\ 0 & 0 & 0 & 0 & 0 \end{pmatrix}$$

is a row echelon matrix because it satisfies conditions (a) and (b) of Definition 1.3.2. On the contrary, the matrix

$$B = \begin{pmatrix} 2 & -1 & -1 & 2 & -4 \\ 0 & 1 & -1 & 3 & 5 \\ 0 & 2 & 0 & 1 & \frac{1}{5} \end{pmatrix}$$

is not in such a form because the first nonzero element of the third row is not located to the right of the first nonzero element of the second row (but below it).

Definition 1.3.4 Let A be a row echelon matrix (by rows). We call *pivot* of A the first nonzero element of each nonzero row of A. We call *row rank* of A, denoted by $\mathrm{rr}(A)$, the number of nonzero rows, or equivalently, the number of its pivots.

Example 1.3.5 Given

$$A = \begin{pmatrix} 1 & -1 & -1 & 2 & -4 \\ 0 & 0 & -1 & 3 & 5 \\ 0 & 0 & 0 & \frac{1}{3} & 1 \\ 0 & 0 & 0 & 0 & 0 \end{pmatrix},$$

the pivots of A are 1, -1, $\frac{1}{3}$, so $\mathrm{rr}(A) = 3$.

Observation 1.3.6 Let $A \in \mathrm{M}_{m,n}(\mathbb{R})$ be a row echelon matrix. By definition

$$\mathrm{rr}(A) \leq m. \tag{1.3}$$

However, we also have the inequality

$$\mathrm{rr}(A) \leq n. \tag{1.4}$$

If $m \leq n$, (1.4) follows obviously from (1.3). If $m > n$, it is easy to understand, by writing a row echelon matrix with a number of rows greater than the number of columns, that property (b) of Definition 1.3.2 implies that the maximum number of pivots of A is n.

Definition 1.3.7 The linear system $A\underline{x} = \underline{b}$ is called *row echelon* if the matrix A is row echelon.

We will explain how to quickly solve a linear system whose matrix is in row echelon form.

Example 1.3.8 Consider the following linear system in the unknowns x_1, x_2, x_3, x_4:

$$\begin{cases} 4x_1 + 2x_2 + 3x_3 + 4x_4 = 1 \\ x_2 - 2x_3 = 2 \\ x_3 - x_4 = 0 \\ x_4 = 1 \end{cases}$$

The complete matrix associated with the system is:

$$(A|\underline{b}) = \left(\begin{array}{cccc|c} 4 & 2 & 3 & 4 & 1 \\ 0 & 1 & -2 & 0 & 2 \\ 0 & 0 & 1 & -1 & 0 \\ 0 & 0 & 0 & 1 & 1 \end{array} \right),$$

which is in row echelon form and has row rank 4. Obviously also the incomplete matrix A is in row echelon form, and we note that it also has row rank 4. The fact that the matrix A is in row echelon form means that if you choose any of the sytem equations, there exists one unknown wich appears in that equation but not in the following ones. The linear system can therefore be easily solved starting from the bottom and proceeding with successive replacements, that is, from the last equation and going back to the first. From the fourth equation we get $x_4 = 1$; replacing $x_4 = 1$ in the third equation we get $x_3 = x_4 = 1$. Replacing $x_3 = 1$ in the second equation we get $x_2 = 2 + 2 = 4$. Finally, replacing $x_2 = 4$ and $x_3 = x_4 = 1$ in the first equation, we obtain $x_1 = \frac{1}{4}(1 - 2x_2 - 3x_3 - 4x_4) = \frac{1}{4}(1 - 8 - 3 - 4) = -\frac{7}{2}$. The system has therefore only one solution: $(-\frac{7}{2}, 4, 1, 1)$.

Example 1.3.9 Consider the linear system in the unknowns x_1, x_2, x_3, x_4 obtained from that of the previous example by deleting the last equation:

$$\begin{cases} 4x_1 + 2x_2 + 3x_3 + 4x_4 = 1 \\ x_2 - 2x_3 = 2 \\ x_3 - x_4 = 0. \end{cases}$$

The complete matrix associated with the system is:

$$(A|\underline{b}) = \left(\begin{array}{cccc|c} 4 & 2 & 3 & 4 & 1 \\ 0 & 1 & -2 & 0 & 2 \\ 0 & 0 & 1 & -1 & 0 \end{array} \right).$$

It is in row echelon form and has row rank 3. The incomplete matrix A is in echelon form and also has row rank 3. Of course, the solution $(-\frac{7}{2}, 4, 1, 1)$, found in the previous example, continues to be a solution of the system, so that the system is certainly compatible. However, how many solutions of the system do we have? Also in this case, we can proceed from the bottom with subsequent replacements because, as before, for each equation there exists one unknown which appears in that equation but not in the following ones. From the last equation, we get $x_3 = x_4$. Replacing $x_3 = x_4$ in the second equation, we get $x_2 = 2 + 2x_3 = 2 + 2x_4$. Replacing x_2 and x_3 in the first equation we obtain $x_1 = \frac{1}{4}(1 - 2x_2 - 3x_3 - 4x_4) = \frac{1}{4}(1 - 4 - 4x_4 - 3x_4 - $

$4x_4) = \frac{1}{4}(-3 - 11x_4)$. The system therefore has infinitely many solutions of the form $(\frac{1}{4}(-3 - 11x_4), 2 + 2x_4, x_4, x_4)$ where the variable x_4 is allowed to vary in the set of real numbers.

What we have illustrated in Examples 1.3.8, 1.3.9 is a general fact, and we have the following proposition.

Proposition 1.3.10 *Let $A\underline{x} = \underline{b}$ be a row echelon linear system in the n unknowns $x_1, \ldots, x_n$. Then:*

(a) *the system admits solutions if and only if* $\mathrm{rr}(A) = \mathrm{rr}(A|\underline{b})$;

(b) *if* $\mathrm{rr}(A) = \mathrm{rr}(A|\underline{b}) = n$, *the system admits only one solution;*

(c) *if* $\mathrm{rr}(A) = \mathrm{rr}(A|\underline{b}) = k < n$, *the system admits infinitely many solutions, which depend on $n - k$ free variables.*

Proof. First we observe that by deleting the column $\underline{b}$ from the matrix $(A|\underline{b})$ we still have a matrix in row echelon form, thus also the incomplete matrix A associated with the system is a matrix in such a form. Also, by deleting the column $(\underline{b})$ from the matrix $(A|\underline{b})$, the number of pivots may decrease at most 1 unit. More precisely this happens if and only if the matrix A has at least one zero row, let us say the i-th, and the element b_i is different from 0. Going back to the corresponding equation, we obtain that this is equivalent to the condition $0 = b_i \neq 0$, which cannot evidently be satisfied. So if $\mathrm{rr}(A) \neq \mathrm{rr}(A|\underline{b})$, the system does not admit solutions. Now suppose that $\mathrm{rr}(A) = \mathrm{rr}(A|\underline{b}) = n$. This means that the number of pivots, i.e. the number of "steps" , coincides with the number of unknowns, so the system consists of exactly n equations, the unknown x_1 appears only in the first equation, x_2 appears only in the first two equations, x_3 appears only in the first three and so on. In particular the last equation of the system contains only the unknown x_n and establishes its value. By substituting this value in the next-to-last equation, we obtain the value of the variable x_{n-1} and so on, proceeding by subsequent replacements from below as in Example 1.3.8, we get the solution of the system, which is unique. If, instead, $\mathrm{rr}(A) = \mathrm{rr}(A|\underline{b}) = k < n$ it is possible, by proceeding from the bottom with subsequent replacements, to express the k variables corresponding to the pivots of nonzero rows as a function of the other $n - k$ variables, which remain free to vary in the set of real numbers. In this way, we get infinitely many solutions. ■

Example 1.3.11 We want to solve the linear system:

$$\begin{cases} x_1 - x_2 + x_3 - x_4 = 1 \\ x_3 + \frac{1}{2}x_4 = 0 \end{cases}$$

The complete matrix associated with the system is:

$$\begin{cases} x_1 - x_2 + x_3 - x_4 = 1 \\ x_3 + \frac{1}{2}x_4 = 0 \end{cases}$$

We note that $\mathrm{rr}(A) = \mathrm{rr}(A|\underline{b}) = 2$ so, by Proposition 1.3.10 (a), the system has solutions. Since the number of variables is $4 > 2$, by Proposition 1.3.10 (c), the system admits infinitely many solutions. Basically, we have four variables and two conditions on them, thus two variables remain free to vary in the set of real numbers, and we can express two variables as a function of the other two. Proceeding with subsequent substitutions from the bottom we have:

$$x_3 = -\frac{1}{2}x_4$$

$$x_1 = x_2 - x_3 + x_4 + 1 = x_2 + \frac{1}{2}x_4 + x_4 + 1 = x_2 + \frac{3}{2}x_4 + 1.$$

The infinitely many solutions are of the form: $(x_2 + \frac{3}{2}x_4 + 1, \ x_2, \ -\frac{1}{2}x_4, \ x_4)$, with $x_2, x_4 \in \mathbb{R}$.

We observe that we could choose to express the variable x_4 as depending on the variable x_3 and, for example, the variable x_2 as depending on the variables x_1 and x_3 $(x_2 = x_1 + 3x_3 - 1)$. In other words, the choice of the free variables is not forced. However, we can always choose as free those variables corresponding to the columns of the matrix A containing no pivots and express the unknowns corresponding to the columns that contain the pivots as a function of the others. For example, in this case the pivots, both equal to 1, are located on the first and third column of the matrix A and, as a first choice, we have left the variables x_2 and x_4 to be free and we have expressed x_1 and x_3 as a function of x_2 and x_4.

1.4   THE GAUSSIAN ALGORITHM

We have now established how to solve a linear system in row echelon form. What happens for a generic linear system $A\underline{x} = \underline{b}$? It would be convenient to be able to get a new linear system $A'\underline{x} = \underline{b}'$, this time in row echelon form, equivalent to the initial system, i.e. having the same solutions. In this way, we could calculate the solutions of $A\underline{x} = \underline{b}$ by solving the system $A'\underline{x} = \underline{b}'$ in row echelon form. This is exactly what we will do.

Example 1.4.1 The following linear systems in the unknowns x_1, x_2:

$$\begin{cases} x_1 - x_2 = 1 \\ x_1 + x_2 = 2 \end{cases} \qquad \begin{cases} x_1 - x_2 = 1 \\ 2x_2 = 1 \end{cases}$$

are equivalent. In fact, we can easily see that in both cases the solution is: $(\frac{3}{2}, \frac{1}{2})$. We note that the first equation is the same in the two systems and that the second system can be obtained by substituting the second equation with the difference between the second equation itself and the first equation:

$$2^{nd}\text{equation} \rightarrow 2^{nd}\text{equation} - 1^{st}\text{equation}.$$

How can we switch from one system to another one equivalent to it? For example by doing the following:

(a) exchange of two equations;

(b) multiplication of an equation by a real number other than 0;

(c) substitution of the i-th equation with the sum of the i-th equation and the j-th equation multiplied by a real number α. In summary:

$$i\text{-th equation} \quad \longrightarrow \quad i\text{-th equation} + \alpha(j\text{-th equation}).$$

It is straightforward to verify that operations (a) and (b) do not alter the system solutions. As for operation (c), it is enough to observe that it involves only the i-th and j-th equation of the system, so just observe that the systems

$$\begin{cases} j\text{-th equation} \\ i\text{-th equation} \end{cases} \qquad \begin{cases} j\text{-th equation} \\ i\text{-th equation} + \alpha(j\text{-th equation}) \end{cases}$$

are equivalent, i.e. they have the same solutions.

Now we translate the operations (a), (b) and (c) in matrix terms:

- Exchanging two equations of the system is equivalent to exchanging two rows of the complete matrix associated with the system.

- Multiplying an equation by a real number other than 0 is equivalent to multiplying a row of the complete matrix associated with the system by the same real number other than 0; that is, to multiplying each element of the row by this number.

- Operation (c) is equivalent to replacing the i-th row of the complete matrix associated to the system, with the sum of the i-th row and the j-th row multiplied by a real number α.

Let us see a little better what we mean. Let $(a_{i1} \ldots a_{in}\ b_i)$ and $(a_{j1} \ldots a_{jn}\ b_j)$ be, respectively, the i-th and j-th row of the matrix $(A|\underline{b})$. Adding the i-th row with the j-th one multiplied by a number α it means to take the sum:

$$(a_{i1} \ldots a_{in}\ b_i) + \alpha(a_{j1} \ldots a_{jn}\ b_j) = (a_{i1} + \alpha a_{j1} \ldots a_{in} + \alpha a_{jn}\ b_i + \alpha b_j).$$

Because of the importance of these operations we will then give them a name.

Definition 1.4.2 Given a matrix A we call *elementary operations* on the rows of A the following ones:

(a) exchange of two rows;

(b) multiplication of a row by a real number other than 0;

(c) replacing the i-th row with the sum of the i-th row and the j-th row multiplied by a real number α.

Observation 1.4.3 We observe that the elementary operation (c) does not require that the number α is not zero. In fact, if $\alpha = 0$ the operation (c) amounts to leaving the i-th row unchanged.

Given any matrix $A = (a_{ij})$ we can turn it into a row echelon matrix by elementary operations on the rows of A. This process is known as *Gaussian reduction,* and the algorithm that is used is called *Gaussian algorithm* and operates as follows:

1. If $a_{11} = 0$ exchange the first row of A with a row in which the first element is nonzero. We denote by a such nonzero element. If the first element of each row of A is zero, consider the matrix that is obtained by deleting the first column and start again.

2. Check all the rows except the first, one after the other. If the first element of a row is zero, leave the row unchanged. If the first element of a row, say the i-th $(i > 1)$, is equal to $b \neq 0$, replace the i-th row with the sum of the i-th row and the first row multiplied by $-\frac{b}{a}$.

3. At this point all the elements of the first column, except possibly the first, are zero. Consider the matrix that is obtained by deleting the first row and the first column of the matrix and start again from step one.

Example 1.4.4 Consider

$$A = \begin{pmatrix} 0 & 1 & -1 & 0 \\ 1 & 2 & 0 & 1 \\ 2 & -1 & 1 & 2 \end{pmatrix}.$$

Let us use the Gaussian algorithm to reduce A in row echelon form.

Since the entry in place $(1,1)$ is zero, we exchange the first with the second row, obtaining the matrix:

$$\begin{pmatrix} 1 & 2 & 0 & 1 \\ 0 & 1 & -1 & 0 \\ 2 & -1 & 1 & 2 \end{pmatrix}.$$

The first entry of the second row is zero, hence we leave this row unchanged. The first element of the third row is 2, hence we substitute the third row with the sum of the third row and the first one multiplied by -2. We thus obtain:

$$\begin{pmatrix} 1 & 2 & 0 & 1 \\ 0 & 1 & -1 & 0 \\ 2 & -1 & 1 & 2 \end{pmatrix} \rightarrow \begin{pmatrix} 1 & 2 & 0 & 1 \\ 0 & 1 & -1 & 0 \\ 0 & -5 & 1 & 0 \end{pmatrix}.$$

Every entry of the first column except for the first one is zero. We then consider the matrix obtained by deleting the first row and first column:

$$\begin{pmatrix} \cancel{1} & \cancel{2} & \cancel{0} & \cancel{1} \\ \cancel{0} & 1 & -1 & 0 \\ \cancel{0} & -5 & 1 & 0 \end{pmatrix}.$$

We apply again the Gaussian algorithm. The first entry of the first row is nonzero, hence we leave the first row as it is. We substitute the second row with the sum of the second row with the first multiplied by 5. We obtain:

$$
\begin{pmatrix} 1 & 2 & 0 & 1 \\ 0 & 1 & -1 & 0 \\ 0 & -5 & 1 & 0 \end{pmatrix} \rightarrow \begin{pmatrix} 1 & 2 & 0 & 1 \\ 0 & 1 & -1 & 0 \\ 0 & 0 & -4 & 0 \end{pmatrix}.
$$

We thus have obtained a row echelon matrix.

$$
B = \begin{pmatrix} 1 & 2 & 0 & 1 \\ 0 & 1 & -1 & 0 \\ 0 & 0 & -4 & 0 \end{pmatrix}.
$$

At this point, we are able to solve any linear system $A\underline{x} = \underline{b}$. The complete matrix associated with the system is $(A|\underline{b})$. Using the Gaussian algorithm we can "reduce" $(A|\underline{b})$ to a row echelon matrix obtaining a matrix $(A'|\underline{b}')$. The linear system $A'\underline{x} = \underline{b}'$ is equivalent to the system $A\underline{x} = \underline{b}$, since each elementary operation on the rows of $(A|\underline{b})$ is equivalent to an operation on the equations of the system that preserves the solutions. So, to find the solutions of our system $A\underline{x} = \underline{b}$, we need to solve the row echelon system $A'\underline{x} = \underline{b}'$, taking into account Proposition 1.3.10. We note in particular that, as a result of the reasoning just explained and the content of Proposition 1.3.10, given any linear system $A\underline{x} = \underline{b}$ with real coefficients only one of the following situations can happen:

1. the system has no solutions;

2. the system has a single solution;

3. the system has infinitely many solutions.

This means that there is no linear system with real coefficients with a finite number of solutions greater than 1. When a linear system with real coefficients has 2 solutions, then, it has infinitely many ones.

Observation 1.4.5 In the Gaussian algorithm the operations are not "forced". In Example 1.4.4, for example, instead of exchanging the first with the second row, we could exchange the first with the third row. In this way, completing the algorithm, we would have obtained a different row echelon form of the matrix. From the point of view of linear systems, this simply means that we get different row echelon systems, but all equivalent to the initial system (and therefore equivalent to each other).

Example 1.4.6 We want to solve the following linear system of four equations in five unknowns u, v, w, x, y:

$$
\begin{cases} u + 2v + 3w + x + y = 4 \\ u + 2v + 3w + 2x + 3y = -2 \\ u + v + w + x + y = -2 \\ -3u - 5v - 7w - 4x - 5y = 0. \end{cases}
$$

The complete matrix associated with the system is:

$$(A|\underline{b}) = \begin{pmatrix} 1 & 2 & 3 & 1 & 1 & | & 4 \\ 1 & 2 & 3 & 2 & 3 & | & -2 \\ 1 & 1 & 1 & 1 & 1 & | & -2 \\ -3 & -5 & -7 & -4 & -5 & | & 0 \end{pmatrix}.$$

We first reduce the matrix $(A|\underline{b})$ to a row echelon form using the Gaussian algorithm; then we solve the linear system associated with the reduced matrix.

In this first example, we describe the steps of the Gaussian algorithm and at the same time we describe the operations on the equations. The advantage of the Gaussian algorithm is that we can forget the equations and the unknowns, focusing only on matrices, so our present description is merely explanatory.

The entry $(1,1)$ is not zero, so we leave the first row unchanged. Then we perform the following elementary row operations on $(A|\underline{b})$:

- 2nd row $\rightarrow$ 2nd row $-$ 1st row;

- 3rd row $\rightarrow$ 3rd row $-$ 1st row;

- 4th row $\rightarrow$ $4^t h$ row $+ 3$ (1st row).

We get the following matrix (and the equivalent linear system):

$$\begin{pmatrix} 1 & 2 & 3 & 1 & 1 & | & 4 \\ 0 & 0 & 0 & 1 & 2 & | & -6 \\ 0 & -1 & -2 & 0 & 0 & | & -6 \\ 0 & 1 & 2 & -1 & -2 & | & 12 \end{pmatrix} \Longleftrightarrow \begin{cases} u + 2v + 3w + x + y = 4 \\ x + 2y = -6 \\ -v - 2w = -6 \\ v + 2w - x - 2y = 12 \end{cases}$$

Now we exchange the second with the fourth row:

$$\begin{pmatrix} 1 & 2 & 3 & 1 & 1 & | & 4 \\ 0 & 1 & 2 & -1 & -2 & | & 12 \\ 0 & -1 & -2 & 0 & 0 & | & -6 \\ 0 & 0 & 0 & 1 & 2 & | & -6 \end{pmatrix} \Longleftrightarrow \begin{cases} u + 2v + 3w + x + y = 4 \\ v + 2w - x - 2y = 12 \\ -v - 2w = -6 \\ x + 2y = -6 \end{cases}$$

Now we replace the third row with the sum of the third row and the second one:

$$\begin{pmatrix} 1 & 2 & 3 & 1 & 1 & | & 4 \\ 0 & 1 & 2 & -1 & -2 & | & 12 \\ 0 & 0 & 0 & -1 & -2 & | & 6 \\ 0 & 0 & 0 & 1 & 2 & | & -6 \end{pmatrix} \Longleftrightarrow \begin{cases} u + 2v + 3w + x + y = 4 \\ v + 2w - x - 2y = 12 \\ -x - 2y = 6 \\ x + 2y = -6 \end{cases}$$

Finally, we substitute the fourth row with the sum of the fourth row and the third one:

$$\begin{pmatrix} 1 & 2 & 3 & 1 & 1 & | & 4 \\ 0 & 1 & 2 & -1 & -2 & | & 12 \\ 0 & 0 & 0 & -1 & -2 & | & 6 \\ 0 & 0 & 0 & 0 & 0 & | & 0 \end{pmatrix} \Longleftrightarrow \begin{cases} u + 2v + 3w + x + y = 4 \\ v + 2w - x - 2y = 12 \\ -x - 2y = 6 \\ 0 = 0 \end{cases}$$

The initial system is equivalent to the row echelon system that we have obtained, in which the last equation has become an identity. The rank of the incomplete matrix and the rank of the complete matrix of the row echelon matrix obtained coincide and are equal to 3. The number of unknowns of the system is 5, then the system admits infinitely many solutions that depend on $5 - 3 = 2$ free variables. We solve the system from the bottom using subsequent substitutions. Using the third equation we can express the variable x as a function of y:

$$x = -2y - 6.$$

In the second equation, we replace x with its expression in terms of y and we obtain v in terms of w and y:

$$v = -2w + 6.$$

Finally, in the first equation, we substitute x with its expression depending on y, v with its expression depending on w and we obtain u as a function of w and y:

$$u = -2v - 3w - x - y + 4 = -2(-2w + 6) - 3w - (-2y - 6) - y + 4 = w + y - 2.$$

So the system has infinitely many solutions of the type $(w + y - 2, -2w + 6, w, -2y - 6, y)$, that depend on two free variables, $w, y \in \mathbb{R}$.

1.5 EXERCISES WITH SOLUTIONS

1.5.1 Solve the following linear system in the four unknowns x, y, z, t:

$$\begin{cases} x - 2y = 5 \\ -x + 2y - 3z = -2 \\ -2y + 3z - 4t = -11 \\ -3z + 4t = 15 \end{cases}$$

Solution. The complete matrix associated with the system is:

$$(A|\underline{b}) = \left(\begin{array}{cccc|c} 1 & -2 & 0 & 0 & 5 \\ -1 & 2 & -3 & 0 & -2 \\ 0 & -2 & 3 & -4 & -11 \\ 0 & 0 & -3 & 4 & 15 \end{array} \right).$$

We reduce the matrix $(A|\underline{b})$ to row echelon form using the Gaussian algorithm:

$$\left(\begin{array}{cccc|c} 1 & -2 & 0 & 0 & 5 \\ 0 & 0 & -3 & 0 & 3 \\ 0 & -2 & 3 & -4 & -11 \\ 0 & 0 & -3 & 4 & 15 \end{array} \right) \rightarrow \left(\begin{array}{cccc|c} 1 & -2 & 0 & 0 & 5 \\ 0 & -2 & 3 & -4 & -11 \\ 0 & 0 & -3 & 0 & 3 \\ 0 & 0 & -3 & 4 & 15 \end{array} \right)$$

$$\rightarrow \left(\begin{array}{cccc|c} 1 & -2 & 0 & 0 & 5 \\ 0 & -2 & 3 & -4 & -11 \\ 0 & 0 & -3 & 0 & 3 \\ 0 & 0 & 0 & 4 & 12 \end{array} \right) = (A'|\underline{b}').$$

The matrix in row echelon form is the complete matrix associated to the linear system:

$$\begin{cases} x - 2y = 5 \\ -2y + 3z - 4t = -11 \\ -3z = 3 \\ 4t = 12 \end{cases}$$

Note that $\mathrm{rr}(A') = \mathrm{rr}(A'|\underline{b}') = 4$. The system, therefore, admits a unique solution that we can calculate by proceeding with subsequent substitutions from the bottom. From the fourth equation we have

$$t = 3;$$

and from the third equation we have

$$z = -1;$$

replacing these values of t and z in the second equation we get

$$y = -2;$$

finally, by replacing the values of t, z, y in the first equation we get

$$x = 1.$$

So the system has only one solution: $(1, -2, -1, 3)$.

1.5.2 Determine the solutions of the following linear system in the unknowns x, y, z, t, depending on the real parameter α:

$$\begin{cases} x + y + z + t = 0 \\ x - z - t = -1 \\ x + 2y + (2\alpha + 1)z + 3t = 2\alpha - 1 \\ 3x + 4y + (3\alpha + 2)z + (\alpha + 5)t = 3\alpha - 1. \end{cases}$$

Solution. In this exercise, we are dealing with a linear system in which a real parameter α appears. This means that as α varies in $\mathbb{R}$ we get infinitely many different linear systems that we will solve by treating them as much as possible as one. The procedure is always the same, we behave as if the parameter were a fixed real number. First of all, then, let us write the complete matrix associated with the system:

$$(A|\underline{b}) = \left(\begin{array}{cccc|c} 1 & 1 & 1 & 1 & 0 \\ 1 & 0 & -1 & -1 & -1 \\ 1 & 2 & 2\alpha + 1 & 3 & 2\alpha - 1 \\ 3 & 4 & 3\alpha + 2 & \alpha + 5 & 3\alpha - 1 \end{array} \right)$$

and reduce it to row echelon form using the Gaussian algorithm:

$$\rightarrow \begin{pmatrix} 1 & 1 & 1 & 1 & | & 0 \\ 0 & -1 & -2 & -2 & | & -1 \\ 0 & 1 & 2\alpha & 2 & | & 2\alpha - 1 \\ 0 & 1 & 3\alpha - 1 & \alpha + 2 & | & 3\alpha - 1 \end{pmatrix}$$

$$\rightarrow \begin{pmatrix} 1 & 1 & 1 & 1 & | & 0 \\ 0 & -1 & -2 & -2 & | & -1 \\ 0 & 0 & 2\alpha - 2 & 0 & | & 2\alpha - 2 \\ 0 & 0 & 3\alpha - 3 & \alpha & | & 3\alpha - 2 \end{pmatrix}$$

$$\rightarrow \begin{pmatrix} 1 & 1 & 1 & 1 & | & 0 \\ 0 & -1 & -2 & -2 & | & -1 \\ 0 & 0 & \alpha - 1 & 0 & | & \alpha - 1 \\ 0 & 0 & 3\alpha - 3 & \alpha & | & 3\alpha - 2 \end{pmatrix}$$

$$\rightarrow \begin{pmatrix} 1 & 1 & 1 & 1 & | & 0 \\ 0 & -1 & -2 & -2 & | & -1 \\ 0 & 0 & \alpha - 1 & 0 & | & \alpha - 1 \\ 0 & 0 & 0 & \alpha & | & 1 \end{pmatrix} = (A'|\underline{b}')$$

We now have to determine what happens when the parameter α varies in the set of real numbers. We must therefore answer the following questions:

1. For which values of α is the system compatible?

2. For the α values for which the system is compatible, how many solutions do we have and can we determine them explicitly?

As we know the answer is given by Proposition 1.3.10 (a): we must compare the rank of A' with the rank of $(A'|\underline{b}')$. We note that these ranks depend on the value of α. More precisely: $\mathrm{rr}(A') = \mathrm{rr}(A'|\underline{b}') = 4$ for $\alpha \neq 0, 1$. In this case, the system has a unique solution that we can compute by proceeding by substitutions starting from the bottom: the solution is $(\frac{1}{\alpha}, -\frac{\alpha+2}{\alpha}, 1\frac{1}{\alpha})$.

For $\alpha = 0$ we have

$$(A'|\underline{b}') = \begin{pmatrix} 1 & 1 & 1 & 1 & | & 0 \\ 0 & -1 & -2 & -2 & | & -1 \\ 0 & 0 & -1 & 0 & | & -1 \\ 0 & 0 & 0 & 0 & | & 1 \end{pmatrix},$$

therefore $\mathrm{rr}(A') = 3$ and $\mathrm{rr}(A'|\underline{b}') = 4$, so the system is not solvable.

For $\alpha = 1$ we have

$$(A'|\underline{b}') = \begin{pmatrix} 1 & 1 & 1 & 1 & | & 0 \\ 0 & -1 & -2 & -2 & | & -1 \\ 0 & 0 & 0 & 0 & | & 0 \\ 0 & 0 & 0 & 1 & | & 1 \end{pmatrix}$$

therefore $\mathrm{rr}(A') = 3 = \mathrm{rr}(A'|\underline{b}')$, so the system has infinitely many solutions depending on a free variable. As usual we can determine such solutions proceeding with subsequent substitutions: $(x_3, -1 - 2x_3, x_3, 1)$, $x_3 \in \mathbb{R}$.

1.5.3 Consider the linear system Σ_α in the unknowns x_1, x_2, x_3 depending on the real parameter α:

$$\Sigma_\alpha : \begin{cases} \alpha x_1 + (\alpha + 3)x_2 + 2\alpha x_3 = \alpha + 2 \\ \alpha x_1 + (2\alpha + 2)x_2 + 3\alpha x_3 = 2\alpha + 2 \\ 2\alpha x_1 + (\alpha + 7)x_2 + 4\alpha x_3 = 2\alpha + 4 \end{cases}$$

1. Determine the solutions of the linear system Σ_α as the parameter $\alpha \in \mathbb{R}$ varies.

2. Determine the solutions of the linear system Σ_α now interpreted as a linear system in 4 unknowns x_1, x_2, x_3, x_4.

Solution.

1. Consider the complete matrix $(A|\underline{b})$ associated with the linear system Σ_α:

$$\begin{pmatrix} \alpha & \alpha + 3 & 2\alpha & \alpha + 2 \\ \alpha & 2\alpha + 2 & 3\alpha & 2\alpha + 2 \\ 2\alpha & \alpha + 7 & 4\alpha & 2\alpha + 4 \end{pmatrix}$$

and reduce it to row echelon form with the Gaussian algorithm:

$$\begin{pmatrix} \alpha & \alpha + 3 & 2\alpha & \alpha + 2 \\ \alpha & 2\alpha + 2 & 3\alpha & 2\alpha + 2 \\ 2\alpha & \alpha + 7 & 4\alpha & 2\alpha + 4 \end{pmatrix} \to \begin{pmatrix} \alpha & \alpha + 3 & 2\alpha & \alpha + 2 \\ 0 & \alpha - 1 & \alpha & \alpha \\ 0 & -\alpha + 1 & 0 & 0 \end{pmatrix} \to$$

$$\begin{pmatrix} \alpha & \alpha + 3 & 2\alpha & \alpha + 2 \\ 0 & \alpha - 1 & \alpha & \alpha \\ 0 & 0 & \alpha & \alpha \end{pmatrix} = (A'|\underline{b}').$$

If $\alpha \neq 0$ and $\alpha - 1 \neq 0$, i.e. for each $\alpha \in \mathbb{R} \setminus \{0.1\}$, we have $\mathrm{rr}(A') = \mathrm{rr}(A'|b') = 3$, so by Proposition 1.3.10, the system Σ_α admits one solution: $(\frac{2-\alpha}{\alpha}, 0, 1)$.

If $\alpha = 0$ we get the matrix:

$$(A'|\underline{b}') = \begin{pmatrix} 0 & 3 & 0 & 2 \\ 0 & -1 & 0 & 0 \\ 0 & 0 & 0 & 0 \end{pmatrix},$$

which is not in row echelon form, but this can be fixed by replacing the second line with the sum of the second row and first row multiplied by $\frac{1}{3}$:

$$\begin{pmatrix} 0 & 3 & 0 & | & 2 \\ 0 & 0 & 0 & | & 2/3 \\ 0 & 0 & 0 & | & 0 \end{pmatrix}.$$

The given system is therefore equivalent to the linear system

$$\begin{cases} 3x_2 = 2 \\ 0 = 2/3 \end{cases}$$

which obviously has no solutions.

Finally, if $\alpha = 1$ we have:

$$(A'|\underline{b}') = \begin{pmatrix} 1 & 4 & 2 & | & 3 \\ 0 & 0 & 1 & | & 1 \\ 0 & 0 & 1 & | & 1 \end{pmatrix}$$

which we can reduce in row echelon form to obtain the matrix

$$(A''|\underline{b}'') = \begin{pmatrix} 1 & 4 & 2 & | & 3 \\ 0 & 0 & 1 & | & 1 \\ 0 & 0 & 0 & | & 0 \end{pmatrix}$$

so $rr(A'') = rr(A''|\underline{b}'') = 2$, hence the system Σ_1 is equivalent to the linear system of 2 equations in 3 unknowns:

$$\begin{cases} x_1 + 4x_2 + 2x_3 = 3 \\ x_3 = 1 \end{cases}.$$

This system has infinitely many solutions depending on a parameter and the set of solutions is: $\{(1 - 4x_2, x_2, 1) \mid x_2 \in \mathbb{R}\}$.

2. Adding the unknown x_4 means to add to the complete matrix $(A|\underline{b})$ associated with the system a column of zeros corresponding to the coefficients of x_4. Therefore, by reducing $(A|\underline{b})$ to row echelon form, we get the matrix:

$$(A'|\underline{b}') = \begin{pmatrix} \alpha & \alpha + 3 & 2\alpha & 0 & | & \alpha + 2 \\ 0 & \alpha - 1 & \alpha & 0 & | & \alpha \\ 0 & 0 & \alpha & 0 & | & \alpha \end{pmatrix}.$$

Therefore, reasoning as above, but taking into account that in this case the number of variables is 4, we obtain that:

for $\alpha \in \mathbb{R}\setminus\{0, 1\}$ the system has infinitely many solutions, and they are of the form: $(\frac{2-\alpha}{\alpha}, 0.1, x_4)$ with $x_4 \in \mathbb{R}$;

for $\alpha = 0$ the system has no solutions;

for $\alpha = 1$ the system has infinitely many solutions, and they are of the form: $(1 - 4x_2, x_2, 1, x_4)$, with $x_2, x_4 \in \mathbb{R}$.

1.5.4 Determine if there are values of the real parameter k such that the linear system

$$\Sigma : \begin{cases} 2x_1 + x_2 - x_3 = 0 \\ 4x_1 - x_2 = 0 \\ x_1 + \frac{1}{2}x_2 - x_3 = -\frac{3}{2} \end{cases}$$

is equivalent to the linear system

$$\Pi_k : \begin{cases} x_1 + x_2 - \frac{1}{2}x_3 = 1 \\ 2x_1 - x_2 + x_3 = 2 \\ kx_1 - 4x_2 + 3x_3 = k \end{cases}$$

Solution. Two systems are equivalent if they have the same solutions. First we solve the linear system Σ. The complete matrix associated with the system is:

$$(A|\underline{b}) = \begin{pmatrix} 2 & 1 & -1 & 0 \\ 4 & -1 & 0 & 0 \\ 1 & \frac{1}{2} & -1 & -\frac{3}{2} \end{pmatrix}.$$

Using the Gaussian algorithm we can reduce $(A|\underline{b})$ to row echelon form, obtaining the matrix

$$(A'|\underline{b}') = \begin{pmatrix} 2 & 1 & -1 & 0 \\ 0 & -3 & 2 & 0 \\ 0 & 0 & -\frac{1}{2} & -\frac{3}{2} \end{pmatrix}.$$

We have: $\mathrm{rr}(A') = \mathrm{rr}(A'|\underline{b}') = 3$, so the system Σ admits a single solution that we can determine by proceeding by susequent substitutions from the bottom:

$$x_3 = 3; \quad x_2 = 2; \quad x_1 = \frac{1}{2}.$$

The only solution of the system is therefore $\left(\frac{1}{2}, 2, 3\right)$.

For Σ to be equivalent to Π_k it is therefore necessary that $\left(\frac{1}{2}, 2, 3\right)$ simultaneously satisfies all the equations of Π_k. So we replace $x_1 = \frac{1}{2}$, $x_2 = 2$ and $x_3 = 3$ in the equations of Π_k:

$$\begin{cases} \frac{1}{2} + 2 - \frac{3}{2} = 1 \\ 1 - 2 + 3 = 2 \\ \frac{k}{2} - 8 + 9 = k \end{cases}$$

We thus obtained two identities and the necessary condition $k = 2$. Therefore, $\left(\frac{1}{2}, 2, 3\right)$ is a solution of the system Π_k only if $k = 2$. We can therefore say that for $k \neq 2$ the Π_k and Σ systems are not equivalent, but we do not know yet if the systems Σ and Π_2 are equivalent. In fact, this happens if and only if $\left(\frac{1}{2}, 2, 3\right)$ is the only solution of the system Π_2. Consider then, the complete matrix associated with the system Π_k for $k = 2$:

$$\begin{pmatrix} 1 & 1 & -\frac{1}{2} & | & 1 \\ 2 & -1 & 1 & | & 2 \\ 2 & -4 & 3 & | & 2 \end{pmatrix}.$$

By reducing this matrix to row echelon form, we obtain the matrix:

$$(A''|\underline{b}'') = \begin{pmatrix} 1 & 1 & -\frac{1}{2} & | & 1 \\ 0 & -3 & 2 & | & 0 \\ 0 & 0 & 0 & | & 0 \end{pmatrix}.$$

So we have $\mathrm{rr}(A'') = \mathrm{rr}(A''|\underline{b}'') = 2 < 3$, so the Π_2 system has infinitely many solutions. We can therefore conclude that there are no values of k such that the systems Σ and Π_k are equivalent to each other.

1.6 SUGGESTED EXERCISES

1.6.1 Solve the following linear systems in the unknowns x, y, z:

1.

$$\begin{cases} x + y + z = 1 \\ 2x + 2y + z = 1 \\ 3y + z = 1 \end{cases}$$

2.

$$\begin{cases} x - y + 4z = 10 \\ 3x + y + 5z = 15 \\ x + 3y - 3z = 6 \end{cases}$$

1.6.2 Solve the following linear systems in the unknowns x, y, z, w:

1.

$$\begin{cases} x - y + 2z - 3w = 0 \\ 2x + y - w = 3 \\ 2y + z + w = -3 \\ 2x + z = 0 \end{cases}$$

2.

$$\begin{cases} x + y - z + w = 0 \\ 2x - z - w = 0 \\ x - y - 2w = 0 \\ 3x + y - 2z = 0 \end{cases}$$

3.

$$\begin{cases} x + z = 7 \\ x + y = 2 \\ 4x + 12y + z = 1 \\ 5x + 6y + 2z = -1 \end{cases}$$

1.6.3 Consider the following linear system in the unknowns x, y, z, depending on the real parameter k:

$$\begin{cases} x + 2y + kz = 0 \\ x + y = -1 \\ x + ky = -2. \end{cases}$$

Determine for which value of k the system admits solutions and, when possible, determine such solutions.

1.6.4 Determine for which values of $a \in \mathbb{R}$ the following linear system, in the unknowns x, y, z, t, admits solutions and, when possible, determine them:

$$\begin{cases} 2x + y - z = 1 \\ -2x + 3z + t = 1 \\ 2x + 3y + (a^2 + 2a + 3)z + (a^2 - 2)t = a + 6 \\ y + 2(a^2 + 2a + 1)z + (3a^2 - 2a - 7)t = 3a + 4. \end{cases}$$

1.6.5 Given the linear system in the unknowns x, y, z:

$$\Sigma_{a,b} : \begin{cases} x + (2 + a)y = b \\ (2 + 2a)x + 3y - (b + 1)z = 1 + b \\ bx + by - (b + 4)z = b^2 + 3b. \end{cases}$$

1. Determine for which values of $a, b \in \mathbb{R}$ the homogeneous system associated to it admits the solution $(a, -a, 0)$. (We call homogeneous system associated with the linear system $A\underline{x} = \underline{b}$ the system $A\underline{x} = \mathbf{0}$, where $\mathbf{0}$ is the null column.)

2. Determine for which among the values of a, b found in the previous point the system $\Sigma_{a,b}$ is solvable and determine its solutions.

1.6.6 Determine for which values of the real parameter k the following system in the unknowns x_1, x_2, x_3, x_4 is compatible. Determine the system's solutions when possible.

$$\begin{cases} x_1 + 3x_2 + kx_3 + 2x_4 = k \\ x_1 + 6x_2 + kx_3 + 3x_4 = 2k + 1 \\ -x_1 - 3x_2 + (k - 2)x_4 = 1 - k \\ kx_3 + (2 - k)x_4 = 1 \end{cases}$$

CHAPTER 2

Vector Spaces

In this chapter, we want to introduce the main character of linear algebra: the *vector space*. It is a generalization of concepts that we already know very well. The Cartesian plane, the set of functions studied in calculus, the set of $m \times n$ matrices introduced in the previous chapter, the set of polynomials, the set of real numbers are all examples of sets that have a natural vector space structure. The vector space will also be the right environment in which to read and interpret the results obtained in the previous chapter. Before giving its precise definition, we see some concrete examples.

2.1   INTRODUCTION: THE SET OF REAL NUMBERS

Let us briefly recall the main properties of the operations that we are usually performed on numbers, in particular real numbers.

The sum of two real numbers is an operation that associates to each pair of real numbers a and b another real number, denoted by $a + b$. So the sum is a function whose domain is $\mathbb{R} \times \mathbb{R}$ and codomain is $\mathbb{R}$:

$$
\begin{aligned}
+ : \quad \mathbb{R} \times \mathbb{R} \quad &\to \quad \mathbb{R} \\
(a, b) \quad &\mapsto \quad a + b.
\end{aligned}
$$

The sum of real numbers is:

- commutative: $a + b = b + a$ for each $a, b \in \mathbb{R}$;

- associative: $(a + b) + c = a + (b + c)$ for each $a, b, c \in \mathbb{R}$;

- admits a *neutral element*, i.e. there exists a number, 0, such that $0 + a = a + 0 = a$ for every $a \in \mathbb{R}$;

- every real number a admits *an opposite*, that is, there is another number, which we denote by $-a$, such that $a + (-a) = 0$.

The product of two real numbers is an operation that associates to each pair of real numbers a and b another real number, denoted by ab. Therefore, the product is a function whose domain is $\mathbb{R} \times \mathbb{R}$ and codomain is $\mathbb{R}$:

$$
\begin{aligned}
\cdot : \quad \mathbb{R} \times \mathbb{R} \quad &\to \quad \mathbb{R} \\
(a, b) \quad &\mapsto \quad ab.
\end{aligned}
$$

The product of real numbers is:

- commutative: $ab = ba$ for each $a, b \in \mathbb{R}$;

- associative: $(ab)c = a(bc)$ for every $a, b, c \in \mathbb{R}$;

- admits neutral element, i.e. there exists a number, 1, such that $1a = a1 = a$ for every $a \in \mathbb{R}$;

- distributive with respect to the sum: $a(b + c) = ab + ac$ for every $a, b, c \in \mathbb{R}$.

One of the most important properties of real numbers, which distinguishes them from other sets of numbers, is their continuity. Geometrically this means that we think of real numbers as distributed along a straight line. More precisely, given a line, a fixed point on it (origin) and a unit of measure, there is a correspondence between the points on the line and the set of real numbers. In other words, every real number uniquely identifies one and only one point on the line.

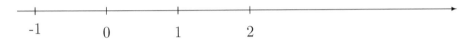

$$-1 \qquad 0 \qquad 1 \qquad 2$$

2.2   THE VECTOR SPACE $\mathbb{R}^N$ AND THE VECTOR SPACE OF MATRICES

We denote by the symbol $\mathbb{R}^2$ the set of ordered pairs of real numbers:

$$\mathbb{R}^2 = \{(x, y) \mid x, y \in \mathbb{R}\}.$$

The fact that the pairs are ordered means, for example, that the element $(1, 2)$ is different from the element $(2, 1)$.

Once we fix a Cartesian coordinate system in a plane, there is a correspondence between $\mathbb{R}^2$ and the set of points in the plane. Attaching a Cartesian reference to the plane means fixing two oriented perpendicular lines r and s and a unit of measure. The point of intersection between the two straight lines is called the origin of the reference system. Each point of the plane is then uniquely identified by a pair of real numbers, called coordinates of the point, which indicate the distance of the point from the line s and its distance from the line r, respectively. The student who is not familiar with the Cartesian plane can think of the boardgame Battleship.

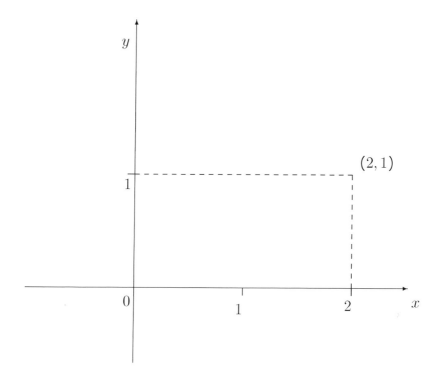

It is natural to try to extend the operations that we perform with numbers to the pairs of real numbers. We then define the following:

- Sum:

$$+ : \quad \begin{matrix} \mathbb{R}^2 \times \mathbb{R}^2 & \to & \mathbb{R}^2 \\ ((x,y),(x',y')) & \mapsto & (x,y)+(x',y') = (x+x',y+y'). \end{matrix}$$

- Multiplication by a real number:

$$\cdot : \quad \begin{matrix} \mathbb{R} \times \mathbb{R}^2 & \to & \mathbb{R}^2 \\ (\lambda,(x,y)) & \mapsto & \lambda(x,y) = (\lambda x, \lambda y). \end{matrix}$$

Note that in the product $\lambda(x,y)$, we have omitted the symbol for the multiplication, just like we usually do when we multiply two real numbers.

We try to interpret geometrically the operations defined in the case of $\mathbb{R}^2$. To this aim, we think of each element (a,b) of $\mathbb{R}^2$ as the endpoint of a *vector* applied at the origin, that is, as an outgoing-oriented segment from the origin with the arrow pointing to the point of coordinates (a,b). In this case, the way to add two elements of $\mathbb{R}^2$ coincides with the well-known *rule of the parallelogram* used to add up forces in physics. This rule states that the sum of two vectors $\vec{u}$ and $\vec{v}$ applied at a point is a vector applied at the same point with the direction and length of the diagonal of the parallelogram having as sides $\vec{u}$ and $\vec{v}$.

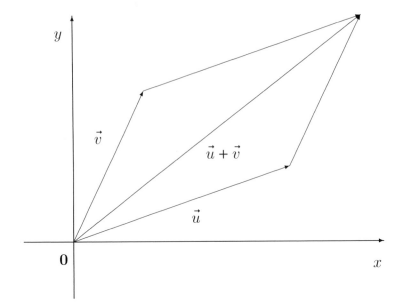

The multiplication of a vector $\vec{v}$ by a real number α is a vector having the same direction as $\vec{v}$ and the length multiplied by the absolute value of α, pointing in the same or the opposite direction as $\vec{v}$, depending on whether α is positive or negative.

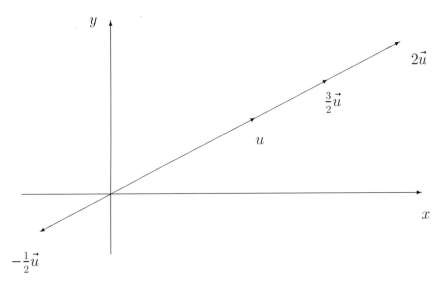

Some students will remember from physics that there are other operations that can be performed with vectors (the dot product, cross product, etc.), but now we are not interested in them and we will not take them into account.

Almost immediately we can verify that the sum of elements of $\mathbb{R}^2$ satisfies the following properties:

1. *commutative:* $(x, y) + (x', y') = (x', y') + (x, y)$ for every $(x, y), (x', y') \in \mathbb{R}^2$;

2. *associative:* $((x, y) + (x', y')) + (x'', y'') = (x, y) + ((x', y') + (x'', y''))$ for every $(x, y), (x', y'), (x'', y'') \in \mathbb{R}^2$;

3. *existence of neutral element* $(0,0)$: $(x,y) + (0,0) = (0,0) + (x,y) = (x,y)$ for each $(x,y) \in \mathbb{R}^2$;

4. *existence of opposite*: for every $(x,y) \in \mathbb{R}^2$ there exists an element (a,b), called *opposite* of (x,y), such that $(a,b) + (x,y) = (x,y) + (a,b) = (0,0)$. Obviously we have: $(a,b) = (-x,-y)$;

5. *distributive property*: $\lambda((x,y) + (x',y')) = \lambda(x,y) + \lambda(x',y')$, for every (x,y), $(x',y') \in \mathbb{R}^2$ and for every $a \in \mathbb{R}$.

6. $(\lambda + \mu)(x,y) = \lambda(x,y) + \mu(x,y)$, for every $(x,y) \in \mathbb{R}^2$ and for all $\lambda, \mu \in \mathbb{R}$.

7. $(\lambda\mu)(x,y) = \lambda(\mu(x,y))$, for every $(x,y) \in \mathbb{R}^2$ and for every $\lambda, \mu \in \mathbb{R}$.

8. $1(x,y) = (x,y)$, for every $(x,y) \in \mathbb{R}^2$.

Of course we can generalize what has been done for $\mathbb{R}^2$ to the set of ordered n-tuples of real numbers, for every $n \in \mathbb{N}$:

$$\mathbb{R}^n = \{(x_1, \ldots, x_n) \mid x_1, \ldots, x_n \in \mathbb{R}\}.$$

We define the sum + and the product · by real numbers:

$$+ : \mathbb{R}^n \times \mathbb{R}^n \to \mathbb{R}^n$$

$$(x_1, \ldots, x_n) + (x'_1, \ldots, x'_n) = (x_1 + x'_1, \ldots, x_n + x'_n);$$

$$\cdot : \mathbb{R} \times \mathbb{R}^n \to \mathbb{R}^n$$

$$\lambda(x_1, \ldots, x_n) = (\lambda x_1, \ldots, \lambda x_n).$$

With a little patience we can check the properties 1 through 8 listed above for the sum and product real numbers in $\mathbb{R}^n$.

Let us examine another example. Consider the set of matrices 2×2 with real coefficients:

$$\mathrm{M}_2(\mathbb{R}) = \left\{ \begin{pmatrix} a & b \\ c & d \end{pmatrix} \mid a, b, c, d \in \mathbb{R} \right\}$$

introduced in Chapter 1. We define in $\mathrm{M}_2(\mathbb{R})$ the following sum + and product · by real number operations:

$$+ : \mathrm{M}_2(\mathbb{R}) \times \mathrm{M}_2(\mathbb{R}) \to \mathrm{M}_2(\mathbb{R})$$

$$\begin{pmatrix} a & b \\ c & d \end{pmatrix} + \begin{pmatrix} a' & b' \\ c' & d' \end{pmatrix} = \begin{pmatrix} a + a' & b + b' \\ c + c' & d + d' \end{pmatrix}.$$

$$\cdot : \mathbb{R} \times \mathrm{M}_2(\mathbb{R}) \to \mathrm{M}_2(\mathbb{R})$$

$$\lambda \begin{pmatrix} a & b \\ c & d \end{pmatrix} = \begin{pmatrix} \lambda a & \lambda b \\ \lambda c & \lambda d \end{pmatrix}.$$

Also in this case, with patience, it is possible to verify properties 1 through 8. Students are strongly encouraged to do so. For example, we prove the commutativity of +:

$$\begin{pmatrix} a & b \\ c & d \end{pmatrix} + \begin{pmatrix} a' & b' \\ c' & d' \end{pmatrix} = \begin{pmatrix} a + a' & b + b' \\ c + c' & d + d' \end{pmatrix}.$$

Since the sum of real numbers is commutative, we can write:

$$\begin{pmatrix} a + a' & b + b' \\ c + c' & d + d' \end{pmatrix} = \begin{pmatrix} a' + a & b' + b \\ c' + c & d' + d \end{pmatrix} = \begin{pmatrix} a' & b' \\ c' & d' \end{pmatrix} + \begin{pmatrix} a & b \\ c & d \end{pmatrix}.$$

Once again it all depends on the properties of the real number operations. This is precisely the strategy to verify the properties 1 through 8 in $M_2(\mathbb{R})$ and more in general in any vector space.

It is clear that, similarly to what was done for 2×2 matrices, it is possible to define a sum and a product by real numbers also in the set of $m \times n$ matrices, and with some patience one can show that such operations satisfy all the properties listed above. So we give the definition of the operations of sum and product by real numbers in $M_{m,n}(\mathbb{R})$:

Sum:

$$\begin{pmatrix} a_{11} & a_{12} & \cdots & a_{1n} \\ a_{21} & a_{22} & \cdots & a_{2n} \\ \vdots & \vdots & \ddots & \vdots \\ a_{m1} & a_{m2} & \cdots & a_{mn} \end{pmatrix} + \begin{pmatrix} a'_{11} & a'_{12} & \cdots & a'_{1n} \\ a'_{21} & a'_{22} & \cdots & a'_{2n} \\ \vdots & \vdots & \ddots & \vdots \\ a'_{m1} & a'_{m2} & \cdots & a'_{mn} \end{pmatrix}$$

$$= \begin{pmatrix} a_{11} + a'_{11} & a_{12} + a'_{12} & \cdots & a_{1n} + a'_{1n} \\ a_{21} + a'_{21} & a_{22} + a'_{22} & \cdots & a_{2n} + a'_{2n} \\ \vdots & \vdots & \ddots & \vdots \\ a_{m1} + a'_{m1} & a_{m2} + a'_{2m} & \cdots & a_{mn} + a'_{mn} \end{pmatrix}.$$

Product by a real number:

$$\lambda \begin{pmatrix} a_{11} & a_{12} & \cdots & a_{1n} \\ a_{21} & a_{22} & \cdots & a_{2n} \\ \vdots & \vdots & \ddots & \vdots \\ a_{m1} & a_{m2} & \cdots & a_{mn} \end{pmatrix} = \begin{pmatrix} \lambda a_{11} & \lambda a_{12} & \cdots & \lambda a_{1n} \\ \lambda a_{21} & \lambda a_{22} & \cdots & \lambda a_{2n} \\ \vdots & \vdots & \ddots & \vdots \\ \lambda a_{m1} & \lambda a_{m2} & \cdots & \lambda a_{mn} \end{pmatrix}.$$

Note that the neutral element of the sum in $M_{m,n}(\mathbb{R})$ is the *zero* matrix

$$\begin{pmatrix} 0 & 0 & \cdots & 0 \\ 0 & 0 & \cdots & 0 \\ \vdots & \vdots & \ddots & \vdots \\ 0 & 0 & \cdots & 0 \end{pmatrix}.$$

The sets described so far with the sum and product operations are all examples of vector spaces.

2.3 VECTOR SPACES

In this section, we give the definition of a vector space, that is we formally define the structure which we introduced through the examples in the previous section.[1]

Definition 2.3.1 A *real vector space* is a set V equipped with two operations called, respectively, *sum* and *multiplication by scalars*:

$$+ : V \times V \longrightarrow V \qquad \cdot : \mathbb{R} \times V \longrightarrow V$$

$$(\mathbf{u}, \mathbf{v}) \mapsto \mathbf{u} + \mathbf{v} \qquad (\lambda, \mathbf{u}) \mapsto \lambda \mathbf{u}$$

satisfying the following properties:

1. *commutative*, i.e. $\mathbf{u} + \mathbf{v} = \mathbf{v} + \mathbf{u}$, for every $\mathbf{u}, \mathbf{v} \in V$;

2. *associative*, that is, $(\mathbf{u} + \mathbf{v}) + \mathbf{w} = \mathbf{u} + (\mathbf{v} + \mathbf{w})$ for every $\mathbf{u}, \mathbf{v}, \mathbf{w} \in V$;

3. there exists a *neutral element* for the sum, i.e. there is $\mathbf{0} \in V$ such that $\mathbf{0} + \mathbf{u} = \mathbf{u} + \mathbf{0} = \mathbf{u}$ for each $\mathbf{u}$ in V;

4. each element of V has an *opposite*, that is, for every $\mathbf{u} \in V$ there exists a vector $\mathbf{a}$ such that $\mathbf{a} + \mathbf{u} = \mathbf{u} + \mathbf{a} = \mathbf{0}$;

5. $1\mathbf{u} = \mathbf{u}$;

6. $(\lambda\mu)\mathbf{u} = \lambda(\mu\mathbf{u})$, for every $\mathbf{u} \in V$ and for every $\lambda, \mu \in \mathbb{R}$;

7. $\lambda(\mathbf{u} + \mathbf{v}) = \lambda\mathbf{u} + \lambda\mathbf{v}$, for every $\mathbf{u}, \mathbf{v} \in V$ and for every $\lambda \in \mathbb{R}$;

8. $(\lambda + \mu)\mathbf{u} = \lambda\mathbf{u} + \mu\mathbf{u}$, for every $\mathbf{u} \in V$ and for every $\lambda, \mu \in \mathbb{R}$.

The elements of a vector space are called *vectors*, while the real numbers are called *scalars*. The neutral element of the sum in V is called *zero vector*. To distinguish vectors from numbers we will indicate the vectors in bold.

In the previous section, we have seen that $\mathbb{R}^n$ and $M_{m,n}(\mathbb{R})$ are real vector spaces. Now let us see other examples.

Example 2.3.2 *The vectors applied at one point.*

Consider the set V of the vectors applied at one point. We define the sum of two such vectors using the rule of the parallelogram. The sum clearly enjoys properties 1 through 4, and the neutral element is the zero-length vector.

Then we can define multiplication by a real number in the following way: if $\alpha \in \mathbb{R}$ and $\vec{v} \in V$ then $\alpha \cdot \vec{v}$ is the vector with the same direction as $\vec{v}$, the length multiplied by the factor $|\alpha|$ (where $|\alpha|$ is the absolute value of α).

---

[1]We will define the concept of the real vector space, in which vectors are multiplied by real numbers we will call *scalars*. It is important to know that one can develop this theory also replacing the real numbers with other sets, such as the rational numbers, or the complex numbers.

Then $\alpha \cdot \vec{v}$ is the vector lying in the same line as $\vec{v}$, its length is multiplied by the factor $|\alpha|$ (where $|\alpha|$ is the absolute value of α) and its direction is the same as that of $\vec{v}$ or opposite to it depending on whether the sign of α is positive or negative.

In this way, the set of vectors of the space applied at a point turns out to be a real vector space.

Example 2.3.3 *The functions.* Let $\mathcal{F}(\mathbb{R})$ be the set of functions $\mathcal{F} : \mathbb{R} \longrightarrow \mathbb{R}$ with two operations:

$$(f + g)(x) = f(x) + g(x), \qquad (\lambda \cdot f)(x) = \lambda f(x).$$

Check by exercise that $\mathcal{F}(\mathbb{R})$ is a real vector space.

Example 2.3.4 *The polynomials.*

Consider the set $\mathbb{R}[x]$ of polynomials with real coefficients in the variable x. With the usual operations of sum and product by a real number, $\mathbb{R}[x]$ is a vector space. Again we leave the verification of properties 1 to 8 by exercise. We observe, for example, that the zero vector in $\mathbb{R}[x]$ is the polynomial identically zero, i.e. 0; the opposite of the polynomial $5x^2 - x$ is the polynomial $-5x^2 + x$.

We now show some useful properties that are valid for any vector space:

Proposition 2.3.5 *Let V be a vector space. Then we have the following properties:*

i) The zero vector is unique and will be denoted by $\mathbf{0}_V$.

ii) If $\mathbf{u}$ is a vector of V, its opposite is unique and it will be denoted by $-\mathbf{u}$.

iii) $\lambda \mathbf{0}_V = \mathbf{0}_V$, for each scalar $\lambda \in \mathbb{R}$.

iv) $0\mathbf{u} = \mathbf{0}_V$ for each $\mathbf{u} \in V$ (note the different sense of zero in the first and second member!).

v) If $\lambda \mathbf{u} = \mathbf{0}_V$, then it is $\lambda = 0$ or $\mathbf{u} = \mathbf{0}_V$.

vi) $(-\lambda)\mathbf{u} = \lambda(-\mathbf{u}) = -\lambda\mathbf{u}$.

Proof. Notice that, while the properties 1 through 8 in Definition 2.3.1 are given, the statements i) through vi), though appear obvious to the student, must be proven and they are a direct consequence of Definition 2.3.1.

i) If $\mathbf{0}'$ is another vector that fulfills property 3 of Definition 2.3.1, we have that $\mathbf{0}' + \mathbf{0}_V = \mathbf{0}_V$ (here we take $\mathbf{u} = \mathbf{0}_V$). Furthermore, using the fact that $\mathbf{0}_V$ satisfies property 3 and taking $\mathbf{u} = \mathbf{0}'$ we also have $\mathbf{0}' + \mathbf{0}_V = \mathbf{0}'$. It follows that $\mathbf{0}_V = \mathbf{0}' + \mathbf{0}_V = \mathbf{0}'$.

ii) If $\mathbf{a}$ and $\mathbf{a}'$ are both opposite of $\mathbf{u}$, by property 4 we have in particular that $\mathbf{a}+\mathbf{u} = \mathbf{0}_V$ and $\mathbf{u}+\mathbf{a}' = \mathbf{0}_V$. Then $\mathbf{a} = \mathbf{a}+\mathbf{0}_V = \mathbf{a}+(\mathbf{u}+\mathbf{a}') = (\mathbf{a}+\mathbf{u})+\mathbf{a}' = \mathbf{0}_V+\mathbf{a}' = \mathbf{a}'$, where we have used associativity of the sum.

iii) We have that $\lambda\mathbf{0}_V = \lambda(\mathbf{0}_V + \mathbf{0}_V)$ by property 3. In addition $\lambda(\mathbf{0}_V + \mathbf{0}_V) = \lambda\mathbf{0}_V + \lambda\mathbf{0}_V$ by property 7, therefore, $\lambda\mathbf{0}_V = \lambda\mathbf{0}_V + \lambda\mathbf{0}_V$. We add $-\lambda\mathbf{0}_V$ to both members, and we get, by property 4, that $\mathbf{0}_V = \lambda\mathbf{0}_V$.

iv) We have that $0\mathbf{u} = (0+0)\mathbf{u} = 0\mathbf{u} + 0\mathbf{u}$ by property 8, thus $0\mathbf{u} = 0\mathbf{u} + 0\mathbf{u}$. We add $-0\mathbf{u}$ to both members, and we get, by property 4, that $\mathbf{0}_V = 0\mathbf{u}$.

v) We show that if $\lambda\mathbf{u} = \mathbf{0}_V$ and $\lambda \neq 0$, then it must be $\mathbf{u} = \mathbf{0}_V$. Let $\frac{1}{\lambda}$ be the inverse of λ, which exists because $\lambda \neq 0$. Multiplying both members of the equality $\lambda\mathbf{u} = \mathbf{0}_V$ by $\frac{1}{\lambda}$, we have: $\mathbf{u} = (\frac{1}{\lambda}\lambda)\mathbf{u} = \frac{1}{\lambda}\mathbf{0}_V = \mathbf{0}_V$.

vi) We have that $(-\lambda)\mathbf{u} + \lambda\mathbf{u} = (-\lambda + \lambda)\mathbf{u} = 0\mathbf{u} = \mathbf{0}_V$ by properties 8 and iv). Similarly $\lambda\mathbf{u} + (-\lambda)\mathbf{u} = \mathbf{0}_V$. Then $(-\lambda)\mathbf{u}$ is the opposite of $\lambda\mathbf{u}$, i.e. $(-\lambda)\mathbf{u} = -\lambda\mathbf{u}$. Also $\lambda(-\mathbf{u}) + \lambda\mathbf{u} = \lambda(-\mathbf{u} + \mathbf{u}) = \lambda\mathbf{0}_V = \mathbf{0}_V$ by properties 7, 4 and iii). Similarly $\lambda\mathbf{u} + (-\lambda)\mathbf{u} = \mathbf{0}_V$, hence also $\lambda(-\mathbf{u})$ is the opposite of $\lambda\mathbf{u}$, i.e. $\lambda(-\mathbf{u}) = -\lambda\mathbf{u}$. ■

Definition 2.3.6 The *trivial vector space*, denoted with $\{\mathbf{0_V}\}$, is a vector space consisting only of the zero vector.

Observation 2.3.7 By definition of zero vector and by property *iii*) of Proposition 2.3.5, the operations of sum and product by a scalar are well defined and of course trivial in the trivial vector space:

$$\mathbf{0}_V + \mathbf{0}_V = \mathbf{0}_V$$

$$\lambda\mathbf{0}_V = \mathbf{0}_V, \quad \text{for all } \lambda \in \mathbb{R}.$$

Observation 2.3.8 Let us think about the definition of $\mathbb{R}$-vector space. First of all we observe that, by definition, *a real vector space can never be empty*. In fact, it must contain at least the zero vector that is the neutral element of the sum. It could happen that a vector space contains *only* the zero vector; In this case, it is called *trivial*.

Now suppose that V is a *nontrivial* vector space, i.e. that it contains at least one vector $\mathbf{v} \neq \mathbf{0}_V$. How many elements does V contain? As we can multiply by real numbers, which are infinitely many, V will contain all the infinitely many multiples of $\mathbf{v}$, that is, all the vectors of the form $\lambda\mathbf{v}$, for every $\lambda \in \mathbb{R}$.

Since λ varies in the set of real numbers we get infinitely many different elements of V. To rigorously prove this statement, we have to show that if λ and μ are distinct real numbers, that is, $\lambda \neq \mu$, and $\mathbf{v} \in V$ is a nonzero vector, then $\lambda\mathbf{v} \neq \mu\mathbf{v}$. In fact, if not, we would have:

$$\lambda\mathbf{v} = \mu\mathbf{v} \leftrightarrow (\lambda - \mu)\mathbf{v} = \mathbf{0}_V$$

with $\lambda - \mu \neq 0$ and $\mathbf{v} \neq \mathbf{0}_V$, which would contradict property (v) of Proposition 2.3.5.

2.4   SUBSPACES

How can we recognize and describe a vector space? How can we single out a subset of a vector space with the same characteristics? To answer to these questions is necessary to introduce the definition of subspace.

Definition 2.4.1 Let W be a subset of the space vector V. We say that W is a *subspace of V* if it satisfies the following properties:

1) W is different from the empty set;

2) W is closed with respect to the sum, that is, for every $\mathbf{u}, \mathbf{v} \in W$ we have that $\mathbf{u} + \mathbf{v} \in W$;

3) W is closed with respect to the product by scalars, that is, for every $\mathbf{u} \in W$ and every $\lambda \in \mathbb{R}$ we have that $\lambda \mathbf{u} \in W$.

We note that, since W is not empty, and $\lambda \mathbf{u} \in W$ for each $\lambda \in \mathbb{R}$, then $\mathbf{0}_V \in W$, because we can take any vector of W and multiply it by $\lambda = 0$. In fact, property (1) can effectively be replaced by the property:

1) $\mathbf{0}_V \in W$,

and we obtain an equivalent definition.

It is important to note that a subspace W of V is a vector space with the operations of V restricted to W. In fact, property 2) of Definition 2.4.1 ensures that the restriction to W of the sum defined in V gives as a result a vector of W:

$$+_V|_{W \times W} : W \times W \to W.$$

Similarly, property 3) of Definition 2.4.1 ensures that the restriction to $\mathbb{R} \times W$ of the product by scalars defined on $\mathbb{R} \times V$ gives as a result a vector of W. Then properties 1 through 8 of Definition 2.3.1 continue to hold because they are true in V.

In particular, therefore, every vector space V has always at least two subspaces: V itself and the zero subspace, consisting of only the zero vector $\mathbf{0}_V$. Because of Observation 2.3.8, if V itself is not trivial, every nontrivial subspace of V contains infinitely many elements.

We now want to clarify the concept of subspace with some examples and counterexamples.

Example 2.4.2 The set $X = \{(x, y) \in \mathbb{R}^2 \mid y = 0\}$ is a subspace of $\mathbb{R}^2$. X is in fact:

1) not empty: it contains infinitely many pairs of real numbers $(x, 0)$;

2) closed under the sum: given two elements $(x_1, 0)$, $(x_2, 0)$ in X, the sum of $(x_1, 0) + (x_2, 0) = (x_1 + x_2, 0)$ still belongs to X;

3) closed with respect to the product by scalars: given any real number α and any element $(x, 0) \in X$, the product $\alpha(x, 0) = (\alpha x, 0)$ belongs to X.

Geometrically, after setting a Cartesian coordinate system in $\mathbb{R}^2$, we can identify the set X with the x-axis. Then adding two vectors lying on the x-axis or multiplying by a scalar one of them, we still get a vector that lies on the x-axis.

More generally, if a is a real number set $W_a = \{(x, y) \in \mathbb{R}^2 \mid y = ax\}$. We observe first that $(0, 0) \in W$. Furthermore, given two elements (x_1, ax_1) and (x_2, ax_2) in W_a, their sum

$$(x_1, ax_1) + (x_2, ax_2) = (x_1 + x_2, a(x_1 + x_2))$$

belongs to W_a, i.e. W_a is closed with respect to the sum. Likewise, given $\lambda \in \mathbb{R}$ and $(x_1, ax_1) \in W_a$, we have:

$$\lambda(x_1, ax_1) = (\lambda x_1, \lambda a x_1) = (\lambda x_1, a(\lambda x_1)),$$

which belongs to W_a, i.e. W_a is closed with respect to the product by scalars. So W_a is a subspace of $\mathbb{R}^2$.

Again, after setting a Cartesian coordinate system in the plane, W_a is identified with a line passing through the origin of the coordinate system.

Example 2.4.3 Consider the set $S = \{(x, y) \in \mathbb{R}^2 \,|\, y = x + 1\}$. We can immediately say that S is not a subspace of $\mathbb{R}^2$ as it does not contain $\mathbf{0}_{\mathbb{R}^2} = (0, 0)$. Therefore, not all the straight lines of the plane give subspaces of $\mathbb{R}^2$, but only those through $(0, 0)$.

Example 2.4.4 The set $X = \left\{ \begin{pmatrix} a & b \\ c & d \end{pmatrix} \in M_2(\mathbb{R}) \,\middle|\, b = 1 \right\} \subseteq M_2(\mathbb{R})$ is not a subspace

of $M_2(\mathbb{R})$, because the zero matrix $\begin{pmatrix} 0 & 0 \\ 0 & 0 \end{pmatrix}$ does not belong to X.

Observation 2.4.5 If S is a subspace of a vector space V, the condition $\mathbf{0}_V \in S$ is necessary, but not sufficient, for S to be a subspace of V. We give a counterexample.

Example 2.4.6 Let $S = \{(x, y, z) \in \mathbb{R}^3 \,|\, xy = z\}$. Despite the fact the set S contains the zero vector $(0, 0, 0)$ of $\mathbb{R}^3$, S is not a subspace of $\mathbb{R}^3$ because it is not closed with respect to the sum. Indeed the vectors $\mathbf{v} = (1, 1, 1)$ and $\mathbf{w} = (-1, -1, 1)$ belong to S, since they satisfy the equation $xy = z$, but their sum $\mathbf{v} + \mathbf{w} = (1, 1, 1) + (-1, -1, 1) = (0, 0, 2)$ does not belong to S since $0 \cdot 0 \neq 2$.

Observation 2.4.7 To say that a subset S of a vector space V is closed with respect to the product by scalars is equivalent to saying that if S contains a nonzero vector, then *it must also contain all of its multiples*. If we think of the geometric interpretation of the product by scalars, which we gave in the case of $\mathbb{R}^2$, this means that if a subspace contains a nonzero vector $\mathbf{v}$, then it contains the whole line of the plane passing through the origin and determined by $\mathbf{v}$. This geometrical reasoning allows us to say immediately that certain known subsets of the plane are not vector spaces, for example

$$C = \{(x, y) \in \mathbb{R}^2 \,|\, x^2 + y^2 = 1\}$$

(a circumference),

$$P = \{(x, y) \in \mathbb{R}^2 \,|\, y = x^2\}$$

(a parabola) and so on on.

In a sense, a vector space can not be *curved*, from which the name *linear* algebra.

Observation 2.4.8 The examples given so far highlight two different types of reasoning. To prove that a subset S of a vector space V is a subspace of V, we must prove it satisfies properties 1), 2) and 3) of Definition 2.4.1. These properties must apply *always*, that is, for each pair of vectors in S (property 2) and for all real numbers (property 3).

On the contrary, to prove that a subset of a vector space V is not a subspace of V, it is enough to show that one of properties 1), 2) and 3) of Definition 2.4.1 fails, i.e. if $S \neq \varnothing$, or if there exists a pair of vectors of S whose sum is not in S, or if there is a vector of S and a scalar whose product is not in S.

Example 2.4.9 In the vector space $\mathbb{R}[x]$ of polynomials with real coefficients in a variable x, consider the subset $\mathbb{R}_2[x]$ consisting of polynomials of degree less than or equal to 2:

$$\mathbb{R}_2[x] = \{p(x) = a + bx + cx^2 \mid a, b, c \in \mathbb{R}\}.$$

Then $\mathbb{R}_2[x]$ is a subspace of $\mathbb{R}[x]$. In fact, adding two polynomials of degrees less than or equal to 2, we obtain a polynomial of degree less than or equal to 2:

$$(a + bx + cx^2) + (a' + b'x + c'x^2) = (a + a') + (b + b')x + (c + c')x^2.$$

Similarly, multiplying a polynomial of degree less than or equal to 2 by a real number λ, we obtain a polynomial of degree less than or equal to 2:

$$\lambda(a + bx + cx^2) = (\lambda a) + (\lambda b)x + (\lambda c)x^2.$$

In other words, $\mathbb{R}_2[x]$ (which is certainly not empty) is closed with respect to the sum and the product by scalars, therefore it is a subspace of $\mathbb{R}[x]$.

Example 2.4.10 In the vector space $\mathbb{R}[x]$ of polynomials with real coefficients in a variable x, we consider the subset $S = \{p(x) = a + bx + cx^2 \in \mathbb{R}[x] \mid ac = 0\}$. The subset S contains, for example, the polynomial identically zero and the monomial x, therefore, it is different from the empty set. However, it is not closed with respect to the sum defined in $\mathbb{R}[x]$. In fact, S contains the polynomials $p(x) = 1 + x$ and $q(x) = x + x^2$ but does not contain their sum: $p(x) + q(x) = 1 + 2x + x^2$.

Since a subspace of a vector space V is primarily a subset of V, it is natural to ask what happens when carrying out the operation of set theoretic union and intersection of two (or more) subspaces of V.

Example 2.4.11 Consider the set $W = X \cup Y$ with

$$X = \{(x, y) \in \mathbb{R}^2 \mid y = 0\} \quad \text{and} \quad Y = \{(x, y) \in \mathbb{R}^2 \mid x = 0\}.$$

In Example, 2.4.2 we showed that X is a subspace of $\mathbb{R}^2$. In a similar way, one can show that Y is a subspace of $\mathbb{R}^2$. However, their union W is not a subspace of $\mathbb{R}^2$, because it is not closed with respect to the sum: in fact, the vector $(1, 0)$ belongs to W because it is an element of X, and the vector $(0, 1)$ belongs to W because it is an element of Y. However their sum $(1, 0) + (0, 1) = (1, 1)$ belongs neither to X nor to Y.

The geometric reasoning is also simple: we can think of X and Y, respectively, as the x-axis and the y-axis in a Cartesian reference in the plane, and W as the union of the two axes. It is clear from the parallelogram rule that the sum of a vector that lies on the x-axis and a vector that lies on the y-axis will be *outside* of the two lines, hence W is not a subspace.

We observe that the set W can be described as follows:

$$W = \{(x, y) \in \mathbb{R}^2 \mid xy = 0\}.$$

In fact, since x and y real numbers, their product is zero if and only if at least one of the two factors is zero.

The example above shows that, in general, the union of two subspaces of a vector space V is *not* a subspace. More precisely, we have the following proposition.

Proposition 2.4.12 *Let W_1, W_2 be two subspaces of a vector space V. Then $W_1 \cup W_2$ is a subspace if and only if $W_1 \subseteq W_2$ or $W_2 \subseteq W_1$.*

Proof. "$\Leftarrow$" If $W_1 \subseteq W_2$ (resp. if $W_2 \subseteq W_1$) then $W_1 \cup W_2 = W_2$ (resp. $W_1 \cup W_2 = W_1$), which is a subspace by hypothesis.

"$\Rightarrow$" To prove this implication, we show that, if $W_1 \nsubseteq W_2$ and $W_2 \nsubseteq W_1$, then $W_1 \cup W_2$ is not a subspace of V. As $W_1 \nsubseteq W_2$, there exists a vector $\mathbf{v}_1 \in W_1 \setminus W_2$; similarly with $W_2 \nsubseteq W_1$, there exists a vector $\mathbf{v}_2 \in W_2 \setminus W_1$. If $W_1 \cup W_2$ were a subspace, then $\mathbf{v} = \mathbf{v}_1 + \mathbf{v}_2$ should be an element of $W_1 \cup W_2$ as it is the sum of an element of W_1 and one of W_2. If $\mathbf{v}$ were in W_1, then also $\mathbf{v}_2 = \mathbf{v} - \mathbf{v}_1$ would belong to W_1, but we had chosen $\mathbf{v}_2 \in W_2 \setminus W_1$. Similarly, if $\mathbf{v}$ were in W_2, then also $\mathbf{v}_1 = \mathbf{v} - \mathbf{v}_2$ would belong to W_2, but we had chosen $\mathbf{v}_1 \in W_1 \setminus W_2$. So $\mathbf{v} \notin W_1 \cup W_2$. ■

With the intersection of two subspaces we have less problems.

Proposition 2.4.13 *The intersection $S_1 \cap S_2$ of two subspaces S_1 and S_2 of a vector space V is a subspace of V.*

Proof. We have to show that $S_1 \cap S_2$ is a subspace of V: we observe first that this intersection is not empty since $\mathbf{0}_V$ belongs both to S_1 and S_2, so it belongs to $S_1 \cap S_2$. Now we show that $S_1 \cap S_2$ is closed with respect to the sum of V: so let $\mathbf{v}_1, \mathbf{v}_2 \in S_1 \cap S_2$. This means, in particular, that $\mathbf{v}_1, \mathbf{v}_2 \in S_1$, which is a subspace of V, so $\mathbf{v}_1 + \mathbf{v}_2 \in S_1$. Similarly since S_2 is a subspace, we have that $\mathbf{v}_1 + \mathbf{v}_2 \in S_2$. Then $\mathbf{v}_1 + \mathbf{v}_2 \in S_1 \cap S_2$.

Similarly, since we show that $S_1 \cap S_2$ is closed with respect to the product by scalars. Let $\mathbf{v} \in S_1 \cap S_2$ and $\lambda \in \mathbb{R}$. In particular, $\mathbf{v} \in S_1$, which is a subspace, so $\lambda \mathbf{v} \in S_1$; similarly, $\mathbf{v} \in S_2$, which is a subspace, so $\lambda \mathbf{v} \in S_2$. Thus $\lambda \mathbf{v}$ belongs both to S_1 and to S_2, so that it belongs to their intersection. ■

Example 2.4.14 Consider the subspaces:

$$S = \left\{ \begin{pmatrix} a & b \\ c & d \end{pmatrix} \in M_2(\mathbb{R}) \mid b = -c \right\} \quad \text{and}$$

$$T = \left\{ \begin{pmatrix} a & b \\ c & d \end{pmatrix} \in M_2(\mathbb{R}) \mid a + b + c + d = 0 \right\}$$

of $M_2(\mathbb{R})$.

What is $S \cap T$? The subspace $S \cap T$ consists of the elements of $M_2(\mathbb{R})$ belonging to S and T, that is:

$$S \cap T = \left\{ \begin{pmatrix} a & b \\ c & d \end{pmatrix} \in M_2(\mathbb{R}) \mid b = -c, a + b + c + d = 0 \right\}.$$

Hence:

$$S \cap T = \left\{ \begin{pmatrix} a & -c \\ c & -a \end{pmatrix} \in M_2(\mathbb{R}) \right\}.$$

It is easy to verify that this subset of $M_2(\mathbb{R})$ is closed with respect to the sum and the product by scalars, as guaranteed by Proposition 2.4.13.

2.5 EXERCISES WITH SOLUTIONS

2.5.1 Determine if the set $X = \{(r, s, r - s) \in \mathbb{R}^3\}$ is a subspace of $\mathbb{R}^3$.
Solution. First of all, we observe that X is not the empty set because $(0, 0, 0) \in X$ (just take $r = s = 0$).

Let us now consider two generic elements of X: $(r_1, s_1, r_1 - s_1)$ and $(r_2, s_2, r_2 - s_2)$. Their sum is: $(r_1, s_1, r_1 - s_1) + (r_2, s_2, r_2 - s_2) = (r_1 + r_2, s_1 + s_2, r_1 - s_1 + r_2 - s_2) = (r_1 + r_2, s_1 + s_2, r_1 + r_2 - (s_1 + s_2))$, and it still belongs to X as it is of the type $(r, s, r - s)$, with $r = r_1 + r_2$ and $s = s_1 + s_2$.

Consider $(r_1, s_1, r_1 - s_1) \in X$ and $\lambda \in \mathbb{R}$. Then $\lambda(r_1, s_1, r_1 - s_1) = (\lambda r_1, \lambda s_1, \lambda(r_1 - s_1)) = (\lambda r_1, \lambda s_1, \lambda r_1 - \lambda s_1)$ still belongs to X as it is of the type $(r, s, r - s)$, with $r = \lambda r_1$ and $s = \lambda s_1$. So X is a subspace of $\mathbb{R}^3$.

2.5.2 Determine whether the set $W = \{(x, y, z) \in \mathbb{R}^3 \mid 2x + z^2 = 0\} \subseteq \mathbb{R}^3$ is a subspace of $\mathbb{R}^3$.
Solution. W is not the empty set because $(0, 0, 0) \in W$.

Consider now two generic elements of W, (x_1, y_1, z_1) and (x_2, y_2, z_2), with $2x_1 + z_1^2 = 0$ and $2x_2 + z_2^2 = 0$. We have that $(x_1, y_1, z_1) + (x_2, y_2, z_2) = (x_1 + x_2, y_1 + y_2, z_1 + z_2)$, and this sum belongs to W if and only if $2(x_1 + x_2) + (z_1 + z_2)^2 = 0$. But $2(x_1 + x_2) + (z_1 + z_2)^2 = 2x_1 + 2x_2 + z_1^2 + z_2^2 + 2z_1z_2 = (2x_1 + z_1^2) + (2x_2Z_2 +^2) + 2z_1z_2 = 0 + 0 + 2z_1z_2 = 2z_1z_2$, and it is not true that $2z_1z_2$ is always equal to zero.

For example, the elements $(-2, 1, 2)$ and $(-8, 3, 4)$ belong to W but $(-2, 1, 2) + (-8, 3, 4) = (-10, 4, 6) \notin W$, because $2 \cdot (-10) + (6)^2 \neq 0$. (Note that these W elements were not chosen randomly but so to satisfy the request $2z_1z_2 \neq 0$).

So W is not a subspace of $\mathbb{R}^3$.

2.5.3 Determine a non-empty subset of $\mathbb{R}^3$ closed with respect to the sum but not with respect to the product by scalars.
Solution. The set $X = \{(x, y, z) \mid x, y, z \in \mathbb{R}, x \geq 0\}$ has this property. In fact, X is not empty because, for example, $(0, 0, 0) \in X$. Let us check if X is closed with respect to the sum. Let $(x_1, y_1, z_1), (x_2, y_2, z_2) \in X$, with $x_1, x_2 \geq 0$. Then $(x_1, y_1, z_1) + (x_2, y_2, z_2) = (x_1 + x_2, y_1 + y_2, z_1 + z_2) \in X$ because $x_1 + x_2 \geq 0$ (the sum of two non-negative real numbers is a non-negative real number). Now let $(x_1, y_1, z_1) \in X$ and $\lambda \in \mathbb{R}$. We have that $\lambda(x_1, y_1, z_1) = (\lambda x_1, \lambda y_1, \lambda z_1)$ belongs to X if and only if $\lambda x_1 \geq 0$. But if we choose λ negative and $x_1 > 0$, for example $\lambda = -1$ and $(x_1, y_1, z_1) = (3, -2, 1)$, this condition it is not verified. So X is not closed with respect to the product by scalars.

2.5.4 Determine for which values of the parameter k the set

$$X_k = \left\{ \begin{pmatrix} r & s \\ r + k & k^2 - k \end{pmatrix} \mid r, s \in \mathbb{R} \right\} \subseteq \mathrm{M}_2(\mathbb{R})$$

is a subspace of $\mathrm{M}_2(\mathbb{R})$.
Solution. We know that in order for X_k to be a subspace of $\mathrm{M}_2(\mathbb{R})$, the null matrix must belong to X_k, that is to say that $\begin{pmatrix} 0 & 0 \\ 0 & 0 \end{pmatrix}$ is of type $\begin{pmatrix} r & s \\ r + k & k^2 - k \end{pmatrix}$ for some

$r, s \in \mathbb{R}$. This happens if

$$
\begin{cases}
r = 0 \\
s = 0 \\
k = 0 \\
k^2 - k = 0,
\end{cases}
$$

that is, $k = 0$.

Let us see now if $X_0 = \left\{ \begin{pmatrix} r & s \\ r & 0 \end{pmatrix} \mid r, s \in \mathbb{R} \right\}$ is a subspace of $M_2(\mathbb{R})$. Certainly X_0 is not empty since it contains the null matrix.

Let $\begin{pmatrix} r_1 & s_1 \\ R_1 & 0 \end{pmatrix} \begin{pmatrix} r_2 & s_2 \\ r_2 & 0 \end{pmatrix} \in X_0$. We have:

$$
\begin{pmatrix} r_1 & s_1 \\ r_1 & 0 \end{pmatrix} + \begin{pmatrix} r_2 & s_2 \\ r_2 & 0 \end{pmatrix} = \begin{pmatrix} r_1 + r_2 & s_1 + s_2 \\ r_1 + r_2 & 0 \end{pmatrix}.
$$

The matrix obtained therefore belongs to X_0. Similarly, if $\lambda \in \mathbb{R}$ also $\lambda \begin{pmatrix} r_1 & s_1 \\ r_1 & 0 \end{pmatrix} = \begin{pmatrix} \lambda r_1 & \lambda s_1 \\ r_1 & 0 \end{pmatrix}$ belongs to X_0, so X_0 is closed with respect to the sum and with respect to the product by scalars, so it is a vector subspace of $M_2(\mathbb{R})$.

In conclusion, X_k is a subspace of $M_2(\mathbb{R})$ for $k = 0$.

2.6   SUGGESTED EXERCISES

2.6.1 Determine which of the following subsets of vector spaces are subspaces:

i) $S = \{(x, y, z) \in \mathbb{R}^3 \mid x = 0\}$.

ii) $T = \{(x, y, z) \in \mathbb{R}^3 \mid x = y\}$.

iii) $W_n = \{p(x) \in \mathbb{R}[x] \mid \deg(p(x)) = n\}$, $N \in \mathbb{N}$. (Here deg $(p(x))$ indicates the degree of the polynomial $p(x)$.

iv) $D = \left\{ \begin{pmatrix} a & 0 \\ 0 & d \end{pmatrix} \in M_2(\mathbb{R}) \right\}$.

v) $T = \left\{ \begin{pmatrix} a & b \\ 0 & d \end{pmatrix} \in M_2(\mathbb{R}) \right\}$.

vi) $A = \{(a_{ij})_{\substack{i=1,\dots,3 \\ j=1,\dots,3}} \in M_3(\mathbb{R}) \mid a_{11} + a_{22} + a_{33} = 0\}$.

vii) $X = \{(x, y, z) \in \mathbb{R}^3 \mid x^2 + y = 0\}$.

viii) $X = \{(x, y, z) \in \mathbb{R}^3 \mid x + y + z = -1\}$.

ix) $X = \{p(x) = 3x^2 + rx + s \mid r, s \in \mathbb{R}\} \subseteq \mathbb{R}_2[x]$.

x) $X = \left\{ \begin{pmatrix} 0 & r \\ 2r & s \end{pmatrix} \mid r, s \in \mathbb{R} \right\} \subseteq M_2(\mathbb{R})$.

xi) $X = \left\{ \begin{pmatrix} r & 2r \\ r & r^2 \end{pmatrix} \mid r \in \mathbb{R} \right\} \subseteq M_2(\mathbb{R}).$

2.6.2 Show that the set of solutions of the homogeneous linear system

$$\begin{cases} x_1 + 3x_2 - x_3 = 0 \\ 2x_1 + 4x_2 - 4x_3 - x_4 = 0 \\ x_2 + x_3 + 2x_4 = 0 \end{cases}$$

in the unknowns x_1, x_2, x_3, x_4 is a subspace of $\mathbb{R}^4$.

2.6.3 Establish if the set $X = \{p(x) \in \mathbb{R}_3[x] \mid p(-1) = 0\}$ is a subspace of $\mathbb{R}_3[x]$, where $p(-1)$ means the value of the polynomial calculated in -1.

2.6.4 Determine whether the set $X = \{p(x) \in \mathbb{R}_2[x] \mid p(1) = -1\}$ is a subspace of $\mathbb{R}_2[x]$.

2.6.5 Determine if the set $X = \{g : \mathbb{R} \to \mathbb{R} \mid g$ is continuous and differentiable in $x = 2\}$ is a subspace of the vector space of continuous functions $f : \mathbb{R} \to \mathbb{R}$.

2.6.6 Determine if the set $X = \{g : \mathbb{R} \to \mathbb{R} \mid g$ is continuous but not differentiable in $x = 0\}$ and is a subspace of the vector space of continuous functions $f : \mathbb{R} \to \mathbb{R}$.

2.6.7 Write, if possible, the set $S = \{(x, y) \in \mathbb{R}^2 \mid x^2 + xy - 2y^2 = 0\}$ as a union of two subspaces of $\mathbb{R}^2$ and say if S is a subspace of $\mathbb{R}^2$.

2.6.8 We call *sequence* of elements in $\mathbb{R}$ any function $s : \mathbb{N} \to \mathbb{R}$. If $s(n) = a_n$, the sequence is also indicated with (a_n). On the set $\mathcal{S}_\mathbb{R}$ of all sequences with elements in $\mathbb{R}$, we define the following operations:

$$(a_n) + (b_n) = (a_n + b_n), \qquad k(a_n) = (ka_n)$$

for every $(a_n), (b_n) \in \mathcal{S}_\mathbb{R}$ and $k \in \mathbb{R}$. Show that with these operations $\mathcal{S}_\mathbb{R}$ is a vector space over $\mathbb{R}$.

2.6.9 Let $\mathcal{C}(\mathbb{R}; \mathbb{R})$ be the set of continuous functions from $\mathbb{R}$ to $\mathbb{R}$. Consider the operation of sum of functions and the operation of product of any function by a real number defined as in Example 2.3.3. Show that with these operations $\mathcal{C}(\mathbb{R}; \mathbb{R})$ is a vector space over $\mathbb{R}$.

CHAPTER 3

Linear Combination and Linear Independence

In the previous chapter, we have seen the definition of vector space and subspace. We now want to describe these objects in a more efficient way. We introduce for this purpose the concept of *linear combination* of a set of vectors and the concept of *linearly independent* vectors. These are two fundamental definitions within the theory of vector spaces, whose understanding is necessary to get to the key concepts of *basis* and *linear transformation*, which we we will treat later.

3.1   LINEAR COMBINATIONS AND GENERATORS

Every vector space $V \neq \{0\}$ contains infinitely many vectors; for if V contains a vector $\mathbf{v}$, it immediately must also contain all its multiples, i.e. $\lambda \mathbf{v} \in V$ for each $\lambda \in \mathbb{R}$. Let us see an example to better understand this fact.

Consider the subspace $W = \{(x, ax) | x \in \mathbb{R}\}$ in $\mathbb{R}^2$ discussed in the previous chapter. It is represented, in the Cartesian plane, by a line whose equation is $y = ax$. We can describe it, in an alternative way, as the set of all multiples of the vector $(1, a)$

$$W = \{x(1, a) \mid x \in \mathbb{R}\}.$$

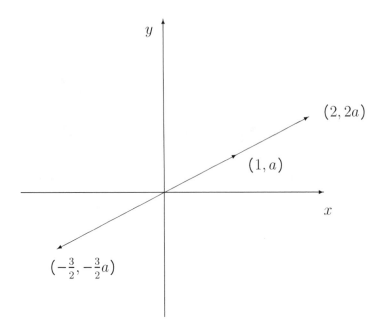

We say that the vector $(1, a)$ generates the subspace W represented by the line $y = x$. The word "generate" is not accidental since, in fact, all vectors of the subspace W are multiples of $(1, a)$. We also note that the choice of the vector $(1, a)$, as a generator of W, is arbitrary, we could as well have choosen any of its multiples, like $(2, 2a)$ or $(-\frac{3}{2}, -\frac{3}{2}a)$.

Graphically it is clear that if we know a point of a straight line (in the plane, but also in three-dimensional space) different from the origin, then we can immediately draw the line passing through it and the origin. We will see later that the fact of knowing the generators of a vector space allows us to determine it uniquely.

Now let us see another example. In $\mathbb{R}^2$, we consider the two vectors $(1, 0)$ and $(0, 1)$. We ask ourselves: what is the smallest subspace W of $\mathbb{R}^2$ that contains both of these vectors? From the previous reasoning, we know that this subspace must contain the two subspaces W_1 and W_2 generated by $(1, 0)$ and $(0, 1)$:

$$W_1 = \{\lambda(1, 0) | \lambda \in \mathbb{R}\} \qquad \text{represented by the } x\text{-axis}$$

$$W_2 = \{\mu(0, 1) | \mu \in \mathbb{R}\} \qquad \text{represented by the } y\text{-axis}$$

We also know that the sum of two vectors of W still belongs to W (by the definition of subspace). For instance $(1, 0) + (0, 1) = (1, 1) \in W$, but also $(1, 2) + (3, 4) = (4, 6) \in W$. The student is invited to draw vectors sums in $\mathbb{R}^2$ considering the points of the plan associated with them and using the parallelogram rule. In this way, we can convince ourselves that actually $W = \mathbb{R}^2$. But the graphic construction is not sufficient to prove this fact, as it is not possible draw all the vectors of the plane, so let us look at an algebraic proof. We take the generic vector $(\lambda, 0)$ in W_1 and the generic vector $(0, \mu)$ in W_2, and we take their sum: $(\lambda, 0) + (0, \mu) = (\lambda, \mu)$. It is clear that *all* vectors (x, y) in $\mathbb{R}^2$ can be written in this way choosing $\lambda = x$ and $\mu = y$. So we found that the smallest subspace of $\mathbb{R}^2$ containing the vectors $(1, 0)$ and $(0, 1)$ is all $\mathbb{R}^2$.

Now we formalize the concept of *generation* of subspace, which we have described with the previous examples.

Definition 3.1.1 Assume that V is a vector space, $\mathbf{v}_1, \ldots, \mathbf{v}_n$ are vectors of V and $\lambda_1, \ldots, \lambda_n \in \mathbb{R}$. The vector $\mathbf{w} = \lambda_1 \mathbf{v}_1 + \cdots + \lambda_n \mathbf{v}_n$ is said to be a *linear combination* of $\mathbf{v}_1, \ldots, \mathbf{v}_n$ with scalars $\lambda_1, \ldots, \lambda_n$.

For example, $(1, 1)$ is a linear combination of $(1, 0)$ and $(0, 1)$ with scalars $\lambda_1 = 1$ and $\lambda_2 = 1$, but also a linear combination of $(2, 1)$ and $(1, 0)$ with scalars $\lambda_1 = 1$ and $\lambda_2 = -1$.

We now come to the concept of vector space generated by some vectors, the main concept of this chapter along with that of linear independence.

Definition 3.1.2 Let V be a vector space and let $\{\mathbf{v}_1, \ldots, \mathbf{v}_n\}$ be a set of vectors of V. The *subspace generated* (or *spanned*) by the vectors $\mathbf{v}_1, \ldots, \mathbf{v}_n$ is the set of all their linear combinations, in symbols

$$\langle \mathbf{v}_1, \ldots, \mathbf{v}_n \rangle = \{\lambda_1 \mathbf{v}_1 + \cdots + \lambda_n \mathbf{v}_n \mid \lambda_1, \ldots, \lambda_n \in \mathbb{R}\}.$$

We have seen that, for example, the subspace generated by a nonzero vector in $\mathbb{R}^2$ corresponds to a straight line, while the subspace generated by the two vectors $(1, 0)$ and $(0, 1)$ of $\mathbb{R}^2$ is all $\mathbb{R}^2$.

Observation 3.1.3 If V is a vector space and $\mathbf{v} \in V$, then the subspace generated by $\mathbf{v}$ is the set of multiples of $\mathbf{v}$, i.e. $\langle \mathbf{v} \rangle = \{\lambda \mathbf{v} \mid \lambda \in \mathbb{R}\}$.

Moreover, the subspace generated by the zero vector is the trivial subspace, which contains only the zero vector: $\langle \mathbf{0} \rangle = \{\mathbf{0}\}$.

Definition 3.1.4 Let V be a vector space and let $\{\mathbf{v}_1, \ldots, \mathbf{v}_n\}$ be a set of vectors of V. We say that $\mathbf{v}_1, \ldots, \mathbf{v}_n$ *generate* V, or $\{\mathbf{v}_1, \ldots, \mathbf{v}_n\}$ is a *set of generators of V* if $V = \langle \mathbf{v}_1, \ldots, \mathbf{v}_n \rangle$.

In the example above, we saw that the vectors $(1, 0)$ and $(0, 1)$ generate the vector space $\mathbb{R}^2$, as each vector (a, b) of $\mathbb{R}^2$ can be written as a linear combination of $(1, 0)$ and $(0, 1)$:

$$(a, b) = a(1, 0) + b(0, 1).$$

Proposition 3.1.5 *Let V be a vector space and let $\{\mathbf{v}_1, \ldots, \mathbf{v}_n\}$ be a set of vectors of V. Then we have that $\langle \mathbf{v}_1, \ldots, \mathbf{v}_n \rangle$ is a subspace of V. Moreover, if Z is a subspace of V containing $\mathbf{v}_1, \ldots \mathbf{v}_n$, then $\langle \mathbf{v}_1, \ldots, \mathbf{v}_n \rangle \subseteq Z$, therefore $\langle \mathbf{v}_1, \ldots, \mathbf{v}_n \rangle$ is the smallest subspace of V containing $\{\mathbf{v}_1, \ldots, \mathbf{v}_n\}$.*

Proof. First of all, we note that $\mathbf{0} \in \langle \mathbf{v}_1, \ldots, \mathbf{v}_n \rangle$, as $\mathbf{0} = 0\mathbf{v}_1 + \cdots + 0\mathbf{v}_n$. Now let $\mathbf{v}, \mathbf{w} \in \langle \mathbf{v}_1, \ldots, \mathbf{v}_n \rangle$. Then by definition there exist scalars $\alpha_1, \ldots, \alpha_n$ and $\beta_1, \ldots, \beta_n$ such that:

$$\mathbf{v} = \alpha_1 \mathbf{v}_1 + \cdots + \alpha_n \mathbf{v}_n, \quad \mathbf{w} = \beta_1 \mathbf{v}_1 + \cdots + \beta_n \mathbf{v}_n$$

thus

$$\mathbf{v} + \mathbf{w} = (\alpha_1 + \beta_1)\mathbf{v}_1 + \cdots + (\alpha_n + \beta_n)\mathbf{v}_n \in \langle \mathbf{v}_1, \ldots, \mathbf{v}_n \rangle.$$

Moreover if $k \in \mathbb{R}$

$$k\mathbf{v} = (k\alpha_1)\mathbf{v}_1 + \cdots + (k\alpha_n)\mathbf{v}_n \in \langle \mathbf{v}_1, \ldots, \mathbf{v}_n \rangle.$$

This proves that $\langle \mathbf{v}_1, \ldots, \mathbf{v}_n \rangle$ is a subspace of V.

Now we prove that $\langle \mathbf{v}_1, \ldots, \mathbf{v}_n \rangle$ is the smallest subspace of V containing $\{\mathbf{v}_1, \ldots, \mathbf{v}_n\}$. Let $\mathbf{v} = \lambda_1 \mathbf{v}_1 + \cdots + \lambda_n \mathbf{v}_n \in \langle \mathbf{v}_1 \ldots \mathbf{v}_n \rangle$, and let Z be a subspace of V containing $\mathbf{v}_1, \ldots \mathbf{v}_n$. Then Z contains also $\lambda_1 \mathbf{v}_1, \ldots, \lambda_n \mathbf{v}_n$ because, being a vector space, if Z contains a vector, it also contains all of its multiples. Moreover, as Z is closed with respect to the sum, it also contains $\lambda_1 \mathbf{v}_1 + \cdots + \lambda_n \mathbf{v}_n = \mathbf{v}$. So $\langle \mathbf{v}_1, \ldots, \mathbf{v}_n \rangle \subseteq Z$. ■

Let us look at an example that is linked to what we have seen in Chapter 1 about the solution of linear systems depending on a parameter.

Example 3.1.6 We want to determine the subspace generated by the vectors $(1, 1)$, $(2, k)$, depending on the parameter k.

$$\langle (1, 1), (2, k) \rangle = \{\lambda_1 (1, 1) + \lambda_2 (2, k) | \lambda_1, \lambda_2 \in \mathbb{R}\}$$

$$= \{(\lambda_1 + 2\lambda_2, \lambda_1 + k\lambda_2) | \lambda_1, \lambda_2 \in \mathbb{R}\}.$$

Since we are considering vectors in $\mathbb{R}^2$ we can represent the vectors with points of the Cartesian plane. The following diagram illustrates the vectors $(1, 1)$ and $(2, k)$ for the values of $k = 1$ and $k = 2$.

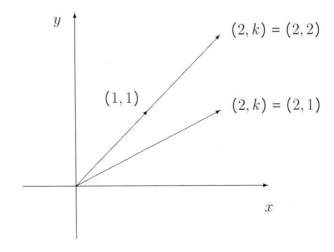

We see at once that if $k = 2$, then the two points lie on the same line through the origin, thus the smallest subspace that contains both of them will be precisely this line, whose equation is $y = x$.

If $k \neq 2$, the two points lie on two distinct lines through the origin, then the smallest subspace that contains both of them must contain such lines, and also the sum of any two points on these lines. Therefore, with a reasoning similar to the one made at the beginning of this chapter, we have that the smallest subspace that contains both points whose coordinates are $(1,1)$, $(2,k)$ is the whole plane, i.e. the vectors $(1,1)$, $(2,k)$ generate $\mathbb{R}^2$.

Let us now see an algebraic proof of this fact. Let (a,b) be a generic vector of $\mathbb{R}^2$. We want to determine when (a,b) belongs to $\langle (1,1),(2,k) \rangle$, that is, when there exist $\lambda_1, \lambda_2 \in \mathbb{R}$ such that

$$(\lambda_1 + 2\lambda_2, \lambda_1 + k\lambda_2) = (a,b).$$

In other words we have to solve the linear system:

$$\begin{cases} \lambda_1 + 2\lambda_2 = a \\ \lambda_1 + k\lambda_2 = b \end{cases}$$

We leave as an exercise to verify that this system in the unknowns λ_1, λ_2 always has a solution if $k \neq 2$. If instead $k = 2$ the complete matrix associated to the system is:

$$\begin{pmatrix} 1 & 2 & a \\ 1 & 2 & b \end{pmatrix},$$

whose echelon form becomes

$$\begin{pmatrix} 1 & 2 & a \\ 0 & 0 & b-a \end{pmatrix}.$$

So if $a \neq b$, the system does not have solutions, i.e. we have that $(a,b) \notin \langle (1,1),(2,2) \rangle$; if instead $a = b$ the system has solutions, that is, $(a,a) \in \langle (1,1),(2,2) \rangle$. Then $\langle (1,1),(2,2) \rangle$ is the set of vectors that have first coordinate equal to the second, i.e. $\langle (1,1),(2,2) \rangle = \{(a,a)|a \in \mathbb{R}\}$.

Example 3.1.7 We now slightly modify the previous example. We want to determine the subspace generated by the vectors: $(1,1)$, $(2,k)$, $(-1,-1)$, depending on the parameter k:

$$\langle (1,1),(2,k),(-1,-1) \rangle =$$

$$\{\lambda_1(1,1) + \lambda_2(2,k) + \lambda_3(-1,-1)|\lambda_1, \lambda_2, \lambda_3 \in \mathbb{R}\} =$$

$$\{(\lambda_1 + 2\lambda_2 - \lambda_3, \lambda_1 + k\lambda_2 - \lambda_3)|\lambda_1, \lambda_2, \lambda_3 \in \mathbb{R}\}.$$

From the previous reasoning, we expect that these vectors almost always generate $\mathbb{R}^2$ (i.e. for almost every value of k), but let us see a rigorous proof of this fact. We want to show that for any fixed vector (a,b), we can always choose λ_1, λ_2, λ_3 such that

$$(\lambda_1 + 2\lambda_2 - \lambda_3, \lambda_1 + k\lambda_2 - \lambda_3) = (a,b).$$

If we solve the linear system, depending on the parameter k, with the Gaussian algorithm, an easy calculation shows that this system always has solution (for every a and b fixed) provided we have $k \neq 2$. But when $k = 2$:

$$(\lambda_1 + 2\lambda_2 - \lambda_3, \lambda_1 + 2\lambda_2 - \lambda_3) = (a, b),$$

therefore, in order for the system to admit solutions, it must necessarily be $a = b$, so the only vectors belonging to the subspace spanned by the two given vectors are those of the type (a, a), namely those corresponding to points lying on the straight line of equation $y = x$. In fact, an accurate analysis could immediately give the answer to this problem without any calculation: it is enough to notice that the vector $(-1, -1)$ is *redundant* in calculating the subspace, as it belongs to the subspace generated by $(1, 1)$. Graphically, it is clear that the point P of coordinates $(-1, -1)$ belongs to the line through the origin and through the point Q of coordinates $(1, 1)$, thus every subspace of the plane containing Q, automatically contains also P. So we could safely ignore $(-1, -1)$ and give right away as an answer the solution of Example 3.1.6.

We have thus seen that, in describing a subspace by using a set of generators, some vectors are *redundant*, i.e. if we eliminate them, the subspace does not change. This happens, for example, when we have that a vector is a multiple of another but also when a vector is the sum of two vectors. For example, we have seen that $\langle (1, 0), (0, 1) \rangle = \mathbb{R}^2$ but also (as the student can verify directly):

$$\langle (1, 0), (0, 1), (1, 1) \rangle = \mathbb{R}^2.$$

This fact is formalized by the following proposition.

Proposition 3.1.8 *Assume that V is a vector space, $\mathbf{v}_1, \ldots, \mathbf{v}_n$ are vectors of V and $\mathbf{w}$ is a linear combination of them, namely: $\mathbf{w} = \lambda_1 \mathbf{v}_1 + \cdots + \lambda_n \mathbf{v}_n$. Then*

$$\langle \mathbf{v}_1, \ldots, \mathbf{v}_n \rangle = \langle \mathbf{v}_1, \ldots, \mathbf{v}_n, \mathbf{w} \rangle.$$

Conversely if

$$\langle \mathbf{v}_1, \ldots, \mathbf{v}_n \rangle = \langle \mathbf{v}_1, \ldots, \mathbf{v}_n, \mathbf{w} \rangle$$

then $\mathbf{w}$ is a linear combination of $\mathbf{v}_1, \ldots, \mathbf{v}_n$.

Proof. In order to show the first part of the result, it is enough to observe that $\mathbf{w} \in \langle \mathbf{v}_1, \ldots, \mathbf{v}_n \rangle$ by assumption, so it follows from Proposition 3.1.5 that $Z = \langle \mathbf{v}_1, \ldots, \mathbf{v}_n \rangle$ is a subspace containing $\{\mathbf{v}_1, \ldots, \mathbf{v}_n, \mathbf{w}\}$, thus $\langle \mathbf{v}_1, \ldots, \mathbf{v}_n, \mathbf{w} \rangle \subseteq \langle \mathbf{v}_1, \ldots, \mathbf{v}_n \rangle$, again by Proposition 3.1.5. The inclusion $\langle \mathbf{v}_1, \ldots, \mathbf{v}_n \rangle \subseteq \langle \mathbf{v}_1, \ldots, \mathbf{v}_n, \mathbf{w} \rangle$ is obvious,

To show the converse, it is enough to note that since $\langle \mathbf{v}_1, \ldots, \mathbf{v}_n \rangle = \langle \mathbf{v}_1, \ldots, \mathbf{v}_n, \mathbf{w} \rangle$ we have that $\mathbf{w} \in \langle \mathbf{v}_1, \ldots, \mathbf{v}_n \rangle$, i.e. $\mathbf{w}$ is a linear combination of $\mathbf{v}_1, \ldots, \mathbf{v}_n$. ■

3.2 LINEAR INDEPENDENCE

Now we ask: How do we establish what the "redundant vectors" are in the description of the subspace generated by a set of vectors? If we want be efficient in the description of a subspace, we must be able to describe it as the subspace generated by the smallest possible number of vectors. The answer to this question comes from the concept of *linear independence*. This is by far the most difficult concept to understand, and it is the cornerstone of the whole theory. Shortly, the story is: if a set of generators of a subspace is a set of linearly independent vectors, then we are sure that it is the most efficient way to describe the subspace generated by those vectors, namely that we are using the smallest number of vectors. Let us see the definition and then with a series of small steps, we will arrive to prove the assertion above.

Definition 3.2.1 Let V be a vector space. The vectors $\mathbf{v}_1, \ldots, \mathbf{v}_n \in V$ are *linearly independent* if for every linear combination

$$\lambda_1 \mathbf{v}_1 + \cdots + \lambda_n \mathbf{v}_n = \mathbf{0},$$

we have $\lambda_1 = \cdots = \lambda_n = 0$. In other words, the only linear combination of the vectors $\mathbf{v}_1, \ldots, \mathbf{v}_n$ giving the zero vector is the one with all zero scalars. We will say also that the set of vectors $\{\mathbf{v}_1, \ldots, \mathbf{v}_n\}$ is linearly independent.[1]

The vectors $\mathbf{v}_1, \ldots, \mathbf{v}_n$ are *linearly dependent*, if they are not independent. In other words, the vectors of the set $\{\mathbf{v}_1, \ldots, \mathbf{v}_n\}$ are linearly dependent if there exist scalars $\lambda_1, \ldots, \lambda_n$, not all zero, such that $\lambda_1 \mathbf{v}_1 + \cdots + \lambda_n \mathbf{v}_n = \mathbf{0}$.

Let us review the previous examples. The set of vectors $\{(1; 0), (0, 1)\}$ in $\mathbb{R}^2$ is a set of vectors which are linearly independent, as their only linear combination that gives the zero vector is obtained with all zero scalars:

$$\alpha(1, 0) + \beta(0, 1) = (\alpha, \beta) = (0, 0),$$

only if $\alpha = \beta = 0$.

On the other hand, the vectors of the set $\{(1, 0), (0, 1), (1, 1)\}$ are linearly dependent, because there is a linear combination of the given vectors with scalars, not all zero, which is equal to the zero vector.

$$1 \cdot (1, 0) + 1 \cdot (0, 1) - 1 \cdot (1, 1) = (0, 0).$$

Let us see a more elaborate example.

Example 3.2.2 Consider the following set of vectors in $\mathbb{R}_2[x]$: $\{x+1, x^2-1, 2, x-1\}$. Is this a set of linearly independent vectors? If we knew something more about linear algebra the answer would be immediate, for the moment we have to perform the

---

[1]The words "linearly independent" can be used for the vectors $\mathbf{v}_1, \ldots, \mathbf{v}_n$, but also for the set of vectors $\{\mathbf{v}_1, \ldots, \mathbf{v}_n\}$ indifferently, i.e. the two terminologies have the same meaning.

calculations. We write a generic linear combination of these vectors, and we put it equal to the zero vector:

$$\alpha_1(x + 1) + \alpha_2(x^2 - 1) + 2\alpha_3 + \alpha_4(x - 1) = 0.$$

From which:

$$\alpha_2 x^2 + (\alpha_1 + \alpha_4)x + (\alpha_1 - \alpha_2 + 2\alpha_3 - \alpha_4) = 0.$$

A polynomial is zero if and only if all its coefficients are zero, so we obtain the linear system:

$$\begin{cases} \alpha_2 = 0 \\ \alpha_1 + \alpha_4 = 0 \\ \alpha_1 - \alpha_2 + 2\alpha_3 - \alpha_4 = 0 \end{cases}.$$

We leave as an exercise for the student to verify that this system admits infinitely many solutions. For example, it has the solution: $\alpha_1 = 1$, $\alpha_2 = 0$, $\alpha_3 = -1$, $\alpha_4 = -1$. So we can explicitly write a linear combination of the given vectors, which is equal to the zero vector, while the scalars are not all zero:

$$1 \cdot (x + 1) - 1 \cdot 2 - 1 \cdot (x - 1) = 0.$$

So the given vectors are linearly dependent.

When we have a linear combination of this type, we can always write one of the vectors as a linear combination of the others, for example in our case we have:

$$(x + 1) = 2 + (x - 1)$$

Of course, in a set of linearly dependent vectors, it is not true that each of the given vectors can be expressed as a function of the others; for example, we see that there is no way to express $x^2 - 1$ as a linear combination of the others.

The important thing to note is that, in a set of linearly dependent vectors, if we eliminate one vector that is a linear combination of the others, the subspace they generate does not change (see Proposition 3.1.8), and the vectors of the new set thus obtained may have become linearly independent. Be careful, however, that this is not always the case, for example, in the set $\{2x, 3x, 4x\}$ even if we eliminate a vector, the remaining vectors are linearly dependent, as the student may verify. Somehow the linear independence tells us that we have reached the smallest number of vectors to describe the subspace. This concept will be explored very carefully in the next chapter on bases.

Observation 3.2.3 We note that, if a set of vectors contains the zero vector, then it is always a set of linearly dependent vectors. In fact, if we consider the set $\{\mathbf{v}_1 = \mathbf{0}, \mathbf{v}_2 \ldots, \mathbf{v}_n\}$ we have that;

$$1\mathbf{v}_1 + 0\mathbf{v}_2 + \cdots + 0\mathbf{v}_n = \mathbf{0}.$$

So we obtained a linear combination equal to the zero vector, while the first scalar is not zero.

Proposition 3.2.4 *In a vector space V, the vectors $\mathbf{v}_1, \ldots, \mathbf{v}_n$ are linearly dependent if and only if at least one of them is a linear combination of the others.*

Proof. Suppose that $\mathbf{v}_1, \ldots, \mathbf{v}_n$ are linearly dependent. Then, there exist scalars $\alpha_1, \ldots, \alpha_n \in \mathbb{R}$, not all zero, such that

$$\alpha_1 \mathbf{v}_1 + \cdots + \alpha_n \mathbf{v}_n = \mathbf{0}.$$

Since at least one of the scalars is nonzero, we have that $\alpha_k \neq \mathbf{0}$ for some k. Then:

$$\mathbf{v}_k = -\frac{\alpha_1}{\alpha_k}\mathbf{v}_1 - \cdots - \frac{\alpha_{k-1}}{\alpha_k}\mathbf{v}_{k-1} - \frac{\alpha_{k+1}}{\alpha_k}\mathbf{v}_{k+1} - \cdots - \frac{\alpha_n}{\alpha_k}\mathbf{v}_n,$$

and therefore $\mathbf{v}_k$ is a linear combination of the other vectors.

Conversely, suppose that there are scalars $\alpha_1, \ldots, \alpha_{k-1}, \alpha_{k+1}, \ldots, \alpha_n$, such that

$$\mathbf{v}_k = \alpha_1 \mathbf{v}_1 + \cdots + \alpha_{k-1}\mathbf{v}_{k-1} + \alpha_{k+1}\mathbf{v}_{k+1} + \cdots + \alpha_n \mathbf{v}_n,$$

then

$$\alpha_1 \mathbf{v}_1 + \cdots + \alpha_{k-1}\mathbf{v}_{k-1} + (-1)\mathbf{v}_k + \alpha_{k+1}\mathbf{v}_{k+1} + \cdots + \alpha_n \mathbf{v}_n = 0,$$

and at least one of the coefficients is not zero, that of $\mathbf{v}_k$. So the vectors $\mathbf{v}_1, \ldots, \mathbf{v}_n$ are linearly dependent. ∎

We see a particular case of this proposition, very useful in the exercises.

Proposition 3.2.5 *Two vectors are linearly independent if and only if neither is a multiple of the other.*

The proof of this proposition is immediate: just use the previous proposition and remember that a vector is a linear combination of another if and only if it is one of its multiples.

Observation 3.2.6 In order to prove that the vectors of a set are linearly dependent, it is enough to find a vector that is a linear combination of the others. For example if we see that a vector is a multiple of another, then we know that the vectors of the set are linearly dependent. The vectors of the following sets are linearly dependent, and we can verify it without any calculation (but the student should do it if he does not see why and wants to convince himself!).

- In $\mathbb{R}^3$: $\{(1,0,0), (0,1,0), (0,0,1), (0,2,0)\}$.

- In $\mathbb{R}^3$: $\{(1,0,0), (0,1,0), (-1,2,0), (1,2,1)\}$.

- In $\mathbb{R}_3[x]$: $\{2, x+7, x^3 - 3x, 1\}$.

- In $\mathbb{R}_3[x]$: $\{0, x, 1-x, x^3\}$.

- In $M_2(\mathbb{R})$:
$$\left\{ \begin{pmatrix} 1 & 0 \\ 0 & 1 \end{pmatrix}, \begin{pmatrix} 0 & 0 \\ 0 & 3 \end{pmatrix}, \begin{pmatrix} 0 & 0 \\ 0 & \sqrt{2} \end{pmatrix}, \begin{pmatrix} 3 & 0 \\ 0 & 3 \end{pmatrix} \right\}.$$

Observation 3.2.7 Although it is true that in a set of vectors it is enough to find out that a vector is a multiple of another one in order to prove that the vectors are linearly dependent, the vice versa *is not true!* As shown in Example 3.2.2, we can have a set of linearly dependent vectors where *no one is a multiple of another.*

The next proposition shows that removing some vectors from a set of linearly independent vectors, we get a set of vectors that are still linearly independent.

Proposition 3.2.8 *A non-empty subset of a set of linearly independent vectors consists of vectors which are still linearly independent.*

Proof. Suppose, by contradiction, that I is a set of linearly independent vectors and that $\varnothing \neq J \subseteq I$ is a subset of linearly dependent vectors. Then there exists a vector in J which is written as a linear combination of the others. But then it is also expressed as a linear combination of vectors of I and then I is a set of linearly dependent vectors, contradicting the hypothesis. ■

Example 3.2.9 Suppose we want to determine for which values of k the following vectors in $M_2(\mathbb{R})$:

$$\begin{pmatrix} 1 & 0 \\ 1 & -3 \end{pmatrix}, \begin{pmatrix} 6 & 0 \\ k & -18 \end{pmatrix}, \begin{pmatrix} k & 0 \\ 1 & 5 \end{pmatrix},$$

are linearly independent. We proceed as suggested by the definition: we write a linear combination of the given vectors, and we see if there are nonzero scalars that allow us to get the zero vector. For the values of k for which this happens, we will have that the given vectors are linearly dependent.

So we write a generic linear combination of the given vectors and set it equal to the zero vector:

$$\lambda_1 \begin{pmatrix} 1 & 0 \\ 1 & -3 \end{pmatrix} + \lambda_2 \begin{pmatrix} 6 & 0 \\ -18 & k \end{pmatrix} + \lambda_3 \begin{pmatrix} k & 0 \\ 1 & 5 \end{pmatrix} = \begin{pmatrix} 0 & 0 \\ 0 & 0 \end{pmatrix} \tag{3.1}$$

from which:

$$\begin{pmatrix} \lambda_1 + 6\lambda_2 + k\lambda_3 & 0 \\ \lambda_1 + k\lambda_2 + \lambda_3 & -3\lambda_1 - 18\lambda_2 + 5\lambda_3 \end{pmatrix} = \begin{pmatrix} 0 & 0 \\ 0 & 0 \end{pmatrix}.$$

Then equality (3.1) is satisfied if and only if $\lambda_1, \lambda_2, \lambda_3$ are solutions of the homogeneous linear system:

$$\begin{cases} \lambda_1 + 6\lambda_2 + k\lambda_3 = 0 \\ \lambda_1 + k\lambda_2 + \lambda_3 = 0 \\ -3\lambda_1 - 18\lambda_2 + 5\lambda_3 = 0. \end{cases}$$

The complete matrix associated with the system is:

$$(A|b) = \left(\begin{array}{ccc|c} 1 & 6 & k & 0 \\ 1 & k & 1 & 0 \\ -3 & -18 & 5 & 0 \end{array} \right),$$

and, if we reduce it in the echelon form with the Gaussian algorithm, it becomes:

$$(A'|\underline{b}') = \begin{pmatrix} 1 & 6 & k & | & 0 \\ 0 & k-6 & 1-k & | & 0 \\ 0 & 0 & 5+3k & | & 0 \end{pmatrix}.$$

Note that the system always admits the zero solution $\lambda_1 = \lambda_2 = \lambda_3 = 0$. The question is whether there are also nonzero solutions or not. The row echelon form of the matrix is particularly appropriate to understand whether or not we have only the zero solution. We can immediately observe that if the initial vectors were 5, of course at least one of the 5 initial unknowns (the scalars that give the zero linear combination) would be indeterminate, in other words the vectors would definitely be linearly dependent, because we could arbitrarily assign a nonzero value to that unknown. In the next chapter, using the concept of basis, we will formalize this reasoning which, however, even now should be clear intuitively.

Returning to the example in question, if $k \neq 6$ and $k \neq \frac{3}{5}$, we have that $\mathrm{rr}(A') = \mathrm{rr}(A'|\underline{b}') = 3$ is equal to the number of unknowns, thus the system admits a unique solution $\lambda_1 = \lambda_2 = \lambda_3 = 0$, and the given vectors are therefore linearly independent.

If $k = 6$, further reducing the matrix in row echelon form we obtain:

$$(A''|\underline{b}'') = \begin{pmatrix} 1 & 6 & 6 & | & 0 \\ 0 & 0 & -5 & | & 0 \\ 0 & 0 & 0 & | & 0 \end{pmatrix},$$

thus $\mathrm{rr}(A'') = \mathrm{rr}(A''|\underline{b}'') = 2$, and the system admits infinitely many solutions depending on one parameter. In particular, there are nonzero solutions, and the given vectors are linearly dependent.

Finally, if $k = -\frac{3}{5}$ we get:

$$(A'|\underline{b}') = \begin{pmatrix} 1 & 6 & -\frac{3}{5} & | & 0 \\ 0 & -\frac{33}{5} & -\frac{8}{5} & | & 0 \\ 0 & 0 & 0 & | & 0 \end{pmatrix}.$$

Thus $\mathrm{rr}(A') = \mathrm{rr}(A'|\underline{b}') = 2$, and as before the given vectors are linearly dependent.

We conclude this chapter with some exercises that clarify the techniques for verifying linear dependence or independence and the concept of generators.

3.3 EXERCISES WITH SOLUTIONS

3.3.1 Determine for which values of k the vectors $x^2 + 2x + k$, $5x^2 + 2kx + k^2$, $kx^2 + x + 3$ generate $\mathbb{R}_2[x]$.

Solution. Consider the generic vector $ax^2 + bx + c \in \mathbb{R}_2[x]$, and see when $ax^2 + bx + c \in \langle x^2 + 2x + k, 5x^2 + 2kx + k^2, kx^2 + x + 3 \rangle$. This happens if we have

$$ax^2 + bx + c = \lambda_1(x^2 + 2x + k) + \lambda_2(5x^2 + 2kx + k^2) + \lambda_3(kx^2 + x + 3),$$

for some $\lambda_1, \lambda_2, \lambda_3 \in \mathbb{R}$, that is if

$$ax^2 + bx + c = (\lambda_1 + 5\lambda_2 + k\lambda_3)x^2 + (2\lambda_1 + 2k\lambda_2 + \lambda_3)x + (k\lambda_1 + k^2\lambda_2 + 3\lambda_3).$$

So $\lambda_1, \lambda_2, \lambda_3$ must be solutions of the system:

$$\begin{cases} \lambda_1 + 5\lambda_2 + k\lambda_3 = a \\ 2\lambda_1 + 2k\lambda_2 + \lambda_3 = b \\ k\lambda_1 + k^2\lambda_2 + 3\lambda_3 = c. \end{cases}$$

The complete matrix associated with the system is:

$$(A|\underline{b}) = \begin{pmatrix} 1 & 5 & k & a \\ 2 & 2k & 1 & b \\ k & k^2 & 3 & c \end{pmatrix},$$

which reduced to row echelon form with Gaussian algorithm becomes:

$$(A'|\underline{b}') = \begin{pmatrix} 1 & 5 & k & a \\ 0 & 2k-10 & 1-2k & b-2a \\ 0 & 0 & 3-\frac{1}{2}k & c-\frac{b}{2}k \end{pmatrix}.$$

If $k \neq 5$ and $k \neq 6$, we have $\mathrm{rr}(A') = \mathrm{rr}(A'|\underline{b}') = 3$, therefore the system admits solution, independently from the values of a, b, c, therefore each vector of the type $ax^2 + bx + c$ belongs to $\langle x^2 + 2x + k, 5x^2 + 2kx + k^2, kx^2 + x + 3 \rangle$, and the given vectors generate $\mathbb{R}_2[x]$.

If $k = 5$, further reducing the matrix to row echelon form, we obtain:

$$(A''|\underline{b}'') = \begin{pmatrix} 1 & 5 & 5 & a \\ 0 & 0 & 1 & 2c-5b \\ 0 & 0 & 0 & -a-22b+9c \end{pmatrix},$$

and if $-a - 22b + 9c \neq 0$ the system does not admit solution, because $3 = \mathrm{rr}(A'') \neq \mathrm{rr}(A''|\underline{b}'') = 2$. This means that if $-a-22b+9c \neq 0$ it is not possible to write ax^2+bx+c as a linear combination of $x^2+2x+k, 5x^2+2kx+k^2, kx^2+x+3$, thus the given vectors do not generate $\mathbb{R}_2[x]$. For example, $x^2-x+1 \notin \langle x^2+2x+k, 5x^2+2kx+k^2, kx^2+x+3 \rangle$.

If $k = 6$ we obtain

$$(A'|\underline{b}') = \begin{pmatrix} 1 & 5 & 6 & a \\ 0 & 2 & -11 & b-2a \\ 0 & 0 & 0 & c-3b \end{pmatrix},$$

and if $c \neq 3b$ the system does not admit solution, thus the given vectors do not generate $\mathbb{R}_2[x]$. For example $3x^2 + 5x - 8 \notin \langle x^2 + 2x + k, 5x^2 + 2kx + k^2, kx^2 + x + 3 \rangle$.

3.3.2 Let W be the vector subspace of $\mathbb{R}^4$ given by the set of solutions of the homogeneous linear system:

$$\begin{cases} x_1 + x_2 - x_4 = 0 \\ 2x_1 + x_2 - x_3 + 3x_4 = 0 \end{cases}$$

in the unknowns x_1, x_2, x_3, x_4. Determine, if possible, a finite set of generators of W.

Solution. The complete matrix associated to the system is

$$(A|b) = \begin{pmatrix} 1 & 1 & 0 & -1 & | & 0 \\ 2 & 1 & -1 & 3 & | & 0 \end{pmatrix},$$

which reduced to row echelon form with the Gaussian algorithm becomes:

$$(A'|\underline{b}') = \begin{pmatrix} 1 & 1 & 0 & -1 & | & 0 \\ 0 & -1 & -1 & 5 & | & 0 \end{pmatrix}.$$

The system solutions are: $(x_3 - 4x_4, -x_3 + 5x_4, x_3, x_4)$, with $x_3, x_4 \in \mathbb{R}$.

To determine the generators of W we separate the free variables. Then

$$W = \{(x_3 - 4x_4, -x_3 + 5x_4, x_3, x_4) \,|\, x_3, x_4 \in \mathbb{R}\} =$$

$$\{(x_3, -x_3, x_3, 0) + (-4x_4, 5x_4, 0, x_4) \,|\, x_3, x_4 \in \mathbb{R}\} =$$

$$\{x_3(1, -1, 1, 0) + x_4(-4, 5, 0, 1) \,|\, x_3, x_4 \in \mathbb{R}\}.$$

We remember that $x_3, x_4 \in \mathbb{R}$ can take any real value. At this point, it is clear that $W = \langle (1, -1, 1, 0), (-4, 5, 0, 1) \rangle$, i.e. the vectors $(1, -1, 1, 0)$, $(-4, 5, 0, 1)$ generate W.

3.3.3 Determine for which values of k the following vectors of $\mathbb{R}^4$ are linearly dependent:

$$\mathbf{v}_1 = (2, 2k, k^2, 2k + 2), \quad \mathbf{v}_2 = (-1, -k, 2k + 2, -k - 1).$$

Choose one of these values of k and show that $\mathbf{v}_1 \in \langle \mathbf{v}_2 \rangle$.

Solution. We need to see if there are two nonzero scalars λ_1, λ_2, such that $\lambda_1 \mathbf{v}_1 + \lambda_2 \mathbf{v}_2 = \mathbf{0}$. It must happen that:

$$\lambda_1(2, 2k, k^2, 2k + 2) + \lambda_2(-1, -k, 2k + 2, -k - 1) = (0, 0, 0, 0),$$

i.e.

$$(2\lambda_1 - \lambda_2, 2k\lambda_1 - k\lambda_2, k^2\lambda_1 + (2k + 2)\lambda_2, (2k + 2)\lambda_1 - (k + 1)\lambda_2) = (0, 0, 0, 0),$$

that is, λ_1, λ_2 must satisfy the following homogeneous linear system:

$$\begin{cases} 2\lambda_1 - \lambda_2 = 0 \\ 2k\lambda_1 - k\lambda_2 = 0 \\ k^2\lambda_1 + (2k + 2)\lambda_2 = 0 \\ (2k + 2)\lambda_1 - (k + 1)\lambda_2 = 0, \end{cases}$$

The complete matrix associated with the system is:

$$(A|b) = \begin{pmatrix} 2 & -1 & | & 0 \\ 2k & -k & | & 0 \\ k^2 & 2k + 2 & | & 0 \\ 2k + 2 & -k - 1 & | & 0 \end{pmatrix},$$

which reduced to row echelon form with the Gaussian algorithm becomes:

$$(A'|\underline{b}') = \begin{pmatrix} 2 & -1 & 0 \\ 0 & (k+2)^2 & 0 \\ 0 & 0 & 0 \\ 0 & 0 & 0 \end{pmatrix}.$$

If $k \neq -2$, then $\mathrm{rr}(A') = \mathrm{rr}(A'|\underline{b}') = 2$ so the system has only one solution, which is $\lambda_1 = \lambda_2 = 0$, and the given vectors are linearly independent.

For $k = -2$ we get:

$$(A'|\underline{b}') = \begin{pmatrix} 2 & -1 & 0 \\ 0 & 0 & 0 \\ 0 & 0 & 0 \\ 0 & 0 & 0 \end{pmatrix},$$

then $\mathrm{rr}(A') = \mathrm{rr}(A'|\underline{b}') = 1$, the system has infinitely many solutions that depend on one parameter and the given vectors are linearly dependent.

For $k = -2$ we have that:

$$\mathbf{v}_1 = (2, -4, 4, -2), \mathbf{v}_2 = (-1, 2, -2, -1)$$

and $\mathbf{v}_1 = -2\mathbf{v}_2$.

3.4   SUGGESTED EXERCISES

3.4.1 Say if the vectors of the following sets are linearly dependent or independent. If they are linearly dependent write one vector as a linear combination of the others:

(i) $\{(2, 1, 1), (3, 2, 1), (6, 2, 2)\} \subseteq \mathbb{R}^3$.

(ii) $\{(1, 0, 1, 4), (2, 1, 0, 6), (1, -2, 5, 8)\} \subseteq \mathbb{R}^4$.

(iii) The set of polynomials $\{2x^2 + 1, x^2 + 2x, 4x - 1\} \subseteq \mathbb{R}_2[x]$.

(iv) The set of polynomials $\{1, x, x^2, x^3\} \subseteq \mathbb{R}[x]$.

(v) The set of following vectors:

$$\left\{ \begin{pmatrix} 1 & 0 \\ 2 & -1 \end{pmatrix}, \begin{pmatrix} 0 & 3 \\ 1 & -3 \end{pmatrix}, \begin{pmatrix} 1 & 3 \\ -2 & -1 \end{pmatrix} \right\} \subseteq M_2(\mathbb{R}).$$

3.4.2 Given the vectors

$$\mathbf{u} = \begin{pmatrix} 1 \\ 1 \end{pmatrix}, \mathbf{v} = \begin{pmatrix} 0 \\ 2 \end{pmatrix}, \mathbf{w} = \begin{pmatrix} 2 \\ -2 \end{pmatrix},$$

determine if they are linearly independent and determine the subspace generated by them.

3.4.3 Determine if $x^3 - x$ belongs to $\langle x^3 + x^2 + x, x^2 + 2x, x^2 \rangle$.

3.4.4 Determine for which values of k the polynomial $k^2x^2 + x + 1$ belongs to $\langle 2x^2 - x, -x^2 + 3x + 1 \rangle$.

3.4.5 Determine for which values of k we have that

$$\mathbf{w} = \begin{pmatrix} 2 \\ 5 \end{pmatrix} \in \left\langle \begin{pmatrix} k \\ 1 \end{pmatrix}, \begin{pmatrix} 1 \\ -2 \end{pmatrix} \right\rangle.$$

3.4.6 Determine for which values of k the polynomial $x^2 + 2k$ belongs to $\langle x^2 + kx, x^2 - (k+1)x - k \rangle$.

3.4.7 In $\mathbb{R}_2[x]$ give examples of sets with the following properties:

a) A set of generators that are not linearly independent.

b) A set of linearly independent vectors that do not generate the space.

3.4.8 In $M_{2,3}(\mathbb{R})$ consider the following vectors:

$$A = \begin{pmatrix} 1 & 0 & -2 \\ 4 & 1 & 0 \end{pmatrix}, \quad B = \begin{pmatrix} -1 & -1 & 1 \\ 0 & -2 & 1 \end{pmatrix}.$$

Determine a matrix C, such that $C \in \langle A, B \rangle$, $C \notin \langle A \rangle$ and $C \notin \langle B \rangle$, and a vector D such that $D \notin \langle A, B \rangle$. The motivation is requested.

3.4.9 Given the following vectors in $\mathbb{R}^3$:

$$\mathbf{v}_1 = \begin{pmatrix} 1 \\ 0 \\ 1 \end{pmatrix} \quad \mathbf{v}_2 = \begin{pmatrix} 2 \\ k \\ 1 \end{pmatrix} \quad \mathbf{v}_3 = \begin{pmatrix} 1 \\ -2 \\ k \end{pmatrix} \quad \mathbf{w} = \begin{pmatrix} -2 \\ 3 \\ 1 \end{pmatrix}.$$

a) Determine the values of k for which the three vectors $\mathbf{v}_1$, $\mathbf{v}_2$, $\mathbf{v}_3$ are linearly independent.

b) Determine the values of k for which $\mathbf{w} \in \langle \mathbf{v}_1, \mathbf{v}_2, \mathbf{v}_3 \rangle$.

3.4.10 a) Determine the solutions of the following linear system as the parameter k varies:

$$\begin{cases} x - z = 1 \\ kx - ky + 2z = 0 \\ 2x + 3ky - 11z = -1. \end{cases}$$

b) Determine for which values of the parameter k the polynomial $x^2 - 1$ belongs to the subspace generated by the polynomials $x^2 + kx + 2$, $kx - 3k^2$, $x^2 - 2x + 11$.

3.4.11 Let V be a vector space and let $\mathbf{v}_1, \ldots, \mathbf{v}_n, \mathbf{w}_1, \ldots, \mathbf{w}_m \in V$. If $\mathbf{v}_1, \ldots, \mathbf{v}_n$ generate V is it also true that $\mathbf{v}_1, \ldots, \mathbf{v}_n, \mathbf{w}_1, \ldots, \mathbf{w}_m$ generate V? If yes, prove it; otherwise give a counterexample.

3.4.12 Find the values of h and k for which the vectors of the set $\{x^2 + h, kx - h, x^2 + 2kx - h\} \subseteq \mathbb{R}_2[x]$ are linearly independent. If $h = 1$ and $k = 2$ do such vectors generate $\mathbb{R}_2[x]$?

3.4.13 a) Determine for which values of the parameter k the vectors $\mathbf{v}_1 = x + 3$, $\mathbf{v}_2 = kx + 5$, $\mathbf{v}_3 = kx^2 + 5x$, $\mathbf{v}_4 = x^3$ of $\mathbb{R}_3[x]$ are linearly independent.
b) Choose a value of k for which $\mathbf{v}_1, \mathbf{v}_2, \mathbf{v}_3$ are linearly dependent and write one of them as a linear combination of the others.

3.4.14 a) Establish for which values of the parameter k the matrices $\begin{pmatrix} 1 & 2 \\ 0 & 0 \end{pmatrix}, \begin{pmatrix} k & 0 \\ 4 & 0 \end{pmatrix}$, $\begin{pmatrix} -1 & k-2 \\ k & 0 \end{pmatrix}$ are linearly independent.
   b) Establish for which values of the parameter k such vectors generate the subspace $W = \left\{ \begin{pmatrix} r & s \\ t & 0 \end{pmatrix} \mid r, s, t \in \mathbb{R} \right\}$ of $M_2(\mathbb{R})$.

3.4.15 Determine for which values of the parameter k we have that:
$$\begin{pmatrix} 3 & -2 \\ 2 & 2 \end{pmatrix} \in \left\langle \begin{pmatrix} 2 & 0 \\ 2 & 0 \end{pmatrix}, \begin{pmatrix} 1 & k \\ 0 & -k \end{pmatrix}, \begin{pmatrix} k & 6 \\ 1 & -6 \end{pmatrix} \right\rangle.$$

3.4.16 Determine a set of generators of the following vector spaces:

   i) $S = \{(x, y, z) \in \mathbb{R}^3 \mid x + 2y = 0\}$.

   ii) $T = \{hx^3 - kx^2 + 4hx + k \mid h, k \in \mathbb{R}\} \subseteq \mathbb{R}_3[x]$.

CHAPTER 4

Basis and Dimension

The concepts of basis and dimension, which are closely related, are central in the theory of vector spaces.

Let us start with some examples, mainly, but not only, in $\mathbb{R}^2$ and $\mathbb{R}^3$. Thanks to these examples, we will develop geometric intuition, which will be valuable in order to understand what happens in vector spaces that cannot be visualized. Then, we will discuss the theory and state the *Completion Theorem*. This is the most important result in this chapter; starting from the concept of basis it allows us to reach the definition of dimension. At the end, we will revisit the Gaussian algorithm, described in Chapter 1, and we will see how it can be effectively used to answer the main questions regarding a basis or the dimension of a vector space.

4.1 BASIS: DEFINITION AND EXAMPLES

As the word itself suggests, the concept of basis of a vector space contains all the information necessary to rebuild the vector space, starting from very "few" vectors. Let us see some examples to motivate us.

Example 4.1.1 In the previous chapter, we have seen several examples of sets of generators of the vector space $\mathbb{R}^2$. We will mention a few:

$$\mathbb{R}^2 = \langle (1,0), (0,1) \rangle = \langle (1,0), (0,1), (1,1) \rangle.$$

If we add a vector to a set that generates $\mathbb{R}^2$, this set continues to generate $\mathbb{R}^2$ by Proposition 3.1.8. The question that we ask ourselves is: how can we find a minimal set, i.e. a set as small as possible, of generators for the space $\mathbb{R}^2$?

Proposition 3.1.8 comes again to help us: if we remove from the set a vector which is a linear combination of the others, the new set obtained generates the same vector space. In the example we are considering, we can remove the vector $(1,1)$ as it is a linear combination of $(1,0)$ and $(0,1)$: $(1,1) = (1,0) + (0,1)$. If now, however, we try to further decrease the number of generators in the set, the vector space generated by them changes. Indeed $\langle (1,0) \rangle$ is just the x-axis, while $\langle (0,1) \rangle$ is just the y-axis. So, if we remove one of the two vectors $(1,0)$ or $(0,1)$, from the given set, the vector space generated by the set changes, in other words, there are no "redundant generators". The important difference between the two sets: $\{(1,0),(0,1)\}$ and $\{(1,0),(0,1),(1,1)\}$ cannot go unnoticed. The first set consists of linearly independent vectors, while the vectors of the second set are linearly dependent. So we have seen in this example that, starting from a set of generators, we can delete one by one the generators that are linear combination of other vectors in the set, until we obtain a set of linearly independent vectors, that is a set in which no vector is a linear combination of the other vectors (Proposition 3.2.4 of Chapter 3). At this point, we cannot remove any vector from the set, without changing the vector space generated by the vectors in the set.

The next proposition formalizes the conclusions of the previous example and gives us an algorithm to obtain a minimal set of generators; as we will see, this set is called *basis*.

Proposition 4.1.2 *Let* $V = \langle \mathbf{v}_1, \ldots, \mathbf{v}_n \rangle \neq \{\mathbf{0}\}$. *Then there exists a subset of* $\{\mathbf{v}_1, \ldots, \mathbf{v}_n\}$, *consisting of linearly independent vectors, which generates* V.

Proof. We proceed algorithmically by steps.

Step one. We have that $V = \langle \mathbf{v}_1, \ldots, \mathbf{v}_n \rangle$ by assumption. If $\mathbf{v}_1, \ldots, \mathbf{v}_n$ are linearly independent, then we have proved the statement. Otherwise, one of the vectors, suppose $\mathbf{v}_n$, is a linear combination of the others, by Proposition 3.2.4. By Proposition 3.1.8, we have:
$$V = \langle \mathbf{v}_1, \ldots, \mathbf{v}_n \rangle = \langle \mathbf{v}_1, \ldots, \mathbf{v}_{n-1} \rangle.$$

Step two. In step one, we have eliminated the vector $\mathbf{v}_n$ from the set of generators of V, thus $V = \langle \mathbf{v}_1, \ldots, \mathbf{v}_{n-1} \rangle$. If $\mathbf{v}_1, \ldots, \mathbf{v}_{n-1}$ are linearly independent, then we have finished our proof. Otherwise, we go back to step one, that is, one of the vectors, suppose $\mathbf{v}_{n-1}$, is a linear combination of the others. By Proposition 3.1.8 in Chapter 3 we have:
$$V = \langle \mathbf{v}_1, \ldots, \mathbf{v}_n \rangle = \langle \mathbf{v}_1, \ldots, \mathbf{v}_{n-1} \rangle = \langle \mathbf{v}_1, \ldots, \mathbf{v}_{n-2} \rangle.$$

It is clear that, after a finite number of steps, $n-1$ at most, we get a set in which no vector is a linear combination of the others, therefore by Proposition 3.2.4, it is a set of linearly independent vectors. ■

We are ready for the definition of basis.

Definition 4.1.3 Let V be a vector space. The set $\{\mathbf{v}_1, \ldots, \mathbf{v}_n\}$ is called a *basis* if:

1. The vectors $\mathbf{v}_1, \ldots, \mathbf{v}_n$ are linearly independent.

2. The vectors $\mathbf{v}_1, \ldots, \mathbf{v}_n$ generate V.

We say also that V is *finitely generated*, if there exists a finite set of generators of V, i.e. $V = \langle \mathbf{v}_1, \ldots, \mathbf{v}_n \rangle$.

If V admits a basis $\{\mathbf{v}_1, \ldots, \mathbf{v}_n\}$, then it is finitely generated. We will soon see that the converse is also true.

Henceforth, we will say that a set X is *maximal (minimal)* with respect to a certain property if X enjoys that properties, but as soon as we add (remove) an element to (from) X, then X does not enjoy the property anymore.

Theorem 4.1.4 *Let* $\mathbf{v}_1, \ldots \mathbf{v}_n$ *be vectors in a vector space* V.

1. $\{\mathbf{v}_1, \ldots, \mathbf{v}_n\}$ *is a basis of* V *if and only if it is a minimal set of generators of* V.

2. $\{\mathbf{v}_1, \ldots, \mathbf{v}_n\}$ *is a basis of* V *if and only if it is a maximal set of linearly independent vectors.*

Proof. (1). If $\{\mathbf{v}_1, \ldots, \mathbf{v}_n\}$ is a basis of a vector space V, then by definition it is a set of generators. We now see that it is also a minimal set with this property. In fact, if we remove any vector from $\{\mathbf{v}_1, \ldots, \mathbf{v}_n\}$, then the vector space generated by the vectors changes. This happens because otherwise, by Proposition 3.1.8, a vector among $\mathbf{v}_1, \ldots, \mathbf{v}_n$ would be a linear combination of the others, while we know that these vectors are linearly independent by hypothesis. Vice versa, if we consider a minimal set of generators, then it is a basis because it consists of linearly independent vectors. In fact, by minimality, we have that, by removing any of the generators, the remaining vectors do not generate the given vector space anymore, and therefore, by Propositions 3.1.8 and 3.2.4, this means that none of them is a linear combination of the other vectors.

(2) If $\{\mathbf{v}_1, \ldots, \mathbf{v}_n\}$ is a basis of a vector space V, by definition it is a set of linearly independent vectors, and it is also maximal with respect to this property. Indeed, as $\mathbf{v}_1, \ldots, \mathbf{v}_n$ generate V, we will have that if $\mathbf{w} \in V$ then

$$\langle \mathbf{v}_1, \ldots, \mathbf{v}_n \rangle = \langle \mathbf{v}_1, \ldots, \mathbf{v}_n, \mathbf{w} \rangle = V,$$

and thus by Proposition 3.1.8, $\mathbf{w}$ is necessarily a linear combination of $\{\mathbf{v}_1, \ldots, \mathbf{v}_n\}$, therefore the vectors $\mathbf{v}_1, \ldots, \mathbf{v}_n, \mathbf{w}$ are linearly dependent by Proposition 3.2.4.

Conversely, if $\{\mathbf{v}_1, \ldots, \mathbf{v}_n\}$ is a maximal set of linearly independent vectors, if we add any other vector $\mathbf{v}$, we get a linearly dependent set of vectors, that is, there are scalars $\alpha, \alpha_1, \ldots, \alpha_n$, not all equal to zero, such that

$$\alpha \mathbf{v} + \alpha_1 \mathbf{v}_1 + \cdots + \alpha_n \mathbf{v}_n = \mathbf{0}.$$

We note that it must be $\alpha \neq 0$, otherwise the vectors $\{\mathbf{v}_1, \ldots, \mathbf{v}_n\}$ would be linearly dependent. Then we have that

$$\mathbf{v} = -\frac{\alpha_1}{\alpha} \mathbf{v}_1 - \cdots - \frac{\alpha_n}{\alpha} \mathbf{v}_n,$$

thus $\mathbf{v} \in \langle \mathbf{v}_1, \ldots, \mathbf{v}_n \rangle$. As we chose $\mathbf{v}$ arbitrarily, we have that $\{\mathbf{v}_1, \ldots, \mathbf{v}_n\}$ generates V. ∎

Example 4.1.5 Consider the following vectors in $\mathbb{R}_3[x]$:

$$x^3, x^2, 2, 5, x + 2, 3x, -7x, 2x^3.$$

We want to find a basis for the subspace they generate. The procedure we will follow is not the standard one, but only an example of the procedure described in Proposition 4.1.2. First, we see immediately that 5 is a linear combination of 2, as it is a multiple of it: $5 = (2/5)2$. Therefore, we can eliminate the vector 5 (thanks to Proposition 3.1.8). In the same way, we can eliminate $-7x = (-7/3)3x$ and also $2x^3 = 2(x^3)$. So we have:

$$W = \langle x^3, x^2, 2, 5, x + 2, 3x, -7x, 2x^3 \rangle = \langle x^3, x^2, 2, x + 2, 3x \rangle.$$

Now we see that $x + 2 = 1/3 \cdot 3x + 2$, thus:

$$W = \langle x^3, x^2, 2, x + 2, 3x \rangle = \langle x^3, x^2, 2, 3x \rangle.$$

To verify that these vectors are linearly independent and thus form a basis of W, we have to show that the equation:

$$ax^3 + bx^2 + 2c + 3dx = 0$$

is satisfied only for $a = b = c = d = 0$, but this is clear, since a polynomial is identically zero if and only if all its coefficients are equal to zero. Therefore $\{x^3, x^2, 2, 3x\}$ is a basis of W. We leave, as an exercise to the student, to prove that $W = \mathbb{R}_3[x]$. We will see later that the latter statement is obvious, using the concept of dimension.

Now the following question arises: given a vector space V, is there always a basis for V? The answer is yes, although we will see the proof only for finitely generated vector spaces.

Proposition 4.1.6 *If a vector space $V \neq \{\mathbf{0}\}$ is generated by a finite number of vectors $\mathbf{v}_1, \ldots, \mathbf{v}_n$, then there exists a basis of V.*

Proof. By assumption $V = \langle \mathbf{v}_1 \ldots \mathbf{v}_n \rangle$. Then, by Proposition 4.1.2, we have that there is a subset of $\{\mathbf{v}_1 \ldots \mathbf{v}_n\}$, consisting of linearly independent vectors generating V, that is, there is a basis of V. ∎

By convention, we have that the empty set $\varnothing$ is linearly independent, and it is the basis of the vector space $V = \{\mathbf{0}\}$.

Observation 4.1.7 Not all vector spaces are finitely generated. An example of such space is $\mathbb{R}[x]$, the vector space of polynomials with coefficients in $\mathbb{R}$. In fact, suppose we have a basis of $\mathbb{R}[x]$ with a finite number of elements and let N be the highest degree of the polynomials in such a basis. Then, the polynomial x^{N+1} cannot be expressed as a linear combination of elements of the basis and we get a contradiction.

However, even for vector spaces that are not finitely generated, we can always find a basis: the proof is difficult, and we will not see it here. In the special case of $\mathbb{R}[x]$, a basis must necessarily contain an infinite number of elements; for example, the student is invited to check that $\{1, x, x^2, x^3, \ldots\}$ is a basis of $\mathbb{R}[x]$.

Observation 4.1.8 Proposition 4.1.6 ensures that each vector space, different from the zero space and generated by a finite number of vectors, has at least one basis. However, this basis is not unique. For example it is easy to verify that, if $k \neq 0$, then the set $\mathcal{B}_k = \{(k, 1), (0, 1)\}$ is a basis of $\mathbb{R}^2$, so $\mathbb{R}^2$ has infinitely many bases. We invite the student to convince himself that every vector space that admits a basis, actually admits infinitely many.

4.2 THE CONCEPT OF DIMENSION

A vector space, as we already know, can admit different bases, however, as we shall see:

> All the bases have the same number of elements,

> This number is called the *dimension* of the vector space.

It is unlikely that the student now understands the importance of this number uniquely associated to a vector space. The concept of dimension is the key to answer many questions about the linear independence of certain sets of vectors, or to determine if a set of vectors of a vector space V generates or not V. In order to precisely define the dimension, we first need the Completion Theorem, which will be proved in the appendix.

Completion Theorem 4.2.1 *Let $S = \{\mathbf{v}_1, \ldots \mathbf{v}_m\}$ be a set of linearly independent vectors in a finitely generated vector space V. If $\mathcal{B} = \{\mathbf{w}_1, \ldots, \mathbf{w}_n\}$ is a basis of V (we know that there is always at least one) then $m \leq n$, and we can always add to S $n - m$ vectors from the basis $\{\mathbf{w}_1, \ldots, \mathbf{w}_n\}$ in order to obtain a basis of V.*

We will see, at the end of this chapter, that we can put in action very explicitly the Completion Theorem in $\mathbb{R}^n$, to *complete* any given set of linearly independent vectors to obtain a basis of $\mathbb{R}^n$.

Proposition 4.2.2 *All the bases of the same finitely generated vector space have the same number of elements.*

Proof. Let $\mathcal{B}_1 = \{\mathbf{v}_1, \ldots, \mathbf{v}_n\}$ and $\mathcal{B}_2 = \{\mathbf{w}_1, \ldots, \mathbf{w}_m\}$ be two bases of V with $n \leq m$. Since the vectors in $\mathcal{B}_1$ are linearly independent and $\mathcal{B}_2$ is a basis, by the Completion Theorem we have $n \leq m$. Exchanging the roles of $\mathcal{B}_1$ and $\mathcal{B}_2$, we get that $m \leq n$, thus $n = m$. ■

Definition 4.2.3 The number of elements of a basis of a vector space is called *dimension* of the vector space, and it is denoted with $\dim(V)$. When this number is finite, or equivalently, when V is generated by a finite number of vectors, V is said to be *finite dimensional.*

We now describe some bases for the vector spaces we have encountered so far, which are particularly simple; they are called *canonical bases*, and the number of elements each basis contains is the dimension of the corresponding vector space. We leave to the student to verify that these are indeed bases.

- $\mathbb{R}^n$. The canonical basis is given by $\mathcal{C} = \{\mathbf{e}_1, \ldots, \mathbf{e}_n\}$, where $\mathbf{e}_i$ is the vector that has 1 in position i and 0 in the other positions.

  For example the canonical basis of $\mathbb{R}^3$ is $\mathcal{C} = \{(1,0,0), (0,1,0), (0,0,1)\}$, and $\mathbb{R}^3$ has dimension 3, or we can also say that it is a three-dimensional vector space.

- $\mathbb{R}_n[x]$. The canonical basis is given by $\mathcal{C} = \{x^n, x^{n-1}, \ldots, x, 1\}$.

- The canonical basis is given by $\mathcal{C} = \{\mathbf{E}_{1,1}, \ldots, \mathbf{E}_{m,n}\}$, where $\mathbf{E}_{i,j}$ is the matrix that has 1 in position (i, j) and 0 in the other positions. For example the canonical basis of $\mathrm{M}_{2,3}$ is

$$\mathcal{C} = \left\{ \begin{pmatrix} 1 & 0 & 0 \\ 0 & 0 & 0 \end{pmatrix}, \begin{pmatrix} 0 & 1 & 0 \\ 0 & 0 & 0 \end{pmatrix}, \right.$$

$$\left. \begin{pmatrix} 0 & 0 & 1 \\ 0 & 0 & 0 \end{pmatrix}, \begin{pmatrix} 0 & 0 & 0 \\ 1 & 0 & 0 \end{pmatrix}, \begin{pmatrix} 0 & 0 & 0 \\ 0 & 1 & 0 \end{pmatrix}, \begin{pmatrix} 0 & 0 & 0 \\ 0 & 0 & 1 \end{pmatrix} \right\}.$$

The knowledge of the canonical bases tells us immediately the dimension of the vector spaces considered above:

$$\dim(\mathbb{R}^n) = n, \qquad \dim(\mathbb{R}_n[x]) = n + 1, \qquad \dim = mn.$$

If one keeps in mind this fact, many exercices become simple. For example, we ask ourself if the set

$$\{(1, 2, 1), \quad (1, 1, 5), \quad (2, 3, 1), \quad (0, 1, 0)\}$$

is a basis of $\mathbb{R}^3$. The answer is immediate: this set cannot be a basis as we know that all bases in $\mathbb{R}^3$ have precisely 3 elements, while the given set has 4 elements.

The proposition below is particularly useful for solving exercises.

Proposition 4.2.4 *Let V be a vector space of dimension n and let W be a subspace of V. Then:*

a) $\dim(W) \le \dim(V)$;

b) $\dim(W) = \dim(V)$ *if and only if $V = W$.*

Proof. a) Recall that the dimension of a vector space is the number of elements of a basis, which is also a maximal set of linearly independent vectors. Since W is contained in V, we cannot choose in W a larger number of linearly independent vectors than the number of linearly independent vectors in V, therefore the dimension of W cannot be larger than the dimension of V.
b) Since the vectors of a basis of W are linearly independent, by the Completion Theorem, we can add to them $\dim(V) - \dim(W)$ vectors to obtain a basis of V. If $\dim(W) = \dim(V)$, it means that a basis of W is already a basis of V, and then in particular generates V, i.e. $W = V$. ■

Let us see how this theorem greatly simplifies the exercises.

Example 4.2.5 We want to show that $\mathcal{B} = \{(1, -1), (2, 0)\}$ is a basis of $\mathbb{R}^2$. In principle, we have to verify that the two vectors are linearly independent and that they generate $\mathbb{R}^2$. However, while linear independence is obvious, because one vector is not a multiple of the other, for generation we should do the calculations. Now we see that the calculations are not necessary. In fact, we have two linearly independent vectors in $\mathbb{R}^2$, so the subspace generated by them has dimension two. So, by the previous theorem, it must be equal to $\mathbb{R}^2$.

Similarly, in Example 4.1.5 we have shown that the vectors $x^3, x^2, 2, 3x$ are linearly independent. Then, since there are four of them, we can immediately conclude, *without* making calculations, that they are a basis of $\mathbb{R}_3[x]$, therefore $\langle x^3, x^2, 2, 3x \rangle = \mathbb{R}_3[x]$.

In fact, a much stronger property is true. In general, given a set of vectors, the property of being linearly independent or the property of being generators of a certain vector space are not related with each other. But, if we are in a vector space of dimension n and we consider a set with exactly n vectors, then the two properties are equivalent.

Proposition 4.2.6 *Let V be a vector space of dimension n, and let $\{\mathbf{v}_1, \ldots, \mathbf{v}_n\}$ be set of n vectors of V. The following are equivalent:*

a) $\{\mathbf{v}_1, \ldots, \mathbf{v}_n\}$ *is a basis of V;*

b) $\mathbf{v}_1, \ldots, \mathbf{v}_n$ *are linearly independent;*

c) $\mathbf{v}_1, \ldots, \mathbf{v}_n$ *generate V.*

Proof. a) implies *b)* by the definition of basis.
Let us see that *b)* implies *c)*. Set $W = \langle \mathbf{v}_1, \ldots, \mathbf{v}_n \rangle$. We have that $\{\mathbf{v}_1, \ldots, \mathbf{v}_n\}$ is a basis of W, so W has dimension n. Then, by Proposition 4.2.4 (b), we have that $W = V$, so the given vectors generate V.
Now, let us see that *c)* implies *a)*. By Proposition 4.1.2, there is a subset of $\{\mathbf{v}_1, \ldots, \mathbf{v}_n\}$ which is a basis of V, but as V has dimension n, this subset must contain exactly n elements, then it is $\{\mathbf{v}_1, \ldots, \mathbf{v}_n\}$. ∎

We now see that a basis is a very "efficient" way to represent vectors in a vector space. Let us see an example.

Example 4.2.7 In $\mathbb{R}^2$ we know that all the vectors are linear combinations of the two vectors of the canonical basis $\mathbf{e}_1 = (1, 0)$, $\mathbf{e}_2 = (0, 1)$. We can also verify that each vector in $\mathbb{R}^2$ is not only a linear combination of $\mathbf{e}_1$ and $\mathbf{e}_2$, but it is so in a *unique* way. In fact, if we take the vector $(2, 3)$, we can write $(2, 3) = 2(1, 0) + 3(0, 1)$ and the numbers 2 and 3 are the only scalars which give us $(2, 3)$ as a linear combination of $(1, 0)$, $(0, 1)$. But the situation is different if we take the three vectors $(1, 0)$, $(0, 1)$ and $(1, 1)$. Indeed, we already know that the vector $(1, 1)$ is somehow "redundant", that is, we know that: $\langle (1, 0), (0, 1), (1, 1) \rangle = \langle (1, 0), (0, 1) \rangle$, and this is because $(1, 1)$ is a linear combination of $(1, 0)$ and $(0, 1)$. This is reflected by the fact that a vector in $\mathbb{R}^2$ *is no longer a linear combination in a unique way* of these three vectors. Indeed

$$(2, 3) = (1, 0) + 2(0, 1) + (1, 1) = 0(1, 0) + (0, 1) + 2(1, 1).$$

The concept of uniqueness of expression, seen in the previous example, is the content of the following proposition.

Theorem 4.2.8 *Let $\mathcal{B} = \{\mathbf{v}_1, \ldots, \mathbf{v}_n\}$ be an ordered basis for the vector space V (that is, we fixed an order in the set of vectors numbering them) and let $\mathbf{v} \in V$. Then there exists a unique n-tuple of scalars $(\alpha_1, \ldots, \alpha_n)$, such that*

$$\mathbf{v} = \alpha_1 \mathbf{v}_1 + \cdots + \alpha_n \mathbf{v}_n.$$

Proof. Since $\mathcal{B}$ is a system of generators, each $\mathbf{v} \in V$ is written as a linear combination of the elements of $\mathcal{B}$, that is, there are scalars $\alpha_1, \ldots, \alpha_n$ such that:

$$\mathbf{v} = \alpha_1 \mathbf{v}_1 + \cdots + \alpha_n \mathbf{v}_n.$$

We prove the uniqueness of the α_i's. Suppose that it is also

$$\mathbf{v} = \beta_1 \mathbf{v}_1 + \cdots + \beta_n \mathbf{v}_n.$$

Subtracting these two equations, we obtain

$$\mathbf{0} = (\alpha_1 - \beta_1)\mathbf{v}_1 + \cdots + (\alpha_n - \beta_n)\mathbf{v}_n,$$

where $\alpha_1 - \beta_1 = \ldots = \alpha_n - \beta_n = 0$ because of the linear independence of the vectors in $\mathcal{B}$, and therefore $\alpha_1 = \beta_1, \ldots, \alpha_n = \beta_n$. ■

Definition 4.2.9 The scalars $(\alpha_1, \ldots, \alpha_n)$ are called the *components* of $\mathbf{v} \in V$ in the basis $\mathcal{B}$ or also the *coordinates* of $\mathbf{v}$ with respect to the basis $\mathcal{B}$ and will be denoted by $(\mathbf{v})_{\mathcal{B}} = (\alpha_1, \ldots, \alpha_n)$.

Example 4.2.10 As an example, we prove by exercise that $\mathcal{B} = \{(1, -1), (2, 0)\}$ is a basis of $\mathbb{R}^2$, and we determine the coordinates of $\mathbf{v} = (-3, 1)$ with respect to this basis.

Clearly the vectors in $\mathcal{B}$ are linearly independent as $(2, 0)$ is not a multiple of $(1, -1)$. At this point, as $\mathbb{R}^2$ has dimension 2, we already know that $\mathcal{B}$ is a basis. But it is instructive to prove directly that it is a set of generators. Let $(x, y) \in \mathbb{R}^2$ and let $a, b \in \mathbb{R}$, such that

$$(x, y) = a(1, -1) + b(2, 0) = (a + 2b, -a);$$

it must be $a = -y$, and therefore $b = \frac{x+y}{2}$, and this is always possible, thus $\mathcal{B}$ generates $\mathbb{R}^2$. In particular $(-3, 1) = (-1)(1, -1) + (-1)(2.0)$, and thus the coordinates of $\mathbf{v}$ with respect to the basis $\mathcal{B}$ are $(\mathbf{v})_{\mathcal{B}} = (-1, -1)$.

4.3 THE GAUSSIAN ALGORITHM AS A PRACTICAL METHOD FOR SOLVING LINEAR ALGEBRA PROBLEMS

We have already seen that, to determine if some vectors of a given vector space V are linearly independent or if they generate V, we need to solve a linear system. So it is clear that the Gaussian algorithm is of great help in problems related to linear dependence or independence and to the concept of generation or more generally to the concept of basis. What we want to see now is how to use the Gaussian algorithm *directly*, without having to set a linear system.

We observe that, if we have a matrix $A \in M_{m,n}(\mathbb{R})$, we can consider its rows as vectors of $\mathbb{R}^n$, such vectors will be called *row* vectors of A. For example, if $A = \begin{pmatrix} 0 & 1 & -3 \\ 2 & -1 & 1 \end{pmatrix}$, its row vectors are $R_1 = (0, 1, -3)$ and $R_2 = (2, -1, 1)$.

Proposition 4.3.1 *Given a matrix $A \in M_{m,n}(\mathbb{R})$ the elementary row operations do not change the subspace of $\mathbb{R}^n$ generated by the row vectors of A.*

Proof. Recall that the elementary row operations are:

(a) exchange of two rows;

(b) multiplying a row by a real number other than 0;

(c) replacing the i-th row with the sum of the i-th row and j-th multiplied by any real number α.

It is immediate to verify that the statement is true for the type of operations (a) and (b). For operations of type (c), it is sufficient to show that if R_i and R_j are two row vectors of A and $\alpha \in \mathbb{R}$, we have that $\langle R_i, R_j + \alpha R_i \rangle = \langle R_i, R_j \rangle$. We obviously have that $R_i, R_j + \alpha R_i \in \langle R_i, R_j \rangle$, so $\langle R_i, R_j \rangle$ is a subspace of $\mathbb{R}^n$ containing $R_i, R_j + \alpha R_i$. Then by Proposition 3.1.5, we have that $\langle R_i, R_j + \alpha R_i \rangle \subseteq \langle R_i, R_j \rangle$.

The inclusion $\langle R_i, R_j \rangle \subseteq \langle R_i, R_j + \alpha R_i \rangle$ is shown in a similar way taking into account the fact that $R_i = (R_i + \alpha R_j) - \alpha R_j$, thus $R_i \in \langle R_i, R_j + \alpha R_i \rangle$. ■

Observation 4.3.2 The elementary row operations do not change the subspace of $\mathbb{R}^n$ generated by the row vectors of A, but they do change the subspace of $\mathbb{R}^m$ generated by the column vectors of A. We invite the reader to verify this fact with an example in order to convince himself.

Proposition 4.3.3 *If a matrix A is row echelon, its nonzero row vectors are linearly independent.*

Proof. Let $R_1, \ldots, R_k$ be the nonzero rows of A, and let $a_{1j_1}, \ldots, a_{kj_k}$ be the corresponding pivots. Now let $\lambda_1 R_1 + \cdots + \lambda_k R_k = \mathbf{0}$, and we want to prove that $\lambda_1 = \lambda_2 = \cdots = \lambda_k = 0$. In the vector $\lambda_1 R_1 + \cdots + \lambda_k R_k$, the element in the position j_1 is $\lambda_1 a_{1j_1}$, the element in the position j_2 is $\lambda_1 a_{1j_2} + \lambda_2 a_{2j_2}$, and so on, until we reach the element in position j_k, which is $\lambda_1 a_{1j_k} + \lambda_2 a_{2j_k} + \cdots + \lambda_k a_{kj_k}$. So, from the fact that $\lambda_1 R_1 + \cdots + \lambda_k R_k = 0$, it follows that:

$$\begin{cases} \lambda_1 a_{1j_1} = 0 \\ \lambda_1 a_{1j_2} + \lambda_2 a_{2j_2} = 0 \\ \vdots \\ \lambda_1 a_{1j_k} + \lambda_2 a_{2j_k} + \cdots + \lambda_k a_{kj_k} = 0. \end{cases}$$

Since $a_{1j_1} \neq 0$, from the first equation, we get $\lambda_1 = 0$. Substituting $\lambda_1 = 0$ in the second equation, and since $a_{2j_2} \neq 0$, we get that $\lambda_2 = 0$, and so on. After k steps we obtain $\lambda_k = 0$, thus $\lambda_1 = \cdots = \lambda_k = 0$, and this shows that the rows $R_1, \ldots, R_k$ are linearly independent vectors of $\mathbb{R}^n$. ■

Let us now see an example of how these propositions can be applied in the exercises.

Example 4.3.4 Given the following vectors of $\mathbb{R}^3$

$$\mathbf{v}_1 = \begin{pmatrix} 2 \\ 3 \\ -1 \end{pmatrix}, \qquad \mathbf{v}_2 = \begin{pmatrix} 0 \\ -1 \\ 3 \end{pmatrix}, \qquad \mathbf{v}_3 = \begin{pmatrix} 2 \\ 2 \\ 2 \end{pmatrix},$$

we want to establish if they are linearly independent and if they are a basis of $\mathbb{R}^3$. Moreover, we want to find a basis of the subspace W generated by them and to calculate the dimension of W.

The matrix:

$$A = \begin{pmatrix} 2 & 3 & -1 \\ 0 & -1 & 3 \\ 2 & 2 & 2 \end{pmatrix}$$

can be reduced to the following row echelon form:

$$\begin{pmatrix} 1 & 0 & 4 \\ 0 & 1 & -3 \\ 0 & 0 & 0 \end{pmatrix}.$$

By Proposition 4.3.1 we have that

$$W = \left\langle \begin{pmatrix} 2 \\ 3 \\ -1 \end{pmatrix}, \begin{pmatrix} 0 \\ -1 \\ 3 \end{pmatrix}, \begin{pmatrix} 2 \\ 2 \\ 2 \end{pmatrix} \right\rangle =$$

$$= \left\langle \begin{pmatrix} 1 \\ 0 \\ 4 \end{pmatrix}, \begin{pmatrix} 0 \\ -1 \\ 3 \end{pmatrix} \right\rangle.$$

Therefore, we have that the subspace W is generated by the two vectors $\mathbf{u}_1 = \begin{pmatrix} 1 \\ 0 \\ 4 \end{pmatrix}$

and $\mathbf{u}_2 = \begin{pmatrix} -1 \\ 0 \\ 3 \end{pmatrix}$, which are linearly independent by Proposition 4.3.3. Therefore,

$\{\mathbf{u}_1, \mathbf{u}_2\}$ is a basis for W, which consequently has dimension 2. Since by Theorem 4.1.4, the number of vectors in a basis is the maximum number of linearly independent vectors in the vector space, we have that $\mathbf{v}_1, \mathbf{v}_2, \mathbf{v}_3$ are linearly dependent, therefore they cannot be a basis of $\mathbb{R}^3$.

Let us now see in general how to proceed to find a basis of the subspace W generated by vectors $\mathbf{v}_1, \ldots, \mathbf{v}_k \in \mathbb{R}^n$ and how to decide if they are linearly independent, that is, we formalize what we learned from the previous example.

- We write the matrix A, whose rows are the vectors $\mathbf{v}_1, \ldots, \mathbf{v}_k$ (A will be a $k \times n$ matrix).

- Using the Gaussian algorithm, we obtain a matrix A' which is in row echelon form.

- The vectors forming the nonzero rows of A' generate W and are they are linearly independent, so they are a basis of W.

- Let r be the number of nonzero rows of A'; we have that $\dim W = r$. If $k = r$ the vectors $\mathbf{v}_1, \ldots, \mathbf{v}_k$ generate a vector space of dimension k and so they are linearly independent. If $k > r$ the vectors $\mathbf{v}_1, \ldots, \mathbf{v}_k$ are linearly dependent, because a maximal set of vectors of W, which are linearly independent, has $r < k$ elements.

Now let us see a method to get a basis of a subspace $W \subset \mathbb{R}^n$ and then to complete such a basis to a basis of $\mathbb{R}^n$.

Let $W \subset \mathbb{R}^n$ be the vector subspace generated by the vectors $\mathbf{w}_1, \ldots, \mathbf{w}_k$.

- We write the matrix A whose rows are the vectors $\mathbf{w}_1, \ldots, \mathbf{w}_k$ (A will be a $k \times n$ matrix).

- Using the Gaussian algorithm, we obtain a matrix A', which is in row echelon form.

- The r vectors forming the nonzero rows of A' generate W and are linearly independent, so they are a basis of W.

- To complete this basis to a basis of $\mathbb{R}^n$, just add $n - r$ row vectors in the positions corresponding to the missing pivots in the row echelon matrix A', to obtain a row echelon matrix A'' with n pivots.
  In fact, by Proposition 4.3.3, the rows of A'' are linearly independent, and since $\mathbb{R}^n$ has dimension n, by Proposition 4.2.6, the n row vectors of A'' form a basis of $\mathbb{R}^n$.

Example 4.3.5 In Example 4.3.4, to complete the basis $\{\mathbf{u}_1, \mathbf{u}_2\}$ to a basis of $\mathbb{R}^3$, just add the vector $(0, 0, 1)$, or any vector of the type $(0, 0, h)$ with $h \neq 0$.

Observation 4.3.6 Observe at this point that, if we fix an ordered basis $\mathcal{B} = \{\mathbf{v}_1, \ldots, \mathbf{v}_n\}$ of the vector space V, we can consider the function $c : V \to \mathbb{R}^n$ that associates to every vector its coordinates with respect to the basis $\mathcal{B}$. If we write $\mathbf{v} \in V$ as a linear combination of the elements of $\mathcal{B}$, $\mathbf{v} = \alpha_1 \mathbf{v}_1 + \cdots + \alpha_n \mathbf{v}_n$, we have $c(v) = (\mathbf{v})_\mathcal{B} = (\alpha_1, \ldots, \alpha_n)$. By Theorem 4.2.8, c is a bijection. If $\mathbf{v}$ has coordinates $(\alpha_1, \ldots, \alpha_n)$ and $\lambda \in \mathbb{R}$ is a scalar, then the coordinates of $\lambda \mathbf{v}$ are $(\lambda \alpha_1, \ldots, \lambda \alpha_n)$, and if $\mathbf{w}$ is another vector, whose coordinates are $(\beta_1, \ldots, \beta_n)$, then the coordinates of $\mathbf{v} + \mathbf{w}$ are $c(\mathbf{v} + \mathbf{w}) = (\alpha_1 + \beta_1, \ldots, \alpha_n + \beta_n)$. The student is invited to verify these assertions, that we will recall later on, when we talk about isomorphisms of vector spaces. For now, we are just content to note that, thanks to these properties, we have that $c(\lambda_1 \mathbf{v}_1 + \cdots + \lambda_n \mathbf{v}_n) = \lambda_1 c(\mathbf{v}_1) + \cdots + \lambda_n c(\mathbf{v}_n)$ for every $\mathbf{v}_1, \ldots, \mathbf{v}_n \in V$ and for every $\lambda_1, \ldots, \lambda_n \in \mathbb{R}$. From this, it follows that the vectors $\mathbf{v}_1, \ldots, \mathbf{v}_n$ are linearly independent if and only if their coordinates $c(\mathbf{v}_1), \ldots, c(\mathbf{v}_n)$ are linearly independent, seen as vectors in $\mathbb{R}^n$. Similarly, $\mathbf{w}$ is a linear combination of $\mathbf{v}_1, \ldots, \mathbf{v}_n$ if and only if $c(\mathbf{w})$ is a linear combination of $c(\mathbf{v}_1), \ldots, c(\mathbf{v}_n)$, and $\{\mathbf{w}_1, \ldots, \mathbf{w}_k\}$ is a basis of $\langle \mathbf{v}_1, \ldots, \mathbf{v}_n \rangle$ if and only if $\{c(\mathbf{w}_1), \ldots, c(\mathbf{w}_k)\}$ is a basis of $\langle c(\mathbf{v}_1), \ldots, c(\mathbf{v}_n) \rangle$. Then, instead of operating on vectors, we can operate on their coordinates, and then go back again to the vectors. This allows us to use all the techniques that we have seen for $\mathbb{R}^n$ for any other vector space, provided that it is finitely generated. Of course to do this, we must always fix *an ordered basis*, otherwise we cannot talk about coordinates as n-tuples, which are uniquely associated with each vector. In other words, the same vector may have different coordinates with respect to different bases. We will continue this discussion in a later chapter; for now, we just see an example on how we can make calculations in any finite dimensional vector space, using the techniques studied for $\mathbb{R}^n$.

Example 4.3.7 Given the polynomials $2x^2 + 3x - 1$, $-x + 3$, $2x^2 + 2x + 2 \in \mathbb{R}_2[x]$, we want to establish if they are linearly independent and determine a basis of the subspace W they generate. Consider the canonical basis $\mathcal{C} = \{x^2, x, 1\}$ of $\mathbb{R}_2[x]$. With respect to this ordered basis, the coordinates of the polynomials are

$$\mathbf{v}_1 = \begin{pmatrix} 2 \\ 3 \\ -1 \end{pmatrix}, \qquad \mathbf{v}_2 = \begin{pmatrix} 0 \\ -1 \\ 3 \end{pmatrix}, \qquad \mathbf{v}_3 = \begin{pmatrix} 2 \\ 2 \\ 2 \end{pmatrix}.$$

We can then proceed as in Example 4.3.4, making exactly the same calculations, and we obtain a basis of W given by polynomials whose coordinates are $\mathbf{u}_1 = \begin{pmatrix} 1 \\ 0 \\ 4 \end{pmatrix}$, $\mathbf{u}_2 = \begin{pmatrix} -1 \\ 0 \\ 3 \end{pmatrix}$. Returning to polynomials, a basis of W is $\{x^2 + 4, -x + 3\}$.

4.4 EXERCISES WITH SOLUTIONS

4.4.1 Let $W = \langle (1 - k, 1, 1, -k), (2, 2 - k, 2, 0), (1, 1, 1 - k, k) \rangle \subseteq \mathbb{R}^4$. Determine the dimension of W as $k \in \mathbb{R}$ varies. Then choose a suitable value of k and complete the corresponding basis of W to a basis of $\mathbb{R}^4$.

Solution. We write the matrix A that has the given vectors as rows, and we apply the Gaussian algorithm to reduce it to row echelon form. We have:

$$A = \begin{pmatrix} 1 - k & 1 & 1 & -k \\ 2 & 2 - k & 2 & 0 \\ 1 & 1 & 1 - k & k \end{pmatrix},$$

and by reducing in row echelon form we get:

$$A' = \begin{pmatrix} 1 & 1 & 1-k & k \\ 0 & -k & 2k & -2k \\ 0 & 0 & 4k-k^2 & -4k+k^2 \end{pmatrix}.$$

If $k \neq 0$ and $4k - k^2 \neq 0$, that is, if $k \neq 0$ and $k \neq 4$ the matrix A' has 3 nonzero rows, which are linearly independent, therefore as

$$W = \langle (1-k, 1, 1, -k), (2, 2-k, 2, 0), (1, 1, 1-k, k) \rangle =$$

$$\langle (1, 1, 1-k, k), (0, -k, 2k, -2k), (0, 0, 4k-k^2, -4k+k^2) \rangle,$$

W has dimension 3.

If $k = 0$ the matrix A' has only one nonzero row, so $W = \langle (1, 1, 1, 0) \rangle =$ has dimension 1.

If $k = 4$ we get:

$$A' = \begin{pmatrix} 1 & 1 & -3 & 4 \\ 0 & -4 & 8 & -8 \\ 0 & 0 & 0 & 0 \end{pmatrix},$$

so A' has 2 nonzero rows and $W = \langle (1, 1, -3, 4), (0, -4, 8, -8) \rangle$ has dimension 2.

We now choose $k = 4$. To complete $\{(1, 1, -3, 4), (0, -4, 8, -8)\}$ to a basis of $\mathbb{R}^4$ we have to add 2 row vectors having the pivots in the "missing steps", i.e. in the third and fourth place. For example, we can add $(0, 0, -1, 2), (0, 0, 0, 5)$.

So $\{(1, 1, -3, 4), (0, -4, 8, -8), (0, 0, -1, 2), (0, 0, 0, 5)\}$ is a basis of $\mathbb{R}^4$ obtained by completing a basis of W.

4.4.2 Let $\mathbf{v}_1 = (1, 2, 1, 0), \mathbf{v}_2 = (4, 8, k, 5), \mathbf{v}_3 = (-1, -2, 3-k, -k)$. Determine for which values of k the vectors $\mathbf{v}_1, \mathbf{v}_2, \mathbf{v}_3$ are linearly independent. Set $k = 1$ and determine, if possible, a vector $\mathbf{w} \in \mathbb{R}^4$, such that $\mathbf{w} \notin \langle \mathbf{v}_1, \mathbf{v}_2, \mathbf{v}_3 \rangle$.

Solution. We write the matrix A that has the given vectors as rows, and we apply the Gaussian algorithm to reduce it to row echelon form. We have

$$A = \begin{pmatrix} 1 & 2 & 1 & 0 \\ 4 & 8 & k & 5 \\ -1 & -2 & 3-k & -k \end{pmatrix},$$

and by reducing it in row echelon form we get:

$$A' = \begin{pmatrix} 1 & 2 & 1 & 0 \\ 0 & 0 & k-4 & 5 \\ 0 & 0 & 0 & 5-k \end{pmatrix}.$$

Let $W = \langle \mathbf{v}_1, \mathbf{v}_2, \mathbf{v}_3 \rangle$. If $k \neq 4$ and $k \neq 5$, the matrix A' has three nonzero rows, which are linearly independent, so $W = \langle (1, 2, 1, 0), (0, 0, k-4, 5), (0, 0, 0, k-5) \rangle$ has dimension 3. Since $\mathbf{v}_1, \mathbf{v}_2, \mathbf{v}_3$ generate W, by Proposition 4.2.6 they are linearly independent.

If $k = 4$ we get:

$$A' = \begin{pmatrix} 1 & 2 & 1 & 0 \\ 0 & 0 & 0 & 5 \\ 0 & 0 & 0 & 1 \end{pmatrix}.$$

Now $W = \langle (1,2,1,0), (0,0,0,5), (0,0,0,1) \rangle = \langle (1,2,1,0), (0,0,0,1) \rangle$ has dimension 2, and then $\mathbf{v}_1, \mathbf{v}_2, \mathbf{v}_3$ are linearly dependent.

If $k = 5$ we get:

$$A' = \begin{pmatrix} 1 & 2 & 1 & 0 \\ 0 & 0 & 1 & 5 \\ 0 & 0 & 0 & 0 \end{pmatrix},$$

which is a row echelon form matrix with two nonzero rows, so W has dimension 2 and $\mathbf{v}_1, \mathbf{v}_2, \mathbf{v}_3$ are linearly dependent.

Now let $k = 1$. We replace this value in A' to get a basis of W:

$$A' = \begin{pmatrix} 1 & 2 & 1 & 0 \\ 0 & 0 & -3 & 5 \\ 0 & 0 & 0 & 4 \end{pmatrix}.$$

We have that W has dimension 3, and if we choose a row vector that has the second nonzero pivot, for example $\mathbf{w} = (0, -2, 3, -1)$, it follows from Proposition 4.3.3 that the vectors $(1,2,1,0), (0,-2,3,-1), (0,0,-3,5), (0,0,0,4)$ are linearly independent, so by Proposition 3.2.4 we have that $(0,-2,3,-1) \notin \langle (1,2,1,0), (0,0,-3,5), (0,0,0,4) \rangle$, that is $\mathbf{w} \notin W$.

4.4.3 Let W be the subspace of $\mathbb{R}^5$ generated by the set

$$\mathcal{B} = \{(1, 3, -1, 1, 2), (2, 6, -2, 4, 4)\};$$

complete a basis $\mathcal{B}$ of W to a basis of $\mathbb{R}^5$.

Solution. We observe that $\mathcal{B}$ is actually a basis of W, because the two vectors in $\mathcal{B}$ are linearly independent, not being one multiple of the other.

One way to complete $\mathcal{B}$ to a basis of $\mathbb{R}^5$ is to proceed as indicated in the proof of the completion theorem. Another way is the following.

Using the Gaussian algorithm we determine a basis $\mathcal{B}'$ of W such that the matrix A' whose rows are the vectors of $\mathcal{B}'$ is in row echelon form. That is, we consider

$$A = \begin{pmatrix} 1 & 3 & -1 & 1 & 2 \\ 2 & 6 & -2 & 4 & 4 \end{pmatrix},$$

and by applying the Gaussian algorithm we obtain:

$$A' = \begin{pmatrix} 1 & 3 & -1 & 1 & 2 \\ 0 & 0 & 0 & 2 & 0 \end{pmatrix}.$$

At this point, to complete $\{(1, 3, -1, 1, 2), (0, 0, 0, 2, 0)\}$ to a basis of $\mathbb{R}^5$, we need to add 3 row vectors having the pivots in the "missing steps", i.e. the second, third and fifth places. For example, we can add $(0, 1, -1, 0, 1), (0, 0, 2, 1, -3), (0, 0, 0, 0, 1)$. As

$$W = \langle (1, 3, -1, 1, 2), (2, 6, -2, 4, 4) \rangle = \langle (1, 3, -1, 1, 2), (0, 0, 0, 2, 0) \rangle,$$

it is easy to see that

$$\langle (1, 3, -1, 1, 2), (2, 6, -2, 4, 4), (0, 1, -1, 0, 1), (0, 0, 2, 1, -3), (0, 0, 0, 0, 1) \rangle$$
$$= \langle (1, 3, -1, 1, 2), (0, 0, 0, 2, 0), (0, 1, -1, 0, 1), (0, 0, 2, 1, -3), (0, 0, 0, 0, 1) \rangle$$
$$= \mathbb{R}^5.$$

So if $\tilde{\mathcal{B}}$ is the set

$$\{(1, 3, -1, 1, 2), (2, 6, -2, 4, 4), (0, 1, -1, 0, 1), (0, 0, 2, 1, -3), (0, 0, 0, 0, 1)\}$$

the vectors of $\tilde{\mathcal{B}}$ generate $\mathbb{R}^5$, therefore, by Proposition 4.2.6, $\tilde{\mathcal{B}}$ is a basis of $\mathbb{R}^5$, and it was obtained by completing the basis $\mathcal{B}$ of W.

4.4.4 Consider the following vector subspaces of $M_{2,3}(\mathbb{R})$:

$$U = \left\{ \begin{pmatrix} a & 0 & b \\ c & a & d \end{pmatrix} \mid a, b, c, d \in \mathbb{R} \right\},$$

$$W = \left\{ \begin{pmatrix} r & s & t \\ u & x & y \end{pmatrix} \mid r + s + t + u + x + y = 0 \right\},$$

and determine a basis of $U \cap W$.

Solution. We have that:

$$U \cap W = \left\{ \begin{pmatrix} a & 0 & b \\ c & a & d \end{pmatrix} \mid a, b, c, d \in \mathbb{R}, \ a + b + c + a + d = 0 \right\},$$

that is

$$U \cap W = \left\{ \begin{pmatrix} a & 0 & b \\ c & a & d \end{pmatrix} \mid a, b, c, d \in \mathbb{R}, \ d = -2a - b - c \right\} =$$

$$\left\{ \begin{pmatrix} a & 0 & b \\ c & a & -2a - b - c \end{pmatrix} \mid a, b, c \in \mathbb{R} \right\} =$$

$$= \left\{ a \begin{pmatrix} 1 & 0 & 0 \\ 0 & 1 & -2 \end{pmatrix} + b \begin{pmatrix} 0 & 0 & 1 \\ 0 & 0 & -1 \end{pmatrix} + c \begin{pmatrix} 0 & 0 & 0 \\ 1 & 0 & -1 \end{pmatrix} \mid a, b, c \in \mathbb{R} \right\} =$$

$$\left\langle \begin{pmatrix} 1 & 0 & 0 \\ 0 & 1 & -2 \end{pmatrix}, \begin{pmatrix} 0 & 0 & 1 \\ 0 & 0 & -1 \end{pmatrix}, \begin{pmatrix} 0 & 0 & 0 \\ 1 & 0 & -1 \end{pmatrix} \right\rangle.$$

So the vectors $\mathbf{v}_1 = \begin{pmatrix} 1 & 0 & 0 \\ 0 & 1 & -2 \end{pmatrix}, \mathbf{v}_2 = \begin{pmatrix} 0 & 0 & 1 \\ 0 & 0 & -1 \end{pmatrix}, \mathbf{v}_3 = \begin{pmatrix} 0 & 0 & 0 \\ 1 & 0 & -1 \end{pmatrix}$ generate $U \cap V$. To show that they are linearly independent, let us consider their co-ordinates with respect to the canonical basis, that is: $(\mathbf{v}_1)_{\mathcal{C}} = (1, 0, 0, 0, 1, -2)$, $(\mathbf{v}_2)_{\mathcal{C}} = (0, 0, 0, 0, 0, -1)$, $(\mathbf{v}_3)_{\mathcal{C}} = (0, 0, 0, 1, 0, -1)$. We observe that the matrix A whose rows are $(\mathbf{v}_1)_{\mathcal{C}}, (\mathbf{v}_2)_{\mathcal{C}}, (\mathbf{v}_3)_{\mathcal{C}}$ is in row echelon form, so by Proposition 4.3.3 we have that $(\mathbf{v}_1)_{\mathcal{C}}, (\mathbf{v}_2)_{\mathcal{C}}, (\mathbf{v}_3)_{\mathcal{C}}$ are linearly independent. Then also $\mathbf{v}_1, \mathbf{v}_2, \mathbf{v}_3$ are linearly independent, so they are a basis of $U \cap W$.

4.5 SUGGESTED EXERCISES

4.5.1 Determine a basis of $\mathbb{R}_4[x]$ different from the canonical basis, motivating the answer.

4.5.2 Determine for which values of k the following polynomials of $\mathbb{R}_2[x]$ generate $\mathbb{R}_2[x]$: $X^2 + kx + 1.3x^2 + 4x + 1.5x^2 + kx - 3$.

4.5.3 Find a basis of $\langle (1,0,3), (2,3,0), (1,1,1) \rangle$ and complete it to a basis of $\mathbb{R}^3$.

4.5.4 Determine which of the following sets of vectors generate $\mathbb{R}^3$ and which of them are a basis of $\mathbb{R}^3$.

i) $S_1 = \left\{ \begin{pmatrix} 0 \\ 0 \\ -1 \end{pmatrix}, \begin{pmatrix} 2 \\ 0 \\ 0 \end{pmatrix}, \begin{pmatrix} 0 \\ -3 \\ 0 \end{pmatrix} \right\}$.

ii) $S_2 = \left\{ \begin{pmatrix} 1 \\ 1 \\ -1 \end{pmatrix}, \begin{pmatrix} 2 \\ 0 \\ 1 \end{pmatrix}, \begin{pmatrix} 1 \\ -1 \\ 1 \end{pmatrix}, \begin{pmatrix} 0 \\ 6 \\ -3 \end{pmatrix} \right\}$.

iii) $S_3 = \left\{ \begin{pmatrix} 1 \\ 3 \\ 1 \end{pmatrix}, \begin{pmatrix} -2 \\ 1 \\ -9 \end{pmatrix}, \begin{pmatrix} 0 \\ 5 \\ -5 \end{pmatrix} \right\}$.

4.5.5 Determine if the vectors $\mathbf{v}_1 = (1,1), \mathbf{v}_2 = (-1,1), \mathbf{v}_3 = (2,1)$ are linearly independent. Do they generate $\mathbb{R}^2$?

4.5.6 Determine for which values of k the vector $\mathbf{w} = (-1, k, 1)$ belongs to $\langle (1,-1,0), (k,-k,1), (-1, k^2, 1) \rangle$.

4.5.7 Determine for which values of k the vectors $\mathbf{v}_1 = (1, 2k, 0), \mathbf{v}_2 = (1, 0, -3), \mathbf{v}_3 = (-1.0, k+2)$ are linearly independent. Set $k = 1$ and establish if $(1, -2, -6) \in \langle \mathbf{v}_1, \mathbf{v}_2, \mathbf{v}_3 \rangle$.

4.5.8 Let $\mathbf{v}_1 = (1,1,1,0), \mathbf{v}_2 = (2,0,2,-1), \mathbf{v}_3 = (-1,1,-1,1)$. Determine a basis of $\langle \mathbf{v}_1, \mathbf{v}_2, \mathbf{v}_3 \rangle$ and then complete it to a basis of $\mathbb{R}^4$. Also determine for which values of k the vector $\mathbf{w} = (k, -3k, k, -2k)$ belongs to $\langle \mathbf{v}_1, \mathbf{v}_2, \mathbf{v}_3 \rangle$.

4.5.9 Let $X = \{(x,y,z) \in \mathbb{R}^3 \mid x - y + z = 0$ and $x - 2y = 0\}$. Prove that X is a subspace of $\mathbb{R}^3$ and determine a basis for it.

4.5.10 Determine for which values of k the vectors $\mathbf{v}_1 = (0,1,0), \mathbf{v}_2 = (1,k,4), \mathbf{v}_3 = (k,2,3)$ generate $\mathbb{R}^3$. Set $k = 0$ and determine if the vector $(4,-1,6)$ belongs to $\langle \mathbf{v}_1, \mathbf{v}_2, \mathbf{v}_3 \rangle$.

4.5.11 i) Determine for which values of k the vectors $\mathbf{v}_1 = x^2 + 2x - 1, \mathbf{v}_2 = x^2 + kx + 1 - k, \mathbf{v}_3 = 5x + k$ are linearly dependent.
ii) Choose one value of k found in point i) and write one of the 3 vectors as a linear combination of the others. Then find a basis of $\langle \mathbf{v}_1, \mathbf{v}_2, \mathbf{v}_3 \rangle$.

4.5.12 i) Determine for which values of k the vectors $\mathbf{v}_1 = 6x^2 - 6x - k, \mathbf{v}_2 = -kx^2 + kx + 6$ are linearly independent.
ii) Set $k = 0$. Determine the dimension of $\langle \mathbf{v}_1, \mathbf{v}_2 \rangle$, and if possible, find a vector $\mathbf{w}$ such that $\mathbf{w}$ does not belong to $\langle \mathbf{v}_1, \mathbf{v}_2 \rangle$ and $\{\mathbf{v}_1, \mathbf{v}_2, \mathbf{w}\}$ does not generate $\mathbb{R}_2[x]$.

4.5.13 Determine for which values of k the vectors $\mathbf{v}_1 = (1, 2, 0), \mathbf{v}_2 = (-k, 3, 0),$ $\mathbf{v}_3 = (2, 0, 1)$ are a basis $\mathcal{B}$ of $\mathbb{R}^3$. Put $k = 0$ and determine the coordinates of the vector $\mathbf{w} = (-2.0, -1)$ with respect to the basis $\mathcal{B}$.

4.5.14 Determine, if possible, 4 nonzero vectors of $\mathbb{R}_2[x]$ that do not generate $\mathbb{R}_2[x]$. Determine, if possible, two distinct subspaces of $\mathbb{R}_2[x]$ of dimension 2 that both contain the vector $x^2 + x$.

4.5.15 Let $X = \left\{ \begin{pmatrix} a & b \\ -b & a \end{pmatrix} \in M_2(\mathbb{R}) \mid A, b \in \mathbb{R} \right\}$. Prove that X is a subspace of $M_2(\mathbb{R})$ and determine its dimension.

4.5.16 Show that $\mathcal{B} = \left\{ \begin{pmatrix} -1 \\ 2 \end{pmatrix}, \begin{pmatrix} 1 \\ -1 \end{pmatrix} \right\}$ is a basis of $\mathbb{R}^2$ and determine the coordinates of the vectors $\begin{pmatrix} 3 \\ -1 \end{pmatrix}$ and $\begin{pmatrix} 0 \\ 1 \end{pmatrix}$ with respect to this basis.

4.5.17 i) Determine for which values of k the vectors $\mathbf{v}_1 = (-1, 0, 2k), \mathbf{v}_2 = (1, k, 3k), \mathbf{v}_3 = (0, 4k, 2)$ generate $\mathbb{R}^3$.
ii) Choose a value of k and determine if the vector $(2, k, k)$ belongs to $\langle \mathbf{v}_1, \mathbf{v}_2, \mathbf{v}_3 \rangle$.

4.5.18 i) Determine for which values of k the vectors $\mathbf{v}_1 = x^2 + 2x + 2, \mathbf{v}_2 = -x^2 + 2kx + k - 1, \mathbf{v}_3 = kx^2 + (2k + 4)x + 3k$ are linearly independent.
ii) Set $k = 0$ and determine, if possible, a vector $\mathbf{w}$ that does not belong to $\langle \mathbf{v}_1, \mathbf{v}_2, \mathbf{v}_3 \rangle$.

4.5.19 Determine, if possible, 4 vectors $\{\mathbf{v}_1, \mathbf{v}_2, \mathbf{v}_3, \mathbf{v}_4\}$ of $\mathbb{R}_2[x]$ that simultaneusly satisfy the following properties:

   i) none of them is a multiple of another;

   ii) $\{\mathbf{v}_1, \mathbf{v}_2, \mathbf{v}_3, \mathbf{v}_4\}$ generate $\mathbb{R}_2[x]$;

   ii) $\{\mathbf{v}_1, \mathbf{v}_2, \mathbf{v}_3\}$ do not generate $\mathbb{R}_2[x]$.

4.5.20 Determine if $X = \{sx^3 + rx^2 + sx - 2r \mid r, s \in \mathbb{R}\} \subseteq \mathbb{R}^3[x]$ is a vector subspace of $\mathbb{R}_3[x]$, and if so, determine a basis of X.

4.5.21 Determine a basis of the vector subspace of $\mathbb{R}^4$ given by $A \cap B$, where $A = \langle (1, 0, 0, 0), (0, 1, 0, 0) \rangle$ and $B = \{(x, y, z, t) \in \mathbb{R}^4 \mid y = t = 0\}$.

4.5.22 Prove that a nonzero vector space of finite dimension has infinitely many bases.

4.6 APPENDIX: THE COMPLETION THEOREM

We write here, for the interested reader, the proof of the Completion Theorem. We start with a technical lemma.

Lemma 4.6.1 *If* $\mathcal{B} = \{\mathbf{w}_1, \ldots, \mathbf{w}_n\}$ *is a basis of* V *and* $\mathbf{v}$ *is a vector of the type* $\mathbf{v} = \lambda_1 \mathbf{w}_1 + \cdots + \lambda_k \mathbf{w}_k + \cdots + \lambda_n \mathbf{w}_n$ *with* $\lambda_k \neq 0$ *then also* $\mathcal{B}' = \{\mathbf{w}_1, \ldots, \mathbf{w}_{k-1}, \mathbf{v}, \mathbf{w}_{k+1}, \ldots, \mathbf{w}_n\}$ *is a basis of* V.

Proof. We observe that $\mathbf{w}_1, \ldots, \mathbf{w}_{k-1}, \mathbf{w}_{k+1}, \ldots, \mathbf{w}_n \in \langle \mathcal{B} \rangle$, but since

$$\mathbf{v} = \lambda_1 \mathbf{w}_1 + \cdots + \lambda_k \mathbf{w}_k + \cdots + \lambda_n \mathbf{w}_n$$

with $\lambda_k \neq 0$ we also have that:

$$\mathbf{w}_k = -\frac{\lambda_1}{\lambda_k}\mathbf{w}_1 - \cdots + \frac{1}{\lambda_k}\mathbf{v} - \cdots - \frac{\lambda_n}{\lambda_k}\mathbf{w}_n \in \langle \mathcal{B}' \rangle.$$

So $\mathcal{B} \subseteq \langle \mathcal{B}' \rangle$. Since $\mathcal{B}$ generates V, we have that $V = \langle \mathcal{B} \rangle \subseteq \langle \mathcal{B}' \rangle$. The inclusion $\langle \mathcal{B}' \rangle \subseteq V$ is obvious, so we have $\langle \mathcal{B}' \rangle = V$, i.e. $\mathcal{B}'$ generates V.

We now show that the vectors of $\mathcal{B}'$ are linearly independent. Consider $\beta_1, \ldots, \beta_n \in \mathbb{R}$ such that:

$$\beta_1 \mathbf{w}_1 + \cdots + \beta_k \mathbf{v} + \cdots + \beta_n \mathbf{w}_n = \mathbf{0}.$$

By replacing $\mathbf{v}$ with $\lambda_1 \mathbf{w}_1 + \cdots + \lambda_k \mathbf{w}_k + \cdots + \lambda_n \mathbf{w}_n$ and rearranging we obtain that:

$$(\beta_1 + \beta_k \lambda_1)\mathbf{w}_1 + \cdots + \beta_k \lambda_k \mathbf{w}_k + \cdots + (\beta_n + \beta_k \lambda_n)\mathbf{w}_n = \mathbf{0}.$$

Since the $\mathbf{w}_i$ are linearly independent, we have that all coefficients must be zero. In particular, it must happen that $\beta_k \lambda_k = 0$ and $\beta_i + \beta_k \lambda_i = 0$ for every $i \neq k$. From the first equality, being $\lambda_k \neq 0$, it follows that $\beta_k = 0$, and by substituting in the others we get $\beta_i = 0$ for every $i \neq k$. So all β_i are zero, which shows that the vectors of $\mathcal{B}'$ are linearly independent. ■

We can now prove the Completion Theorem 4.2.1.

Theorem 4.6.2 *Let* $\{\mathbf{v}_1, \ldots, \mathbf{v}_m\}$ *be a set of vectors in a finitely generated vector space* V. *If* $\mathcal{B} = \{\mathbf{w}_1, \ldots, \mathbf{w}_n\}$ *is a basis* V *(and we know there always exists at least one basis) then* $m \leq n$ *and we can always add to* $n - m$ *vectors* $\mathbf{v}_1, \ldots, \mathbf{v}_n$ *to* $\mathcal{B}$ *to obtain a basis of* V.

Proof. Since $\mathcal{B}$ generates V, we can write $\mathbf{v}_1$ in the form

$$\mathbf{v}_1 = \alpha_1 \mathbf{w}_1 + \cdots + \alpha_n \mathbf{w}_n,$$

where not all the coefficients are zero. Possibly rearranging the $\mathbf{w}_k$, we can assume $\alpha_1 \neq 0$, and then by Lemma 4.6.1 we have that $\{\mathbf{v}_1, \mathbf{w}_2, \ldots, \mathbf{w}_n\}$ is a basis of V.

Now we consider $\mathbf{v}_2$ and the basis $\{\mathbf{v}_1, \mathbf{w}_2, \ldots, \mathbf{w}_n\}$. We can write $\mathbf{v}_2$ in the form:

$$\mathbf{v}_2 = \beta_1 \mathbf{v}_1 + \beta_2 \mathbf{w}_2 + \cdots + \beta_n \mathbf{w}_n,$$

where at least one of the β_j with $j \geq 2$ is not zero, otherwise $\mathbf{v}_2$ would be a multiple of $\mathbf{v}_1$ contradicting the hypothesis that the vectors $\mathbf{v}_1, \ldots, \mathbf{v}_m$ are linearly independent. It is not restrictive to assume $\beta_2 \neq 0$, and therefore by Lemma 4.6.1 we have that $\{\mathbf{v}_1, \mathbf{v}_2, \mathbf{w}_3, \ldots, \mathbf{w}_n\}$ is basis.

We can then continue in the same way. At the i-th step we can assume that $\{\mathbf{v}_1, \ldots, \mathbf{v}_{i-1}, \mathbf{w}_i, \ldots, \mathbf{w}_n\}$ is a basis. We can write $\mathbf{v}_i$ in the form:

$$\mathbf{v}_i = \lambda_1 \mathbf{v}_1 + \cdots + \lambda_{i-1} \mathbf{v}_{i-1} + \lambda_i \mathbf{w}_i + \cdots + \lambda_n \mathbf{w}_n,$$

where at least one of the λ_j with $j \geq i$ is not zero, otherwise we would have that $\mathbf{v}_i \in \langle \mathbf{v}_1, \ldots, \mathbf{v}_{i-1} \rangle$, contradicting the hypothesis that the vectors $\mathbf{v}_1, \ldots, \mathbf{v}_m$ are linearly independent. It is not restrictive; suppose that it is $\lambda_i \neq 0$, and then by Lemma 4.6.1 we have that $\{\mathbf{v}_1, \ldots, \mathbf{v}_i, \mathbf{w}_{i+1}, \ldots, \mathbf{w}_n\}$ is a basis for V.

If $m \leq n$, possibly rearranging the vectors $\mathbf{w}_k$ appropriately, after m steps we obtain that $\{\mathbf{v}_1, \ldots, \mathbf{v}_m, \mathbf{w}_{m+1}, \ldots, \mathbf{w}_n\}$ is a basis for V, as we wanted.

If $m > n$, after n steps we get that $\{\mathbf{v}_1, \ldots, \mathbf{v}_n\}$ is a basis of V, from which it follows that $\mathbf{v}_{n+1} \in \langle \mathbf{v}_1, \ldots, \mathbf{v}_n \rangle$, but this contradicts the hypothesis that vectors $\mathbf{v}_1, \ldots, \mathbf{v}_m$ are linearly independent. So it must be $m \leq n$, and this ends the proof. ■

CHAPTER 5

Linear Transformations

Linear transformations are functions between vector spaces that preserve their structure, i.e. they are compatible with the operations of sum of vectors and multiplication of a vector by a scalar. As we will see, linear maps are represented very efficiently using matrices. The purpose of this chapter is to introduce the concept of linear transformation and understand how it is possible to uniquely associate a matrix to each linear transformation between $\mathbb{R}^n$ and $\mathbb{R}^m$, once we fix the canonical bases in both spaces. Then we will study the kernel, the image of a linear transformation and the *Rank Nullity Theorem*, which is one of the most important results in the theory of vector spaces of finite dimension.

5.1 LINEAR TRANSFORMATIONS: DEFINITION

In calculus we study the real valued functions. These are laws which associate to a real number, ranging in a certain set called *domain* of the function, another real number, belonging to another subset of the real numbers, namely the *codomain* of the function.

For example, $f : \mathbb{R} \longrightarrow \mathbb{R}$, $f(x) = x^2$ is a well-known function from $\mathbb{R}$ to $\mathbb{R}$; its graph is the parabola of equation $y = x^2$. As we know, $\mathbb{R}$ is also a vector space, then f is a function between the vector space $\mathbb{R}$ (the domain) and vector space $\mathbb{R}$ (the codomain). However, this function does not behave well with respect to the vector space structure of $\mathbb{R}$. In fact, if we take two vectors $\mathbf{u}$ and $\mathbf{v}$, and we consider the sum $\mathbf{u} + \mathbf{v}$, we have immediately that $f(\mathbf{u} + \mathbf{v}) \neq f(\mathbf{u}) + f(\mathbf{v})$ (if we draw a graph, we see it immediately). In the same way, we can see that the multiplication by a scalar is not preserved. For example, we see that $f(2 \cdot 3) = 6^2 \neq 2 \cdot f(3) = 2 \cdot 9$.

On the other hand, there are other functions $f : \mathbb{R} \longrightarrow \mathbb{R}$, that behave well with respect to the vector space structure, i.e. they verify the equalities $f(\mathbf{u} + \mathbf{v}) = f(\mathbf{u}) + f(\mathbf{v})$ and $f(\lambda \mathbf{v}) = \lambda f(\mathbf{v})$. Consider for example, the function $f(x) = 3x$. We see immediately that $f(x_1 + x_2) = 3(x_1 + x_2) = 3x_1 + 3x_2 = f(x_1) + f(x_2)$ and also that $f(\lambda x_1) = 3\lambda x_1 = \lambda(3x_1) = \lambda f(x_1)$.

As we will see, those functions are *linear transformations* between vector spaces, and they preserve the structure, i.e. the sum of vectors has as image, via the function, the sum of the images of the vectors, and the image of the product of a vector by a scalar is the product of the scalar and the image of the vector.

Before the formal definition of linear map, we give the definition of function and image.

Definition 5.1.1 We define a *function* f between two sets A and B as a law which associates to each element of A one and only one element of B and denote this law as $f : A \longrightarrow B$. The set A is called *domain* of the function, while the set B is called *codomain* of the function. We define *image* of an element $a \in A$, the element $f(a) \in B$. The set of images of all the elements of A is called *image* of f and is denoted by $\mathrm{Im}(f)$ or sometimes with $f(A)$.

Not all laws that associate elements of a set to elements of another set are functions. For example, we can define a law that goes from the set A of all human beings, to the set B of all human beings (A and B can be the same set), that associates to every person a brother. This is not a function because someone may have more than one brother.

Another example: We consider the law that goes from the set of natural numbers, to the set of natural numbers, that associates to every number one of its divisors. Also this law is not a function.

Let us define linear transformations.

Definition 5.1.2 Let V and W be two vector spaces and let $F : V \longrightarrow W$ be a function. F is called a *linear transformation* if:
1. $F(\mathbf{u} + \mathbf{v}) = F(\mathbf{u}) + F(\mathbf{v})$, for every $\mathbf{u}, \mathbf{v} \in V$,
2. $F(\lambda \mathbf{u}) = \lambda F(\mathbf{u})$ for every $\lambda \in \mathbb{R}$ and for every $\mathbf{u} \in V$.

***Observation* 5.1.3** It is easy to verify that if V and W are two vector spaces, a function $F : V \longrightarrow W$ is a linear transformation if and only if $F(\lambda \mathbf{u} + \mu \mathbf{v}) = \lambda F(\mathbf{u}) + \mu F(\mathbf{v})$, for every $\mathbf{u}, \mathbf{v} \in V$ and for every $\lambda, \mu \in \mathbb{R}$.

Proposition 5.1.4 *Let $F : V \to W$ be a linear transformation; then $F(\mathbf{0}_V) = \mathbf{0}_W$.*

Proof. Let $\mathbf{v}$ be any vector of V. We have

$$F(\mathbf{0}_V) = F(0\mathbf{v}) = 0F(\mathbf{v}) = \mathbf{0}_W$$

because of Definition 5.1.2 and Proposition 2.3.5(iv). ■

Example 5.1.5   1. Let $F : \mathbb{R}^2 \longrightarrow \mathbb{R}^2$ be the function defined by: $F(x, y) = (x + 1, y)$. We want to determine if F is linear, that is, see if the properties 1 and 2 of Definition 5.1.2 are verified. We see property 1 first:

$$F((x_1, y_1) + (x_2, y_2)) = F(x_1 + x_2, y_1 + y_2) = (x_1 + x_2 + 1, y_1 + y_2).$$

On the other hand:

$$F(x_1, y_1) + F(x_2, y_2) = (x_1 + 1, y_1) + (x_2 + 1, y_2) = (x_1 + x_2 + 2, y_1 + y_2).$$

Therefore, the given function is not a linear transformation.

We could also conclude more quickly observing that $F(0,0) = (1,0) \neq (0,0)$, so by the previous proposition the map is not linear.

2. Let $D : \mathbb{R}[x] \longrightarrow \mathbb{R}[x]$ be defined by: $D(p(x)) = p'(x)$, i.e. D is the function which associates to a polynomial its derivative. From calculus we know that $D(p(x) + q(x)) = D(p(x)) + D(q(x))$, i.e. the derivative of the sum of two polynomials is the sum of their derivatives. In addition, we also know that $D(kp(x)) = kD(p(x))$, for each constant $k \in \mathbb{R}$. So we have shown that the derivative D is a linear transformation. We invite the student to verify, in a similar manner, that also the integral is a linear transformation from $\mathbb{R}[x]$ to $\mathbb{R}[x]$.

3. The identity map id $: V \to V$ defined by: $\mathbf{v} \mapsto \mathbf{v}$ for every $\mathbf{v} \in V$ is a linear transformation.

4. The zero map $T : V \to V$ defined by: $\mathbf{v} \mapsto \mathbf{0}_V$ for every $\mathbf{v} \in V$ is a linear transformation.

5. Given a basis $\mathcal{B} = \{\mathbf{v}_1, \ldots, \mathbf{v}_n\}$ of V, the map that associates to each vector $\mathbf{v} = \lambda_1 \mathbf{v}_1 + \ldots + \lambda_n \mathbf{v}_n$ its coordinates $(\lambda_1, \ldots, \lambda_n)$ with respect to the basis $\mathcal{B}$ is an linear transformation. We leave the easy verification as an exercise.

The following observation is crucially importance to understand the correspondence between linear transformations and matrices. We will return to this concept more in detail later.

Observation 5.1.6 To each matrix

$$A = \begin{pmatrix} a_{11} & a_{12} & \cdots & a_{1n} \\ a_{21} & a_{22} & \cdots & a_{2n} \\ \vdots & \vdots & \ddots & \vdots \\ a_{m1} & a_{m2} & \cdots & a_{mn} \end{pmatrix} \in \mathrm{M}_{m,n}(\mathbb{R}),$$

we can associate the function $L_A : \mathbb{R}^n \to \mathbb{R}^m$ so defined:

$$L_A : \quad \mathbb{R}^n \quad \to \quad \quad \quad \quad \quad \mathbb{R}^m$$

$$\begin{pmatrix} x_1 \\ x_2 \\ \vdots \\ x_n \end{pmatrix} \mapsto \begin{pmatrix} a_{11}x_1 + a_{12}x_2 + \ldots + a_{1n}x_n \\ a_{21}x_1 + a_{22}x_2 + \ldots + a_{2n}x_n \\ \vdots \\ a_{m1}x_1 + a_{m2}x_2 + \ldots + a_{mn}x_n \end{pmatrix}$$

$$= \begin{pmatrix} a_{11} & a_{12} & \cdots & a_{1n} \\ a_{21} & a_{22} & \cdots & a_{2n} \\ \vdots & \vdots & \ddots & \vdots \\ a_{m1} & a_{m2} & \cdots & a_{mn} \end{pmatrix} \begin{pmatrix} x_1 \\ x_2 \\ \vdots \\ x_n \end{pmatrix} = A \begin{pmatrix} x_1 \\ x_2 \\ \vdots \\ x_n \end{pmatrix}$$

where the product of A by the vector $(x_1, \ldots, x_n)$ is the *product rows by columns* defined in Chapter 1.

It is easy to see that L_A is a linear transformation. Property 1 of Definition 5.1.2 comes from Proposition 1.2.1 (distributivity of the product rows by columns with respect to the sum), and property 2 is a simple calculation.

More concisely, we can write:

$$L_A \begin{pmatrix} x_1 \\ x_2 \\ \vdots \\ x_n \end{pmatrix} = A \begin{pmatrix} x_1 \\ x_2 \\ \vdots \\ x_n \end{pmatrix}.$$

Let us see a concrete example. If we consider the matrix:

$$A = \begin{pmatrix} 2 & 1 & 0 \\ -1 & 1 & 3 \end{pmatrix},$$

it follows that the linear transformation $L_A : \mathbb{R}^3 \to \mathbb{R}^2$ is defined by:

$$L_A \begin{pmatrix} x_1 \\ x_2 \\ x_3 \end{pmatrix} = \begin{pmatrix} 2 & 1 & 0 \\ -1 & 1 & 3 \end{pmatrix} \begin{pmatrix} x_1 \\ x_2 \\ x_3 \end{pmatrix} = \begin{pmatrix} 2x_1 + x_2 \\ -x_1 + x_2 + 3x_3 \end{pmatrix}$$

$$= x_1 \begin{pmatrix} 2 \\ -1 \end{pmatrix} + x_2 \begin{pmatrix} 1 \\ 1 \end{pmatrix} + x_3 \begin{pmatrix} 0 \\ 3 \end{pmatrix}.$$

Note that we have:

$$L_A \begin{pmatrix} 1 \\ 0 \\ 0 \end{pmatrix} = \begin{pmatrix} 2 \\ -1 \end{pmatrix}, \qquad L_A \begin{pmatrix} 0 \\ 1 \\ 0 \end{pmatrix} = \begin{pmatrix} 1 \\ 1 \end{pmatrix}, \qquad L_A \begin{pmatrix} 0 \\ 0 \\ 1 \end{pmatrix} = \begin{pmatrix} 0 \\ 3 \end{pmatrix},$$

in other words:

the images of the canonical basis vectors are the columns of the matrix A.

This fact will be crucial for the exercises, when we have to determine the image of a linear transformation.

We now wish to know how many possibilities we have for a linear transformation $F : \mathbb{R} \longrightarrow \mathbb{R}$ such that $F(1) = a \in \mathbb{R}$. We observe that, by property 1 of Definition 5.1.2 we have that $F(x) = xF(1) = ax$. So the only linear transformations from $\mathbb{R}$ to $\mathbb{R}$ correspond to the straight lines passing through the origin (hence the name *linear transformation*). This example is particularly instructive because it showed us that, in order to fully understand a linear transformation from $\mathbb{R}$ to $\mathbb{R}$, it is sufficient to know only one value; we chose $F(1)$, but the student can convince himself that the value of F in any other point (as long as nonzero) would have determined F. This is true in general: a linear transformation is completely determined by knowing only some values and precisely the values corresponding to the vectors of a basis (and in fact in the example considered, any number other than zero is a basis of $\mathbb{R}$). This basic fact is encoded by the following theorem.

Theorem 5.1.7 *Let V and W be two vector spaces. If $\{\mathbf{v}_1, \ldots, \mathbf{v}_n\}$ is a basis of V and $\mathbf{w}_1, \ldots, \mathbf{w}_n$ are arbitrary vectors of W, then there is a unique linear transformation $L : V \to W$, such that $L(\mathbf{v}_1) = \mathbf{w}_1, \ldots, L(\mathbf{v}_n) = \mathbf{w}_n$.*

Proof. As $\{\mathbf{v}_1, \ldots, \mathbf{v}_n\}$ is a basis of V there exist unique scalars $\alpha_1, \ldots, \alpha_n$ (the coordinates of $\mathbf{v}$), such that $\mathbf{v} = \alpha_1 \mathbf{v}_1 + \cdots + \alpha_n \mathbf{v}_n$. We define $L(\mathbf{v})$ as follows:

$$L(\mathbf{v}) = \alpha_1 \mathbf{w}_1 + \cdots + \alpha_n \mathbf{w}_n.$$

To verify that L is linear, we need to verify properties 1 and 2 of Definition 5.1.2. Let $\mathbf{v} = \alpha_1 \mathbf{v}_1 + \cdots + \alpha_n \mathbf{v}_n$, $\mathbf{u} = \beta_1 \mathbf{v}_1 + \cdots + \beta_n \mathbf{v}_n$. We have that $\mathbf{v} + \mathbf{u} = (\alpha_1 + \beta_1) \mathbf{v}_1 + \cdots + (\alpha_n + \beta_n) \mathbf{v}_n$ and, if $\lambda \in \mathbb{R}$, we have $\lambda \mathbf{v} = \lambda \alpha_1 \mathbf{v}_1 + \cdots + \lambda \alpha_n \mathbf{v}_n$. Furthermore:

$$L(\mathbf{v} + \mathbf{u}) \;\; = L((\alpha_1 + \beta_1) \mathbf{v}_1 + \cdots + (\alpha_n + \beta_n) \mathbf{v}_n) =$$

$$= (\alpha_1 + \beta_1) \mathbf{w}_1 + \cdots + (\alpha_n + \beta_n) \mathbf{w}_n =$$

$$= \alpha_1 \mathbf{w}_1 + \cdots + \alpha_n \mathbf{w}_n + \beta_1 \mathbf{w}_1 + \cdots + \beta_n \mathbf{w}_n = L(\mathbf{v}) + L(\mathbf{u}).$$

Now we see that

$$L(\lambda \mathbf{v}) \;\; = L(\lambda \alpha_1 \mathbf{v}_1 + \cdots + \lambda \alpha_n \mathbf{v}_n) =$$

$$= \lambda \alpha_1 \mathbf{w}_1 + \cdots + \lambda \alpha_n \mathbf{w}_n =$$

$$= \lambda(\alpha_1 \mathbf{w}_1 + \cdots + \alpha_n \mathbf{w}_n) = \lambda L(\mathbf{v}).$$

So L is a linear transformation.

Now let us prove uniqueness. Suppose that G is a linear transformation $G : V \to W$, such that $G(\mathbf{v}_1) = \mathbf{w}_1, \ldots, G(\mathbf{v}_n) = \mathbf{w}_n$ and L is the linear transformation defined above. Then:

$$G(\mathbf{v}) \;\; = G(\alpha_1 \mathbf{v}_1 + \cdots + \alpha_n \mathbf{v}_n) = \alpha_1 G(\mathbf{v}_1) + \cdots + \alpha_n G(\mathbf{v}_n) =$$

$$= \alpha_1 \mathbf{w}_1 + \cdots + \alpha_n \mathbf{w}_n = L(\mathbf{v}).$$

So $G = L$, as we wanted. ■

Corollary 5.1.8 *Let V and W be two vector spaces. If two linear transformations, $T, S : V \to W$ coincide on a basis of V, then they coincide on the whole V.*

As we shall see, Theorem 5.1.7 is essential to establish a one-to-one correspondence between linear transformations between vector spaces with fixed bases and matrices. Before proceeding, we pause to consider the extraordinary fact that a linear transformation from $\mathbb{R}^n$ a $\mathbb{R}^m$ is known, once we know its values on a set of n elements! If we recall the functions studied in analysis, this is far from being true, that is for the study of the graph of a function $f : \mathbb{R} \longrightarrow \mathbb{R}$, it is not enough to know one value of f, but one must painstakingly study its maximum and minimum, the asymptotes etc. The difference here is that we are considering linear transformations, which are functions with particular properties and behavior.

5.2 LINEAR MAPS AND MATRICES

In this section, we want to look at different ways to write and represent linear transformations between the two vector spaces $\mathbb{R}^n$ and $\mathbb{R}^m$. By Theorem 5.1.7, we know that a linear transformation is uniquely determined by its values on any basis. We now see what happens in practice, keeping in mind Remark 5.1.6. Let us start by examining an example.

Example 5.2.1 Consider the linear transformation $F : \mathbb{R}^2 \longrightarrow \mathbb{R}^3$ such that $F(\mathbf{e}_1) = \mathbf{e}_2 + \mathbf{e}_3$, $F(\mathbf{e}_2) = 2\mathbf{e}_1 - \mathbf{e}_2 + \mathbf{e}_3$, where $\{\mathbf{e}_1, \mathbf{e}_2\}$ is the the canonical basis of $\mathbb{R}^2$, i.e. $\mathbf{e}_1 = (1,0)$, $\mathbf{e}_2 = (0,1)$, and $\{\mathbf{e}_1, \mathbf{e}_2, \mathbf{e}_3\}$ is the the canonical basis of $\mathbb{R}^3$, i.e. $\mathbf{e}_1 = (1,0,0)$, $\mathbf{e}_2 = (0,1,0)$, $\mathbf{e}_3 = (0,0,1)$. We want to determine $F(x,y)$ for a generic vector $(x,y) \in \mathbb{R}^2$. By properties 1 and 2 of the definition of linear transformation we have that:

$$F(x,y) = F(x(1,0) + y(0,1)) = F(x\mathbf{e}_1 + y\mathbf{e}_2) = xF(\mathbf{e}_1) + yF(\mathbf{e}_2) =$$

$$= x(\mathbf{e}_2 + \mathbf{e}_3) + y(2\mathbf{e}_1 - \mathbf{e}_2 + \mathbf{e}_3) = x(0,1,1) + y(2,-1,1) =$$

$$= (2y, x - y, x + y).$$

Now let us see what happens in general.

Let $F : \mathbb{R}^n \longrightarrow \mathbb{R}^m$ be the linear transformation:

$$\begin{aligned} F(\mathbf{e}_1) &= a_{11}\mathbf{e}_1 + a_{21}\mathbf{e}_1 \cdots + a_{m1}\mathbf{e}_m, \\ F(\mathbf{e}_2) &= a_{12}\mathbf{e}_1 + a_{22}\mathbf{e}_1 \cdots + a_{m2}\mathbf{e}_m, \\ &\vdots \\ F(\mathbf{e}_n) &= a_{1n}\mathbf{e}_1 + a_{2n}\mathbf{e}_1 \cdots + a_{mn}\mathbf{e}_m. \end{aligned}$$

We want to express $F(x_1, \ldots, x_n)$ that is, we want to write the image of any vector $(x_1, \ldots, x_n) \in \mathbb{R}^n$.

We proceed exactly as in the example, the reasoning is the same, only more complicated to write.

$$F(x_1, \ldots, x_n) = F(x_1\mathbf{e}_1 + x_2\mathbf{e}_2 + \cdots + x_n\mathbf{e}_n) =$$

$$= x_1 F(\mathbf{e}_1) + x_2 F(\mathbf{e}_2) \cdots + x_n F(\mathbf{e}_n) =$$

$$= x_1(a_{11}\mathbf{e}_1 + a_{12}\mathbf{e}_2 + \cdots + a_{m1}\mathbf{e}_m) + x_2(a_{12}\mathbf{e}_1 + a_{22}\mathbf{e}_1 \cdots + a_{m2}\mathbf{e}_m) +$$

$$\cdots + x_n(a_{1n}\mathbf{e}_1 + \cdots + a_{mn}\mathbf{e}_m) =$$

$$= (a_{11}x_1 + a_{12}x_2 + \ldots + a_{1n}x_n)\mathbf{e}_1 +$$

$$+ (a_{21}x_1 + a_{22}x_2 + \ldots + a_{2n}x_n)\mathbf{e}_2 + \ldots$$

$$+ (a_{m1}x_1 + a_{m2}x_2 + \ldots + a_{mn}x_n)\mathbf{e}_m =$$

$$= \begin{pmatrix} a_{11}x_1 + a_{12}x_2 + \ldots + a_{1n}x_n \\ a_{21}x_1 + a_{22}x_2 + \ldots + a_{2n}x_n \\ \vdots \\ a_{m1}x_1 + a_{m2}x_2 + \ldots + a_{mn}x_n \end{pmatrix}.$$

Let us take a step further, noting that $F(x_1, \ldots, x_n)$ can also be written in a more compact form, using the notation of multiplication of a matrix by a vector:

$$F\begin{pmatrix} x_1 \\ \vdots \\ x_n \end{pmatrix} = A\begin{pmatrix} x_1 \\ \vdots \\ x_n \end{pmatrix}, \quad \text{where} \quad A = \begin{pmatrix} a_{11} & \ldots & a_{1n} \\ \vdots & & \vdots \\ a_{m1} & \ldots & a_{mn} \end{pmatrix}.$$

In practice, we have shown that F is just the transformation L_A associated with A described in Remark 5.1.6.

We collect all our observations in the following theorem, which is *very important* for the exercises. We will denote the canonical bases of $\mathbb{R}^n$ and $\mathbb{R}^m$, respectively, with $\{\mathbf{e}_1, \ldots, \mathbf{e}_n\}$ and $\{\mathbf{e}_1, \ldots, \mathbf{e}_m\}$ (the notation is unambiguous because each vector is uniquely determined as soon as we know to which vector space it belongs).

Theorem 5.2.2 *Assume we have a linear transformation $F : \mathbb{R}^n \longrightarrow \mathbb{R}^m$ and fix in $\mathbb{R}^n$ and $\mathbb{R}^m$ the respective canonical bases. Then we can equivalently express F in one of three ways:*

1. $F(\mathbf{e}_1) = a_{11}\mathbf{e}_1 + \cdots + a_{m1}\mathbf{e}_m,$

   $\vdots$

   $F(\mathbf{e}_n) = a_{1n}\mathbf{e}_1 + \cdots + a_{mn}\mathbf{e}_m;$

2. $F(x_1, \ldots, x_n) = \begin{pmatrix} a_{11}x_1 + a_{12}x_2 + \ldots + a_{1n}x_n \\ a_{21}x_1 + a_{22}x_2 + \ldots + a_{2n}x_n \\ \vdots \\ a_{m1}x_1 + a_{m2}x_2 + \ldots + a_{mn}x_n \end{pmatrix};$

3. $F(\mathbf{x}) = A\mathbf{x}$, *where*

$$\mathbf{x} = \begin{pmatrix} x_1 \\ \vdots \\ x_n \end{pmatrix} \qquad A = \begin{pmatrix} a_{11} & \cdots & a_{1n} \\ \vdots & & \vdots \\ a_{m1} & \cdots & a_{mn} \end{pmatrix}.$$

Corollary 5.2.3 *There is a one to one correspondence between the $m \times n$ matrices and the linear transformations between the vector spaces $\mathbb{R}^n$ and $\mathbb{R}^m$, where we fix in both spaces the respective canonical bases to represent vectors. More precisely the linear transformation $F(x_1, \ldots, x_n) = (a_{11}x_1 + \cdots + a_{1n}x_n, a_{21}x_1 + \cdots + x_n a_{2n}, \ldots, a_{m1}x_1 + \cdots + a_{mn}x_n)$ is associated with the matrix:*

$$A = \begin{pmatrix} a_{11} & \cdots & a_{1n} \\ \vdots & & \vdots \\ a_{m1} & \cdots & a_{mn} \end{pmatrix}$$

and vice versa.

Observation 5.2.4 We observe that, if the linear transformation $F : \mathbb{R}^n \to \mathbb{R}^m$ is associated with the matrix A, where we fix the canonical bases in the domain and codomain, indicating as usual with $\{\mathbf{e}_1, \ldots, \mathbf{e}_n\}$ the canonical basis of $\mathbb{R}^n$, we have that the i-th column of the matrix A (which we denote with A_i) is $F(\mathbf{e}_i)$.

5.3 THE COMPOSITION OF LINEAR TRANSFORMATIONS

In this section, we deal with the composition of linear transformations. If F and G are linear transformations respectively associated with two matrices A and B (with respect to the canonical bases), we have that the composition $F \circ G$ is a linear transformation and we want to know which matrix is associated with it.

We first recall the definition of composition of functions. If $g : A \to B$ and $f : B \to C$ are functions, and *the domain of f coincides with the codomain of g*, we define the *composition* as $f \circ g : A \to C$, which corresponds to applying first g, then f. Formally:

$$\begin{array}{rccc} f \circ g : & A & \longrightarrow & C \\ & a & \mapsto & f(g(a)). \end{array}$$

We observe that the composition is *associative*, that is, if we have three functions $h : A \to B$, $g : B \to C$, $f : C \to D$, we have that $f \circ (g \circ h) = (f \circ g) \circ h$. In fact, for every $a \in A$:

$$(f \circ (g \circ h))(a) = f((g \circ h)(a)) = f(g(h(a))) = (f \circ g)(h(a)) = ((f \circ g) \circ h)(a).$$

The composition operation is not commutative, i.e. general $f \circ g \neq g \circ f$ and it might even happen that $g \circ f$ is not defined.

Example 5.3.1 Let

$$\begin{array}{rccc} f : & \mathbb{R} & \longrightarrow & \mathbb{R} \\ & y & \mapsto & y^2 - 1 \end{array} \qquad \begin{array}{rccc} g : & \mathbb{R} & \longrightarrow & \mathbb{R} \\ & x & \mapsto & x + 2 \end{array}.$$

We compute $f \circ g : \mathbb{R} \to \mathbb{R}$ and $g \circ f : \mathbb{R} \to \mathbb{R}$,

$$f \circ g : \quad x \mapsto g(x) = x + 2 \mapsto f(x + 2) = (x + 2)^2 - 1 = x^2 + 2x + 3,$$

$$g \circ f : y \mapsto f(y) = y^2 - 1 \mapsto g(y^2 - 1) = (y^2 - 1) + 2 = y^2 + 1.$$

In this case, $f \circ g \neq g \circ f$.

Let us now see an important example in linear algebra.

Example 5.3.2 Consider the two linear transformations $L_A : \mathbb{R}^3 \longrightarrow \mathbb{R}^2$, $L_B : \mathbb{R}^2 \longrightarrow \mathbb{R}^2$ associated with the matrices:

$$A = \begin{pmatrix} -1 & 1 & 2 \\ 3 & 1 & 0 \end{pmatrix}, \qquad B = \begin{pmatrix} 2 & 1 \\ 1 & 3 \end{pmatrix},$$

with respect to the canonical bases in $\mathbb{R}^2$ and $\mathbb{R}^3$. We see immediately that $L_B \circ L_A$ is defined, while $L_A \circ L_B$ it is not defined. This is because L_A must have as argument a vector in $\mathbb{R}^3$, while for every $\mathbf{v} \in \mathbb{R}^2$ we have that $L_B(\mathbf{v}) \in \mathbb{R}^2$, then $L_A(L_B(\mathbf{v}))$ does not make sense.

Let us now see that, whenever we can take the composition of two linear transformations, we still obtain a linear transformation.

Proposition 5.3.3 *Let U, V, W be vector spaces and $G : U \to V$, $F : V \to W$ two linear transformations. Then the function $F \circ G : U \to W$ is a linear transformation.*

Proof. We have to check the two properties of Definition 5.1.2.

1. Let $\mathbf{u}_1, \mathbf{u}_2 \in U$. Then $(F \circ G)(\mathbf{u}_1 + \mathbf{u}_2) = F(G(\mathbf{u}_1 + \mathbf{u}_2)) = F(G(\mathbf{u}_1) + G(\mathbf{u}_2)) = F(G(\mathbf{u}_1)) + F(G(\mathbf{u}_2)) = (F \circ G)(\mathbf{u}_1) + (F \circ G)(\mathbf{u}_2)$, where we used first the definition of composition, then linearity of G, linearity of F and finally again the definition of composition.

2. Let $\mathbf{u} \in U$ and $\lambda \in \mathbb{R}$. Then $(F \circ G)(\lambda \mathbf{u}) = F(G(\lambda \mathbf{u})) = F(\lambda(G(\mathbf{u})) = \lambda F(G(\mathbf{u})) = \lambda(F \circ G)(\mathbf{u})$, where as in the previous case, we have used first the definition of composition, then linearity of G, linearity of F and finally again the definition of composition. ■

From this proposition we can get an easy corollary. It is very useful in applications.

Corollary 5.3.4 *Let $L_A : \mathbb{R}^n \longrightarrow \mathbb{R}^m$, $L_B : \mathbb{R}^m \longrightarrow \mathbb{R}^s$ be two linear transformations associated to the matrices A and B, respectively. Then the linear map $L_B \circ L_A$ is associated with the matrix BA, namely:*

$$L_B \circ L_A = L_{BA}.$$

Proof. The proof is a simple check:

$$(L_B \circ L_A) \begin{pmatrix} x_1 \\ \vdots \\ x_n \end{pmatrix} = L_B \left(A \begin{pmatrix} x_1 \\ \vdots \\ x_n \end{pmatrix} \right) = B \left(A \begin{pmatrix} x_1 \\ \vdots \\ x_n \end{pmatrix} \right) =$$

$$= (BA) \begin{pmatrix} x_1 \\ \vdots \\ x_n \end{pmatrix} = L_{BA} \begin{pmatrix} x_1 \\ \vdots \\ x_n \end{pmatrix}.$$

In this check, we used the associativity of the multiplication rows by columns between matrices. ■

5.4   KERNEL AND IMAGE

We now want to get into the theory of linear transformation and introduce the concepts of *kernel* and *image*, which are respectively subspaces of the domain and codomain of a linear transformation.

Definition 5.4.1 Let V and W be two vector spaces and $L : V \to W$ be a linear transformation. We call *kernel* of L, the set of vectors in V whose image is the zero vector of W. This set is denoted by $\operatorname{Ker} L$.

We call *image* of L, the set of vectors in W which are images of some vectors of V, that is,

$$\operatorname{Im}(L) = \{\mathbf{w} \in W \mid \mathbf{w} = L(\mathbf{v}) \text{ for some } \mathbf{v} \in V\}.$$

Let us see some examples.

Example 5.4.2 1. Consider the derivation $D : \mathbb{R}[x] \longrightarrow \mathbb{R}[x]$, defined by $D(p(x)) = p'(x)$. As we have seen D is a linear transformation. We want to know which are the polynomial $p(x)$ whose image is zero, i.e. such that $D(p(x)) = 0$. From calculus, we know they are all the constant polynomials. So $\operatorname{Ker}(D) = \{c \mid c \in \mathbb{R}\}$. Let us look at the image of D. We ask which polynomials are derivatives of other polynomials. From calculus we know they are all the polynomials (we are in fact integrating), so $\operatorname{Im}(D) = \mathbb{R}[x]$.

2. Consider now the linear transformation $L : \mathbb{R}^3 \longrightarrow \mathbb{R}^2$ defined by: $L(\mathbf{e}_1) = 2\mathbf{e}_1 - \mathbf{e}_2$, $L(\mathbf{e}_2) = \mathbf{e}_1$, $L(\mathbf{e}_3) = \mathbf{e}_1 + 2\mathbf{e}_2$.

As seen in Theorem 5.2.2, we know that $L(x, y, z) = (2x + y + z, -x + 2z)$. We want to determine kernel and image of L. We have that $\operatorname{Ker}(L)$ is the set of vectors whose image is the zero vector, that is,

$$\operatorname{Ker}(L) = \{(x, y, z) \mid (2x + y + z, -x + 2z) = (0, 0)\}$$

$$= \{(x, y, z) \mid 2x + y + z = 0, -x + 2z = 0\}$$

$$= \{(x, y, z) \mid z = \tfrac{1}{2}x, y = -2x - z = -\tfrac{5}{2}x\} = \{(x, -\tfrac{5}{2}x, x) \mid x \in \mathbb{R}\}$$

$$= \langle (2, -5, 1) \rangle.$$

Let us see the image.

$$\text{Im}(L) = \left\{ \mathbf{w} \in \mathbb{R}^2 \mid \mathbf{w} = \begin{pmatrix} 2x + y + z \\ -x + 2z \end{pmatrix} \text{ with } x, y, z \in \mathbb{R} \right\}$$

$$= \left\{ \mathbf{w} \in \mathbb{R}^2 \mid \mathbf{w} = \begin{pmatrix} 2x \\ -x \end{pmatrix} + \begin{pmatrix} y \\ 0 \end{pmatrix} + \begin{pmatrix} z \\ 2z \end{pmatrix} \right.$$

with $x, y, z \in \mathbb{R}$ }

$$= \left\{ \mathbf{w} \in \mathbb{R}^2 \mid \mathbf{w} = x \begin{pmatrix} 2 \\ -1 \end{pmatrix} + y \begin{pmatrix} 1 \\ 0 \end{pmatrix} + z \begin{pmatrix} 1 \\ 2 \end{pmatrix} \right.$$

with $x, y, z \in \mathbb{R}$ }

$$= \left\langle \begin{pmatrix} 2 \\ -1 \end{pmatrix}, \begin{pmatrix} 1 \\ 0 \end{pmatrix}, \begin{pmatrix} 1 \\ 2 \end{pmatrix} \right\rangle.$$

The fact that the linear maps are defined so to preserve both operations of vector spaces, makes both the kernel and the image of a given linear transformation to be linear subspaces.

Proposition 5.4.3 *Let $L : V \longrightarrow W$ be a linear transformation.*
1. The kernel of L is a subspace of the domain V.
2. The image of L is a vector subspace of the codomain W.

Proof. (1) We note first that $\text{Ker}(L)$ is not the empty set, because $\mathbf{0}_V \in \text{Ker}(L)$ by Proposition 5.1.4. We have then to verify that $\text{Ker}(L)$ is closed with respect to the sum of vectors and the multiplication of a vector by a scalar. Let us start with the sum. Let $\mathbf{u}, \mathbf{v} \in \text{Ker}(L)$. Then $L(\mathbf{u}) = L(\mathbf{v}) = \mathbf{0}_W$, so $L(\mathbf{u}+\mathbf{v}) = L(\mathbf{u})+L(\mathbf{v}) = \mathbf{0}_W+\mathbf{0}_W = \mathbf{0}_W$, thus $\mathbf{u}+\mathbf{v} \in \text{Ker}(L)$. Now we verify the closure of L with respect to the product by a scalar. If $\alpha \in \mathbb{R}$ and $\mathbf{u} \in \text{Ker}(L)$ one has $L(\alpha\mathbf{u}) = \alpha L(\mathbf{u}) = \alpha\mathbf{0}_W = \mathbf{0}_W$, so $\alpha\mathbf{u} \in \text{Ker}(L)$.

(2) Let us now see the same two properties for $\text{Im}(L)$. We have that $\mathbf{0}_W \in \text{Im}(L)$ by Proposition 5.1.4. Let now $\mathbf{w}_1, \mathbf{w}_2 \in \text{Im}(L)$. So there exist $\mathbf{v}_1, \mathbf{v}_2 \in V$, such that $L(\mathbf{v}_1) = \mathbf{w}_1$ and $L(\mathbf{v}_2) = \mathbf{w}_2$. Therefore, $\mathbf{w}_1 + \mathbf{w}_2 = L(\mathbf{v}_1) + L(\mathbf{v}_2) = L(\mathbf{v}_1 + \mathbf{v}_2) \in \text{Im}(L)$ and $\alpha\mathbf{w}_1 = \alpha L(\mathbf{v}_1) = L(\alpha\mathbf{v}_1) \in \text{Im}(L)$ for every $\alpha \in \mathbb{R}$. ■

Proposition 5.4.4 *Let $L : V \longrightarrow W$ be a linear transformation. Then the subspace $\text{Im}(L)$ is generated by the image of any basis of V, i.e. if $\{\mathbf{v}_1, \ldots, \mathbf{v}_n\}$ is a basis of V, then:*

$$\text{Im}(L) = \langle L(\mathbf{v}_1), \ldots, L(\mathbf{v}_n) \rangle.$$

Proof. We show first that $\text{Im}(L) \subseteq \langle L(\mathbf{v}_1), \ldots, L(\mathbf{v}_n) \rangle$. $\text{Im}(L)$ consists of all vectors of the type $L(\mathbf{v})$ for all $\mathbf{v} \in V$. Let $\{\mathbf{v}_1, \ldots, \mathbf{v}_n\}$ be a basis of V. If $\mathbf{v} \in V$, then $\mathbf{v} = \lambda_1\mathbf{v}_1 + \cdots + \lambda_n\mathbf{v}_n$, with $\lambda_1, \ldots, \lambda_n \in \mathbb{R}$. So:

$$L(\mathbf{v}) = L(\lambda_1\mathbf{v}_1 + \cdots + \lambda_n\mathbf{v}_n) = \lambda_1 L(\mathbf{v}_1) + \cdots + \lambda_n L(\mathbf{v}_n) \in \langle L(\mathbf{v}_1), \ldots, L(\mathbf{v}_n) \rangle.$$

This proves that $\text{Im}(L) \subseteq \langle L(\mathbf{v}_1), \ldots, L(\mathbf{v}_n)\rangle$.

The inclusion $\langle L(\mathbf{v}_1), \ldots, L(\mathbf{v}_n)\rangle \subseteq \text{Im}(L)$ is true by Proposition 3.1.5, because $\text{Im}(L) \subset V$ is a subspace of V. ■

Consider the earlier example.

Example 5.4.5 We want to determine a basis for the image of the linear transformation $L : \mathbb{R}^3 \longrightarrow \mathbb{R}^2$ defined by: $L(\mathbf{e}_1) = 2\mathbf{e}_1 - \mathbf{e}_2$, $L(\mathbf{e}_2) = \mathbf{e}_1$, $L(\mathbf{e}_3) = \mathbf{e}_1 + 2\mathbf{e}_2$. From the previous proposition we know that

$$\text{Im}(L) = \langle 2\mathbf{e}_1 - \mathbf{e}_2, \mathbf{e}_1, \mathbf{e}_1 + 2\mathbf{e}_2 \rangle.$$

We observe that $\text{Im}L \subseteq \mathbb{R}^2$, then the image has dimension at most 2. We easily see that the two vectors that generate the image are linearly independent. We can therefore conclude that they are a basis of $\mathbb{R}^2$ and then that $\text{Im}(L) = \mathbb{R}^2$.

We shall see in the next section a general method to determine the kernel and the image of a linear transformation.

Kernel and image are related to injectivity and surjectivity of the linear transformation. We recall briefly these basic concepts.

Definition 5.4.6 Given a function between two sets A and B,

$$f : A \longrightarrow B$$

1. We say that f is *injective* if whenever $f(x) = f(y)$ then $x = y$, i.e. two distinct elements x and y can never have the same image.

2. We say that f is *surjective* if every element of B is the image of some element of A, i.e. the codomain of f coincides with the image of f.

3. We say that f is *bijective or a bijection* if it is both injective and surjective.

4. We say that f is *invertible* if there is a function $g : B \longrightarrow A$ called *inverse* of f, such that $f \circ g = \text{id}_B$ and $g \circ f = \text{id}_A$, where in general $\text{id}_X : X \longrightarrow X$ is the identity function, which associates each element to itself, i.e. $\text{id}_X(x) = x$. Very often we denote the inverse of f with f^{-1}.

In the next proposition, we state the fact that a function is bijective if and only if it is invertible; therefore from now on, we will use these two terms interchangeably.

Proposition 5.4.7 *Let $f : A \longrightarrow B$ be a function between two sets A and B; then f is bijective if and only if it is invertible.*

Proof. Suppose that f is bijective, and we want to construct the inverse g of f. Let $b \in B$. Since is f is bijective it is, in particular, surjective, thus there exists $a \in A$ such that $f(a) = b$. In addition to this, a is unique, because f is injective and so if a' is such that $f(a') = b$ then $a = a'$. Define $g : B \longrightarrow A$ by the rule $g(b) = a$. Now by construction $f(g(b)) = f(a) = b$, thus $f \circ g = \text{id}_B$. In addition $g(f(a)) = g(b) = a$, thus $g \circ f = \text{id}_A$. This shows that g is the inverse of f.

Now suppose that f is invertible and $g : B \longrightarrow A$ is the inverse of f. Then $f(g(b)) = b$ for each $b \in B$, and $g(f(a)) = a$ for each $a \in A$. We show that f is injective. Let $a_1, a_2 \in A$ be such that $f(a_1) = f(a_2)$. Then $g(f(a_1)) = g(f(a_2))$, and then $a_1 = g(f(a_1)) = g(f(a_2)) = a_2$. For surjectivity, take $b \in B$, and let $a = g(b)$, then $f(a) = f(g(b)) = b$ and then b is image of a by f. ■

Let us now return to linear transformations.

Proposition 5.4.8 *Let $L : V \longrightarrow W$ be a linear map.*

1. *L is injective if and only if $\mathrm{Ker}\,(L) = \mathbf{0}_V$, that is, its kernel is the zero subspace of the domain V.*

2. *L is surjective if and only if $\mathrm{Im}(L) = W$, i.e. the image of L coincides with the codomain.*

Proof. (1) We show that if L is injective $\mathrm{Ker}\,(L) = \mathbf{0}_V$. If $\mathbf{u} \in \mathrm{Ker}\,(L)$ then $L(\mathbf{u}) = \mathbf{0}_W = L(\mathbf{0}_V)$ and since L is injective, $\mathbf{u} = \mathbf{0}_V$.

Viceversa, let $\mathrm{Ker}\,L = (\mathbf{0}_V)$ and suppose that $L(\mathbf{u}) = L(\mathbf{v})$ for some $\mathbf{u}, \mathbf{v} \in V$. Then $L(\mathbf{u} - \mathbf{v}) = L(\mathbf{u}) - L(\mathbf{v}) = \mathbf{0}_W$. So $\mathbf{u} - \mathbf{v} \in \mathrm{Ker}\,(L) = \mathbf{0}_V$, and we have that $\mathbf{u} - \mathbf{v} = \mathbf{0}_V$, then $\mathbf{u} = \mathbf{v}$, therefore f is an injection.

(2)This is precisely the definition of surjectivity. ■

The next proposition tells us that the injective linear transformations preserve linear independence.

Proposition 5.4.9 *Let V and W be vector spaces, $\mathbf{v}_1, \ldots, \mathbf{v}_r$ linearly indipendent vectors of V and $L : V \to W$ an injective linear map. Then $L(\mathbf{v}_1), \ldots, L(\mathbf{v}_r)$ are linearly indipendent vectors of W.*

Proof. Let $\alpha_1 L(\mathbf{v}_1) + \cdots + \alpha_r L(\mathbf{v}_r) = \mathbf{0}_W$, with $\alpha_1, \ldots, \alpha_r \in \mathbb{R}$. Then by Proposition 5.1.4, we have that $L(\alpha_1 \mathbf{v}_1 + \cdots + \alpha_r \mathbf{v}_r) = \mathbf{0}_W$. Since L is injective, it follows that $\alpha_1 \mathbf{v}_1 + \cdots + \alpha_r \mathbf{v}_r = \mathbf{0}_V$. But since $\mathbf{v}_1, \ldots, \mathbf{v}_r$ are linearly independent, we have $\alpha_1 = \ldots = \alpha_r = 0$, so also $L(\mathbf{v}_1), \ldots, L(\mathbf{v}_r)$ are linearly independent. ■

5.5   THE RANK NULLITY THEOREM

The Rank Nullity Theorem is perhaps the most important result in the theory of linear transformations, and it is a formidable tool to solve exercises.

Theorem 5.5.1 *Let $L : V \longrightarrow W$ be a linear map. Then*

$$\dim V = \dim(\mathrm{Ker}\,L) + \dim(\mathrm{Im}L). \tag{5.1}$$

Proof. Let $\{\mathbf{u}_1, \ldots, \mathbf{u}_r\}$ be a basis for the subspace $\mathrm{Ker}\,L$. By Theorem 4.2.1, we can complete it to a basis $\mathcal{B}$ of V. Let

$$\mathcal{B} = \{\mathbf{u}_1, \ldots, \mathbf{u}_r, \mathbf{w}_{r+1}, \ldots, \mathbf{w}_n\}.$$

If we prove that $\mathcal{B}_1 = \{L(\mathbf{w}_{r+1}), \ldots, L(\mathbf{w}_n)\}$ is a basis for $\mathrm{Im}(L)$ then the theorem is proved, as $\dim(\mathrm{Ker}\,(L)) = r$, $\dim(V) = n$ and $\dim(\mathrm{Im}(L)) = n - r$ (the dimension of $\mathrm{Im}(L)$ is the number of vectors in a basis and $\mathcal{B}_1$ contains $n - r$ vectors).

Certainly $\mathcal{B}_1$ is a system of generators for $\mathrm{Im}(L)$, by Proposition 5.4.4. Now we show that the vectors in $\mathcal{B}_1$ are linearly independent. Let

$$\alpha_{r+1} L(\mathbf{w}_{r+1}) + \cdots + \alpha_n L(\mathbf{w}_n) = \mathbf{0},$$

with $\alpha_{r+1}, \ldots, \alpha_n \in \mathbb{R}$. We want to show that $\alpha_{r+1} = \cdots = \alpha_n = 0$.

We have:

$$\mathbf{0} = \alpha_{r+1} L(\mathbf{w}_{r+1}) + \cdots + \alpha_n L(\mathbf{w}_n) = L(\alpha_{r+1} \mathbf{w}_1 + \cdots + \alpha_n \mathbf{w}_n).$$

and therefore $\mathbf{w} = \alpha_{r+1} \mathbf{w}_1 + \cdots + \alpha_n \mathbf{w}_n$ belongs to the kernel of L. Since $\mathrm{Ker}\,(L) = \langle \mathbf{u}_1, \ldots, \mathbf{u}_r \rangle$ we can write $\mathbf{w}$ in the form $\mathbf{w} = \alpha_1 \mathbf{u}_1 + \cdots + \alpha_r \mathbf{u}_r$, with $\alpha_1, \ldots, \alpha_r \in \mathbb{R}$. Then

$$\alpha_{r+1} \mathbf{w}_1 + \cdots + \alpha_n \mathbf{w}_n = \alpha_1 \mathbf{u}_1 + \cdots + \alpha_r \mathbf{u}_r,$$

from which it follows that

$$\alpha_{r+1} \mathbf{w}_1 + \cdots + \alpha_n \mathbf{w}_n - (\alpha_1 \mathbf{u}_1 + \cdots + \alpha_r \mathbf{u}_r) = \mathbf{0}$$

and being $\mathcal{B}$ a basis for V this implies that $\alpha_1 = \ldots = \alpha_n = 0$, concluding the proof of the theorem. ■

Formula 5.1 places restrictions on the type and the existence of linear maps between two given vector spaces.

Proposition 5.5.2 *Let V and W be two vector spaces.*

1. *If $\dim V > \dim W$ there are no injective linear maps from V to W.*

2. *If $\dim V < \dim W$ there are no surjective linear transformations from V to W.*

Proof. It is a simple application of Theorem 5.5.1.
1. If $L : V \to W$ is a linear injection, then $\dim(\mathrm{Ker}\,L) = 0$ and thus $\dim V = \dim(\mathrm{Im}L)$. Since $\mathrm{Im}L$ is a subspace of W it follows that $\dim V \le \dim W$.
2. Similarly, if $L : V \to W$ is surjective, then $\mathrm{Im}L = W$ and thus $\dim W = \dim V + \dim(\mathrm{Ker}\,L)$, thus $\dim W \le \dim V$. ■

Example 5.5.3 Consider the linear tranformation $F : \mathbb{R}^4 \longrightarrow \mathbb{R}^2$ defined by: $F(\mathbf{e}_1) = \mathbf{e}_1 - \mathbf{e}_2$, $F(\mathbf{e}_2) = 3\mathbf{e}_1 - 4\mathbf{e}_2$, $F(\mathbf{e}_3) = -\mathbf{e}_1 - 5\mathbf{e}_2$, $F(\mathbf{e}_4) = 3\mathbf{e}_1 + \mathbf{e}_2$. We want to determine if the function is injective, surjective, bijective. By the previous theorem, there is no need to make any calculations, because we have that the map cannot be injective as the dimension of $\mathbb{R}^4$ is larger than the dimension of $\mathbb{R}^2$. We now come to surjectivity. The image of F is generated by the vectors: $\mathbf{e}_1 - \mathbf{e}_2$, $3\mathbf{e}_1 - 4\mathbf{e}_2$, $-\mathbf{e}_1 - 5\mathbf{e}_2$, $3\mathbf{e}_1 + \mathbf{e}_2$. Since at least two of them are linearly independent, they form a basis of $\mathbb{R}^2$, and therefore the function is surjective. Since it is not an injection, F is not bijective.

5.6 ISOMORPHISM OF VECTOR SPACES

The concept of isomorphism allows us to identify two vector spaces, and then to treat them in the same way when we have to solve linear algebra problems, such as, for example, to determine if a set of vectors is linearly independent, or for the calculation of the kernel and of the image of a linear transformation.

Definition 5.6.1 A linear map $L : V \longrightarrow W$ is said to be an *isomorphism*, if it is invertible, or equivalently if it is injective and surjective.

Similarly, two vector spaces V and W are called *isomorphic*, if there is an isomorphism $L : V \longrightarrow W$; in this case, we write $V \cong W$.

Example 5.6.2 Consider the linear map $L : \mathbb{R}_2[x] \longrightarrow \mathbb{R}^3$ defined by: $L(x^2) = (1,0,0)$, $L(x) = (0,1,0)$, $L(1) = (0,0,1)$. This linear transformation is invertibile. To show this, we can determine the kernel and see that it is the zero subspace and determine the image and see that it is all $\mathbb{R}^3$. We leave this as an exercise. Alternatively, we can define the linear transformation $T : \mathbb{R}^3 \longrightarrow \mathbb{R}_2[x]$, such that $T(\mathbf{e}_1) = x^2$, $T(\mathbf{e}_2) = x$, $T(\mathbf{e}_3) = 1$ and verify that it is the inverse of L (the student may want to do these verifications as an exercise). Therefore $\mathbb{R}_2[x]$ and $\mathbb{R}^3$ are isomorphic. Somehow, it is as if they were the same space, as we created a *one to one correspondence* that associates to a vector in $\mathbb{R}_2[x]$, one and only one vector in $\mathbb{R}^3$, and vice versa. This correspondence also preserves the operations of sum of vectors and multiplication of a vector by a scalar. In fact, we had already noticed that, once we fix basis in $\mathbb{R}_2[x]$, each vector is written using three coordinates, just like a vector in $\mathbb{R}^3$. If we fix the canonical basis $\{x^2, x, 1\}$, the linear map that associates to each polynomial its coordinates is just the isomorphism $L : \mathbb{R}_2[x] \longrightarrow \mathbb{R}^3$ described above. Once we write the coordinates of a polynomial, we can treat it as an element of $\mathbb{R}^3$. For example, to determine whether some polynomials are linearly independent or to determine a basis of the subspace they generate, we use the Gaussian algorithm as described in Chapter 1.

The next theorem is particularly important, since it tells us that not only $\mathbb{R}_2[x]$, but any vector space of finite dimension is isomorphic to $\mathbb{R}^N$ for a certain N (which of course depends on the vector space we consider). So the calculation methods we have described to solve various problems in the vector space $\mathbb{R}^N$s can be applied to any vector space V using, instead of the N-tuples of real numbers, the coordinates of the vectors of the vector space V with respect to a fixed basis.

Theorem 5.6.3 *Two vector spaces V and W are isomorphic if and only if they have the same dimension.*

Proof. If $\dim V = \dim W = n$, let $\{\mathbf{u}_1, \ldots, \mathbf{u}_n\}$ and $\{\mathbf{w}_1, \ldots, \mathbf{w}_n\}$ be two bases of V and W, respectively. Then the linear map $L : V \to W$ defined by $L(\mathbf{v}_i) = \mathbf{w}_i$ for $i = 1, \ldots, n$ is an isomorphism. In fact, by Proposition 5.4.4, we have that $\mathrm{Im}L = \langle \mathbf{w}_1, \ldots, \mathbf{w}_n \rangle$, and since is the vectors $\mathbf{w}_1, \ldots, \mathbf{w}_n$ generate W, it follows that L is surjective. Then, by the Rank Nullity Theorem 5.5.1, we have that $\dim(\mathrm{Ker}\,L) =$

$\dim V - \dim(\mathrm{Im}L) = \dim V - \dim W = 0$, so Ker L is the zero subspace and by Proposition 5.4.8, we have that L is injection.

Conversely, if two vector spaces are isomorphic, there exists a linear map $L : V \to W$ that is both injective and surjective. Then, by Proposition 5.4.8, we have that $\dim(\mathrm{Ker}\,L) = 0$ and $\dim(\mathrm{Im}L) = \dim W$; applying the Rank Nullity Theorem 5.5.1, we get that $\dim V = \dim(\mathrm{Ker}\,L) + \dim(\mathrm{Im}L) = \dim W$. ■

Since we know the dimensions of the vector spaces $\mathrm{M}_{m,n}(\mathbb{R})$ and $\mathbb{R}_d[x]$, we immediately get the following corollary.

Corollary 5.6.4 $\mathrm{M}_{m,n}(\mathbb{R}) \cong \mathbb{R}^{mn}$, $\mathbb{R}_d[x] \cong \mathbb{R}^{d+1}$.

5.7 CALCULATION OF KERNEL AND IMAGE

This section is extremely important for the exercises as it provides us with practical methods for the calculation of bases for the kernel and the image of a given linear transformation.

We begin with the *calculation of a basis* of the kernel of a linear map.

Suppose we have a linear map $F : \mathbb{R}^n \longrightarrow \mathbb{R}^m$ and we want to determine a basis for the kernel. We endow $\mathbb{R}^n$ and $\mathbb{R}^m$ with the canonical bases; then, by Proposition 5.2.2, we have that $F(\mathbf{x}) = A\mathbf{x}$, for a suitable matrix $A \in \mathrm{M}_{m,n}(\mathbb{R})$. By the definition of kernel, we have:

$$\mathrm{Ker}\,F = \{\mathbf{x} \in \mathbb{R}^n \,|\, A\mathbf{x} = \mathbf{0}\};$$

that is, the kernel of F is *the set of solutions of the homogeneous linear system associated with A.*

Let us see a concrete example.

Example 5.7.1 Consider the linear map $F : \mathbb{R}^4 \longrightarrow \mathbb{R}^2$ defined by: $F(\mathbf{e}_1) = -\mathbf{e}_2$, $F(\mathbf{e}_2) = 3\mathbf{e}_1 - 4\mathbf{e}_2$, $F(\mathbf{e}_3) = -\mathbf{e}_1$, $F(\mathbf{e}_4) = 3\mathbf{e}_1 + \mathbf{e}_2$. We want to determine a basis for the kernel of F. We write the matrix A associated with F with respect to the canonical bases:

$$A = \begin{pmatrix} 0 & 3 & 3 & -1 \\ -4 & -1 & 0 & 1 \end{pmatrix}.$$

Therefore $F(x_1, x_2, x_3, x_4) = (3x_2 - x_3 + 3x_4, -x_1 - 4x_2 + x_4)$, and Ker F is the set of solutions of the homogeneous linear system:

$$\begin{cases} 3x_2 - x_3 + 3x_4 = 0 \\ -x_1 - 4x_2 + x_4 = 0, \end{cases}$$

which is indeed associated with the matrix A.

To reduce A in row echelon form, it is sufficient to exchange its two lines and we get:

$$A' = \begin{pmatrix} -1 & -4 & 0 & 1 \\ 0 & 3 & -1 & 3 \end{pmatrix}.$$

Solving the system, we therefore have that

$$\text{Ker } F = \left\{ \left(-\frac{4}{3} + x_3 5 x_4, \frac{1}{3} x_3 - x_4, x_3, x_4 \right) | x_3, x_4 \in \mathbb{R} \right\} =$$

$$\left\{ \left(-\frac{4}{3} x_3, \frac{1}{3} x_3, x_3, 0 \right) + (5 x_4, -x_4, 0, x_4) | x_3, x_4 \in \mathbb{R} \right\} =$$

$$\left\{ x_3 \left(-\frac{4}{3}, \frac{1}{3}, 1, 0 \right) + x_4 (5, -1, 0, 1) | x_3, x_4 \in \mathbb{R} \right\} =$$

$$\left\langle \left(-\frac{4}{3}, \frac{1}{3}, 1, 0 \right), (5, -1, 0, 1) \right\rangle.$$

We observe that the vectors $\left(-\frac{4}{3}, \frac{1}{3}, 1, 0 \right), (5, -1, 0, 1)$ not only generate $\text{Ker } F$, but they are also linearly independent, because they are not one a multiple of the other, so they are a basis of $\text{Ker } F$.

Another way to understand the above equalities is the following: $\text{Ker } F$ is the set of linear combinations of the vectors $\left(-\frac{4}{3}, \frac{1}{3}, 1, 0 \right), (5, -1, 0, 1)$, obtained by placing, respectively, first $x_3 = 1$, $x_4 = 0$, then $x_3 = 0, x_4 = 1$.

Observation 5.7.2 The phenomenon described in the previous example occurs in general. If the set of solutions of a homogeneous linear system depends on k free variables, then this set is the vector space generated by the k vectors $\mathbf{v}_1, \ldots, \mathbf{v}_k$, where $\mathbf{v}_i$ is obtained by setting the i-th free variable equal to 1 and the other free variables equal to 0, for every $i = 1, \ldots, k$. Furthermore, the vectors so obtained are linearly independent, and therefore they are a basis for this vector space.

Let us formalize what we have just observed. We first need a definition:

Definition 5.7.3 We call *row rank* of a matrix $M \in \text{M}_{m,n}(\mathbb{R})$ the maximum number of linearly independent rows of M, which is the dimension of the subspace of $\mathbb{R}^n$ generated by the rows of M. The row rank of A is denoted by $\text{rr}(A)$.

If $A \in \text{M}_{m,n}(\mathbb{R})$ is a matrix in echelon form the definition of row rank given above coincides with Definition 1.3.4. In fact, by Proposition 4.3.3 the nonzero rows of a matrix A in row echelon form are linearly independent, so the dimension of the subspace generated by the rows of A coincides with the number of nonzero rows of A.

Proposition 5.7.4 *Let* $A\mathbf{x} = \mathbf{0}$ *be a homogeneous linear system, where* $A \in \text{M}_{m,n}(\mathbb{R})$, *and let* W *be the set of its solutions. Then* W *is a vector space of dimension* $n - \text{rr}(A)$.

Proof. Let L_A be the linear transformation $L_A : \mathbb{R}^n \to \mathbb{R}^m$ associated with the matrix A, with respect to the canonical bases of the domain and codomain. Then W is actually the kernel of L_A, and so it is a subspace of the vector space $\mathbb{R}^n$ by Proposition 5.4.3. To determine W, we can use the Gaussian algorithm and reduce the matrix $(A|\mathbf{0})$ in row echelon form, obtaining a matrix $(A'|\mathbf{0})$. Now, by Proposition

4.3.1, we have that $\mathrm{rr}(A) = \mathrm{rr}(A') = r$, and in turn r is equal to the number of non-zero rows of A', that is the number of pivots of A'. Then, as we saw in Chapter 1, we can assign an arbitrary value to each of the $n - r$ variables, and write the r variables corresponding to the pivots in terms of these values. Let $x_{i_1}, \ldots, x_{i_{n-r}}$ be the free variables and let $\mathbf{w}_j$ be the solution of the system obtained by putting $x_{i_j} = 1$ and the other free variables equal to zero. Proceeding exactly as in Example 5.7.1, we obtain that the vectors $\mathbf{w}_1, \ldots, \mathbf{w}_r$ generate W. We now show that they are also linearly independent, from which it follows that they are a basis of W, and so W has dimension $n - r$. Suppose that $\mathbf{w} = \lambda_1 \mathbf{w}_1 + \cdots + \lambda_{n-r} \mathbf{w}_{n-r} = \mathbf{0}$ with $\lambda_1, \ldots, \lambda_{n-r} \in \mathbb{R}$. We observe the element of place i_j of $\mathbf{w}_h$ is 1 if $h = j$, otherwise it is zero, thus the element of place i_j of $\mathbf{w}$ is exactly λ_j. The hypothesis that $\mathbf{w} = \mathbf{0}$ means that all elements of $\mathbf{w}$ are zero, in particular $\lambda_1 = \cdots = \lambda_{n-r} = 0$, and this shows that the vectors $\mathbf{w}_1, \ldots, \mathbf{w}_n$ are linearly independent. ■

We now want to proceed and *calculate a basis for the image* of a linear transformation.

Suppose we have a linear map $F : \mathbb{R}^n \longrightarrow \mathbb{R}^m$, and we want to determine a basis for the image. We endow $\mathbb{R}^n$ and $\mathbb{R}^m$ with the canonical bases; then, by Proposition 5.2.2, we have that $F(\mathbf{x}) = A\mathbf{x}$ for a suitable matrix $A \in \mathrm{M}_{m,n}(\mathbb{R})$. By Proposition 5.4.4 we have:

$$\mathrm{Im}(F) = \langle F(\mathbf{e}_1), \ldots, F(\mathbf{e}_n) \rangle = \langle A_1, \ldots, A_n \rangle,$$

where $A_1, \ldots, A_n$ are the columns of A. At this point, we simply apply the Gaussian algorithm to the vectors that form the columns of A. Recall that, to perform the Gaussian algorithm, we must write the vectors as rows. Let us see an example.

Example 5.7.5 Let $F : \mathbb{R}^3 \longrightarrow \mathbb{R}^4$ be defined by $F(x, y, z) = (x, 2x, x + y + z, y)$.

The matrix associated with F with respect to the canonical bases is:

$$A = \begin{pmatrix} 1 & 0 & 0 \\ 2 & 0 & 0 \\ 1 & 1 & 1 \\ 0 & 1 & 0 \end{pmatrix}.$$

The image of F is generated by the columns of A, i.e. $\mathrm{Im} F = \langle (1, 2, 1, 0), (0, 0, 1, 1), (0, 0, 1, 0) \rangle$. So we apply the Gaussian algorithm to the matrix

$$A^{\mathrm{T}} = \begin{pmatrix} 1 & 2 & 1 & 0 \\ 0 & 0 & 1 & 1 \\ 0 & 0 & 1 & 0 \end{pmatrix},$$

where A^{T} denotes the *transpose* of the matrix A, i.e. it is the matrix which has the columns of the matrix A as rows. Reducing A^T to row echelon form, we get

$$\begin{pmatrix} 1 & 2 & 1 & 0 \\ 0 & 0 & 1 & 1 \\ 0 & 0 & 0 & -1 \end{pmatrix}.$$

We then see that a basis for the image of F is:

$$\{(1, 2, 1, 0), (0, 0, 1, 1), (0, 0, 0, -1)\}.$$

5.8 EXERCISES WITH SOLUTIONS

5.8.1 Let $F_k : \mathbb{R}^4 \to \mathbb{R}^3$ be the linear transformation defined by: $F_k(\mathbf{e}_1) = \mathbf{e}_1 + 2\mathbf{e}_2 + k\mathbf{e}_3$, $F_k(\mathbf{e}_2) = k\mathbf{e}_2 + k\mathbf{e}_3$, $F_k(\mathbf{e}_3) = k\mathbf{e}_1 + k\mathbf{e}_2 + 6\mathbf{e}_3$, $F_k(\mathbf{e}_4) = k + \mathbf{e}_1(6 - k)\mathbf{e}_3$.

a) Determine for which values of k we have that F_k is injective and for which values of k we have that F_k is surjective.

b) Having chosen a value of k for which F_k is not surjective, determine a vector $\mathbf{v} \in \mathbb{R}^3$ such that $\mathbf{v} \notin \mathrm{Im}F_k$.

Solution. By Proposition 5.5.2 F_k is never injective. We now study surjectivity. The matrix associated with F_k with respect to the canonical bases in the domain and in the codomain is:

$$A = \begin{pmatrix} 1 & 0 & k & k \\ 2 & k & k & 0 \\ k & k & 6 & 6-k \end{pmatrix}$$

(see Observation 5.2.4). The image of F_k is the subspace generated by the columns of A. We write these columns as rows (i.e. we consider the matrix A^T, the transpose of A), and we perform the Gaussian algorithm. We obtain:

$$A^T = \begin{pmatrix} 1 & 2 & k \\ 0 & k & k \\ k & k & 6 \\ k & 0 & 6-k \end{pmatrix}$$

and reducing the matrix to row echelon form:

$$\begin{pmatrix} 1 & 2 & k \\ 0 & k & k \\ 0 & 0 & k^2 - k - 6 \\ 0 & 0 & 0 \end{pmatrix}.$$

If $k \neq 0$, $k \neq -2$ and $k \neq 3$, this matrix has three nonzero rows, so $\mathrm{Im}F_k$ has dimension 3 and F_k is surjective.

If $k = 0$, after the exchange of the second with the third row we get:

$$\begin{pmatrix} 1 & 2 & 0 \\ 0 & 0 & -6 \\ 0 & 0 & 0 \\ 0 & 0 & 0 \end{pmatrix},$$

so $\mathrm{Im}F_0$ has dimension 2 and F_0 is not surjective.

If $k = -2$ or $k = 3$ we have a row echelon form matrix with two nonzero rows, then $\mathrm{Im}F_k$ has dimension 2 and F_k is not surjective.

Choosing $k = 0$, we have $\mathrm{Im}F_0 = \langle (1, 2, 0), (0, 0, 6) \rangle$ and $\mathbf{v} = (0, -1, 3) \notin \mathrm{Im}F_0$, since the 3 vectors $(1, 2, 0), (0, -1, 3), (0, 0, 6)$ are linearly independent, because they are the nonzero rows of a row echelon form matrix.

5.8.2 Determine, if possible, a linear transformation $G : \mathbb{R}^4 \to \mathbb{R}^3$ such that $\mathrm{Im}\,G = \langle (1, 1, 0), (0, 3, -1), (3, 0, 1) \rangle$ and $\mathrm{Ker}\,G = \langle (-1, 0, 1, -3) \rangle$.

Solution. Let us see, first of all, if the requests made are compatible with the Rank Nullity Theorem. We first want to find a basis of $\langle (1, 1, 0), (0, 3, -1), (3, 0, 1) \rangle$. To do this, we reduce the matrix to row echelon form:

$$\begin{pmatrix} 1 & 1 & 0 \\ 0 & 3 & -1 \\ 3 & 0 & 1 \end{pmatrix}$$

and we get

$$\begin{pmatrix} 1 & 1 & 0 \\ 0 & 3 & -1 \\ 0 & 0 & 0 \end{pmatrix}.$$

So we would have that $\mathrm{Im}\,G$ has dimension 2 and $\mathrm{Ker}\,G$ has dimension 1, but $\dim \mathbb{R}^4 = 4 \neq 1 + 2 = \dim(\mathrm{Ker}\,G) + \dim(\mathrm{Im}G)$, consequently a linear transformation with the required properties cannot exist.

5.8.3 Determine, if possible, a linear transformation $F : \mathbb{R}^3 \to \mathbb{R}^3$ such that $\mathrm{Ker}\,F = \langle \mathbf{e}_1 - 2\mathbf{e}_3 \rangle$ and $\mathrm{Im}F = \langle 2\mathbf{e}_1 + \mathbf{e}_2 - \mathbf{e}_3, \mathbf{e}_1 - \mathbf{e}_2 - \mathbf{e}_3 \rangle$. Is this transformation unique?

Solution. We observe that the requests are compatible with the Rank Nullity Theorem. In fact, $\langle \mathbf{e}_1 - 2\mathbf{e}_3 \rangle$ has a dimension of 1, $\langle 2\mathbf{e}_1 + \mathbf{e}_2 - \mathbf{e}_3, \mathbf{e}_1 - \mathbf{e}_2 - \mathbf{e}_3 \rangle$ has dimension 2 (because the two vectors are not one multiple of the other) and $\dim \mathbb{R}^3 = 3 = 1 + 2 = \dim(\mathrm{Ker}\,F) + \dim(\mathrm{Im}F)$.

Let us now try to determine the matrix A associated with such F with respect to the canonical basis (in the domain and in the codomain). It must happen that $F(\mathbf{e}_1 - 2\mathbf{e}_3) = \mathbf{0}$, i.e. $F(\mathbf{e}_1) - 2F(\mathbf{e}_3) = \mathbf{0}$ (because F is linear), so $F(\mathbf{e}_1) = 2F(\mathbf{e}_3)$. Since $F(\mathbf{e}_1)$ is represented by the first column of A, and $F(\mathbf{e}_3)$ by third column, we have that the first column of A must be twice the third. Furthermore, the subspace generated by the columns of A, that is $\mathrm{Im}F$ must be equal to $\langle (2, 1, -1), (1, -1, -1) \rangle$. For example, the matrix:

$$A = \begin{pmatrix} 2 & 2 & 1 \\ -2 & 1 & -1 \\ -2 & -1 & -1 \end{pmatrix}$$

meets all of the requirements. In fact by construction $\mathbf{e}_1 - 2\mathbf{e}_3 \in \mathrm{Ker}\,F$ and $\mathrm{Im}F = \langle 2\mathbf{e}_1 + \mathbf{e}_2 - \mathbf{e}_3, \mathbf{e}_1 - \mathbf{e}_2 - \mathbf{e}_3 \rangle$; also from the Rank Nullity Theorem, we get that $\dim(\mathrm{Ker}\,F) = \dim \mathbb{R}^3 - \dim(\mathrm{Im}F) = 3 - 2 = 1$, thus $\mathrm{Ker}\,F = \langle \mathbf{e}_1 - 2\mathbf{e}_3 \rangle$.

This F is not unique; in fact, it can be verified that for example also the matrix

$$A = \begin{pmatrix} 6 & 2 & 3 \\ -6 & 1 & -3 \\ -6 & -1 & -3 \end{pmatrix}$$

meets all the given requirements.

5.8.4 Let $G_k : \mathbb{R}^2 \to \mathbb{R}^3$ be the linear transformation defined by:
$G_k(x, y) = (kx + 5y, 2x + (k + 3)y, (2k - 2) + x(7 - k)y)$. Determine for which values of k we have that G is injective.

Solution. To determine if G_k is injective we need to understand if the kernel of G_k contains only the zero vector or not. By Corollary 5.2.3 the matrix associated with G_k with respect to the canonical bases in the domain and in the codomain is:

$$A = \begin{pmatrix} k & 5 \\ 2 & k + 3 \\ 2k - 2 & 7 - k \end{pmatrix}.$$

To find the kernel of G_k we need to solve the homogeneous linear system associated with A. Reducing the matrix $(A|\underline{0})$ to row echelon form, we get:

$$(A'|\underline{0}) = \begin{pmatrix} 1 & \frac{k+3}{2} & | & 0 \\ 0 & -k^2 - 3k + 10 & | & 0 \\ 0 & 0 & | & 0 \end{pmatrix}.$$

If $k^2 + 3k - 10 \neq 0$, that is, if $k \neq -5$ and $k \neq 2$, we have that $\mathrm{rr}(A'|\underline{0}) = \mathrm{rr}(A') = 2$. Since the unknowns are 2, the system has only one solution, the null one, therefore $\mathrm{Ker}\, G_k = \{(0, 0)\}$ and G is injective. If $k = -5$ or $k = 2$, we easily see that $\mathrm{rr}(A'|\underline{0}) = \mathrm{rr}(A') = 1$, then the system admits infinitely many solutions that depend on one parameter, and G_k is not injective.

An alternative way of proceeding is as follows. By the Rank Nullity Theorem, we have that $2 = \dim \mathbb{R}^2 = \dim(\mathrm{Ker}\, G_k) + \dim(\mathrm{Im} G_k)$, so G_k is injective if and only if the image of G_k has dimension of 2. We then calculate a basis for the image of G_k. We must consider the matrix

$$\begin{pmatrix} k & 2 & 2k - 2 \\ 5 & k + 3 & 7 - k \end{pmatrix},$$

and reduce it with the Gaussian algorithm. We get:

$$\begin{pmatrix} 1 & \frac{k+3}{5} & \frac{7-k}{5} \\ 0 & -k^2 - 3k + 10 & k^2 + 3k - 10 \end{pmatrix}.$$

If $k \neq -5$ and $k \neq 2$, there are two nonzero rows, so the image of G_k has dimension 2 and G_k is injective. If $k = -5$ or $k = 2$ there is only one nonzero row, so the image of G_k has dimension 1 and G_k is not injective. We have thus found once again the result obtained with the previous method.

5.9 SUGGESTED EXERCISES

5.9.1 Consider the function $F : \mathbb{R}^2 \longrightarrow \mathbb{R}^2$ defined by: $F(x, y) = (x + 2ky, x - y)$. Determine the values of k for which F is linear.

5.9.2 Given the function $F : \mathbb{R}_1[x] \longrightarrow \mathbb{R}_2[x]$ defined by: $F(ax + b) = (a - b)x^2 + kb^2 x + 2a$. Determine the values of k for which F is linear.

5.9.3 Given linear transformations $F : \mathbb{R}^3 \to \mathbb{R}^2$ and $G : \mathbb{R}^2 \to \mathbb{R}^3$ defined by: $F(x, y, z) = (x - y, 2x + y + z)$ and $G(x, y) = (3y, -x, 4x + 2y)$, determine if possible, $F \circ G$ and $G \circ F$.

5.9.4 Given the linear transformations $F : \mathbb{R}^2 \to \mathbb{R}^2$ and $G : \mathbb{R}^2 \to \mathbb{R}$ defined by: $F(\mathbf{e}_1) = -\mathbf{e}_1 - \mathbf{e}_2$, $F(\mathbf{e}_2) = \mathbf{e}_1 + \mathbf{e}_2$, $G(\mathbf{e}_1) = 2$, $G(\mathbf{e}_2) = -1$; determine if possible $F \circ G$ and $G \circ F$.

5.9.5 Consider the linear transformation $F : \mathbb{R}^2 \longrightarrow \mathbb{R}^2$ defined by $F(\mathbf{e}_1) = 3\mathbf{e}_1 - 3\mathbf{e}_2$, $F(\mathbf{e}_2) = 2\mathbf{e}_1 - 2\mathbf{e}_2$. Compute a basis of the kernel and a basis for the image of F.

5.9.6 Establish which of the following linear transformations are isomorphisms:

i) $F : \mathbb{R}^3 \longrightarrow \mathbb{R}^3$ defined by $F(x, y, z) = (x + 2z, y + z, z)$;

ii) $F : \mathbb{R}^3 \to \mathbb{R}^2$ defined by $F(x, y, z) = (2x - z, x - y + z)$;

iii) $F : \mathbb{R}^3 \to \mathbb{R}^3$ defined by $F(\mathbf{e}_1) = 2\mathbf{e}_1 + \mathbf{e}_2$, $F(\mathbf{e}_2) = 3\mathbf{e}_1 - \mathbf{e}_3$, $F(\mathbf{e}_3) = \mathbf{e}_1 - \mathbf{e}_2 - \mathbf{e}_3$.

5.9.7 Find a basis for the kernel and one for the image of each of the following linear transformations. Establish if they are injective, surjective and/or bijective.

i) $F : \mathbb{R}^3 \longrightarrow \mathbb{R}^3$ defined by $F(x, y, z) = (x - z, x + 2y - z, x - 4y - z)$.

ii) $F : \mathbb{R}^2 \longrightarrow \mathbb{R}^3$ defined by $F(\mathbf{e}_1) = \mathbf{e}_1 + \mathbf{e}_2 - \mathbf{e}_3$, $F(\mathbf{e}_2) = 2\mathbf{e}_1 - 2\mathbf{e}_2 - \mathbf{e}_3$.

iii) $F : \mathbb{R}^4 \longrightarrow \mathbb{R}^2$ defined by $F(x, y, z, t) = (2x - t, 3y - x + 2z - t)$.

iv) $F : \mathbb{R}^3 \longrightarrow \mathbb{R}^3$ associated with A, with respect to the canonical basis, where

$$A = \begin{pmatrix} -1 & 1 & -1 \\ 0 & 0 & 1 \\ 1 & 0 & 1 \end{pmatrix}.$$

v) $F : \mathbb{R}^3 \longrightarrow \mathbb{R}^4$ defined by $F(x, y, z) = (x + y - z, z, x + y, z)$.

5.9.8 Given the linear transformation $F : \mathbb{R}^4 \to M_{2,2}(\mathbb{R})$ defined by:

$$F(x_1, x_2, x_3, x_4) = \begin{pmatrix} x_1 - x_3 - x_4 & x_1 + 2x_3 \\ x_1 + x_4 & 0 \end{pmatrix},$$

find a basis for Ker F and determine the dimension of ImF.

5.9.9 Given the linear transformation $F : \mathbb{R}^3 \to \mathbb{R}^3$ defined by: $F(\mathbf{e}_1) = \mathbf{e}_1 + \mathbf{e}_2 + \mathbf{e}_3$, $F(\mathbf{e}_2) = 2\mathbf{e}_1 + 2\mathbf{e}_2 + 2\mathbf{e}_3$, $F(\mathbf{e}_3) = \mathbf{e}_1 + \mathbf{e}_2 + \mathbf{e}_3$; find a basis for Ker F and establish if the vector $\mathbf{e}_1 - \mathbf{e}_2 + \mathbf{e}_3$ belongs to ImF.

5.9.10 Determine, if possible, a surjective linear transformation $T : \mathbb{R}^3 \to \mathbb{R}^2$ and an injective linear transformation $F : \mathbb{R}^4 \to \mathbb{R}^3$.

5.9.11 Are there injective transformations $T : \mathbb{R}^3 \to \mathbb{R}^4$? If yes, determine one; if not, give reasons for the answer.

5.9.12 Let $T_k : \mathbb{R}^2 \longrightarrow \mathbb{R}^2$ be the linear transformation associated with the matrix A with respect to the canonical basis, where

$$A = \begin{pmatrix} 1 & 4 \\ k & 0 \end{pmatrix}.$$

Determine $\operatorname{Ker}(T_k)$ and $\operatorname{Im}(T_k)$ as k varies in $\mathbb{R}$.

5.9.13 Let $T_k : \mathbb{R}^3 \longrightarrow \mathbb{R}^3$ be the linear transformation associated with the matrix A with respect to the canonical basis, where

$$A = \begin{pmatrix} 1 & 2 & 0 \\ 1 & 0 & 1 \\ 2 & k & 3 \end{pmatrix}.$$

Say for which values of the parameter k we have that T_k is an isomorphism.

5.9.14 Let : $\mathbb{R}^3 \to \mathbb{R}^4$ be the linear transformation defined by: $T(x, y, z) = (x + 2z, y, 2x + 3y + 4z, 3x - y + 6z)$. Find a basis for $\operatorname{Ker}(T)$ and a basis for $\operatorname{Im}(T)$ and their dimensions. Is T injective?

5.9.15 Determine, if possible, a linear transformation $F : \mathbb{R}^3 \to \mathbb{R}^2$ such that $\operatorname{Ker} F = \langle \mathbf{e}_1 \rangle$ and $\operatorname{Im} F = \langle \mathbf{e}_1 - \mathbf{e}_2 \rangle$.

5.9.16 Determine, if possible, a linear transformation $F : \mathbb{R}^2 \to \mathbb{R}^3$ such that $\operatorname{Ker} F = \langle \mathbf{e}_2 \rangle$ and $\operatorname{Im} F = \langle \mathbf{e}_1 - \mathbf{e}_2 + 2\mathbf{e}_3 \rangle$.

5.9.17 Determine, if possible, a linear transformation $F : \mathbb{R}^2 \to \mathbb{R}^4$ such that $\operatorname{Im} F$ has dimension 1.

5.9.18 Determine, if possible, a linear linear transformation $F : \mathbb{R}^3 \to \mathbb{R}^2$ such that $(1, 1) \notin \operatorname{Im} F$.

5.9.19 Let $F : \mathbb{R}^3 \to \mathbb{R}^2$ be the linear transformation defined by: $F(\mathbf{e}_1) = 2\mathbf{e}_1 + k\mathbf{e}_2$, $F(\mathbf{e}_2) = k\mathbf{e}_1 + 2\mathbf{e}_2$, $F(\mathbf{e}_3) = -2\mathbf{e}_1 - k\mathbf{e}_2$. Determine for which values of k we have that F is not surjective. Set $k = -1$, determine a vector $\mathbf{v}_1$ that belongs to $\operatorname{Ker} F$ and a vector $\mathbf{v}_2$ which does not belong to $\operatorname{Ker} F$.

5.9.20 Given the linear transformation $F_k : \mathbb{R}^3 \to \mathbb{R}^2$ defined by: $F_k(x, y, z) = (kx - 3y + kz, kx + ky - 3z)$, determine for which values of k we have that F_k is injective and for which values of k we have that F_k is surjective.

5.9.21 Given the linear transformation $F_k : \mathbb{R}^2 \to \mathbb{R}^3$ defined by: $F_k(\mathbf{e}_1) = k\mathbf{e}_1 - 4\mathbf{e}_2 + k\mathbf{e}_3$, $F_k(\mathbf{e}_2) = -3\mathbf{e}_1 + k\mathbf{e}_2 + 3\mathbf{e}_3$, determine for which values of k we have that F_k is injective and for which values of k we have that F_k is surjective.

CHAPTER 6

Linear Systems

In this chapter, we want to revisit the theory of linear systems and interpret the results already discussed in Chapter 1 in terms of linear transformations, using the knowledge we have gained in Chapter 5. We will use the notation and terminology introduced in Chapter 1.

6.1   PREIMAGE

The inverse image or preimage of a vector $\mathbf{w} \in W$ under a linear map $f : V \longrightarrow W$ constists of all the vectors in the vector space V, whose image is $\mathbf{w}$. It is a basic concept in mathematics; for us, it will be a useful tool for expressing the solutions of a linear system.

We already know an example of inverse image, namely the kernel of a linear transformation F. In fact, $\mathrm{Ker}\,(F)$ is the inverse image of the zero vector of W, i.e. it consists of all vectors, whose image under F is $\mathbf{0}_W$.

Let us look at the definition.

Definition 6.1.1 Let $F : V \to W$ be a linear transformation and let $\mathbf{w} \in W$. The *inverse image* or *preimage* of $\mathbf{w}$ under F is the set

$$F^{-1}(\mathbf{w}) = \{\mathbf{v} \in V \mid F(\mathbf{v}) = \mathbf{w}\}.$$

Notice that the notation $F^{-1}(\mathbf{w})$, introduced in the previous definition, does not have anything to do with the invertibility of the function. When we speak of inverse image of a vector under a function F, *we are not saying* that F is an invertible function: The notation $F^{-1}(\mathbf{w})$ simply indicates a subset of the domain.

Example 6.1.2 Let $F : \mathbb{R}^2 \to \mathbb{R}^3$ be the linear transformation defined by:

$$F(x, y) = (x + y, x + y, x).$$

What is the inverse image of the vector $(1,1,3)$ under F? By definition we have:

$$F^{-1}(1,1,3) = \{(x,y) \in \mathbb{R}^2 \mid F(x,y) = (1,1,3)\} =$$

$$= \{(x,y) \in \mathbb{R}^2 \mid (x+y, x+y, x) = (1,1,3)\} =$$

$$= \left\{(x,y) \in \mathbb{R}^2 \middle| \begin{cases} x+y = 1, \\ x+y = 1, \\ x = 3 \end{cases} \right\} = \{(3,-2)\}.$$

This means that $F(3,-2) = (1,1,3)$, and there are no other elements of $\mathbb{R}^2$ that have $(1,1,3)$ as image. In particular, we have that $(1,1,3) \in \mathrm{Im} F$.

We now calculate the inverse image of the vector $(1,0,0)$ under F. By definition we have:

$$F^{-1}(1,0,0) = \{(x,y) \in \mathbb{R}^2 \mid F(x,y) = (1,0,0)\} =$$

$$= \{(x,y) \in \mathbb{R}^2 \mid (x+y, x+y, x) = (1,0,0)\} =$$

$$= \left\{(x,y) \in \mathbb{R}^2 \middle| \begin{cases} x+y = 1 \\ x+y = 0 \\ x = 0 \end{cases} \right\} = \varnothing.$$

The vector $(1,0,0)$ is not the image of any vector of $\mathbb{R}^2$, i.e. $(1,0,0) \notin \mathrm{Im} F$.

These examples show that calculating the preimage of a vector under a linear transformation is equivalent to solving a linear system. We will now deepen our understanding on this point.

Observation 6.1.3 The set $F^{-1}(\mathbf{w})$ is a vector subspace of V if and only if $F^{-1}(\mathbf{w}) = \mathrm{Ker}\, F$ and in this case $\mathbf{w} = \mathbf{0}_W$.

In fact, if $F^{-1}(\mathbf{w}) = \mathrm{Ker}\, F$, we have immediately that the inverse image is a subspace. Conversely, reasoning by contradiction, if $\mathbf{w} \neq \mathbf{0}_W$ then the set $F^{-1}(\mathbf{w})$ does not contain $\mathbf{0}_V$. In fact, being F a linear transformation, we have $F(\mathbf{0}_V) = \mathbf{0}_W$, therefore $\mathbf{0}_V$ belongs only to the inverse image of $\mathbf{0}_W$ (the same vector in V cannot belong to the inverse image of different vectors). So $F^{-1}(\mathbf{w})$ cannot be a a subspace of V.

Proposition 6.1.4 Let $F : V \to W$ be a linear transformation and let $\mathbf{w} \in W$. Then $F^{-1}(\mathbf{w})$ is not empty if and only if $\mathbf{w} \in \mathrm{Im} F$. In this case we have:

$$F^{-1}(\mathbf{w}) = \{\mathbf{v} + \mathbf{z} \mid \mathbf{z} \in \mathrm{Ker}\, F\}, \tag{6.1}$$

where $\mathbf{v}$ is any element of V such that $F(\mathbf{v}) = \mathbf{w}$.

Proof. If $\mathbf{w} \notin \text{Im } F$, then, by definition, $F^{-1}(\mathbf{w}) = \emptyset$. We see now the case when $\mathbf{w} \in \text{Im} F$. By definition of image, there exists an element $\mathbf{v} \in V$, such that $F(\mathbf{v}) = \mathbf{w}$. It might not be the case that $\mathbf{v}$ is the only element in $F^{-1}(\mathbf{w})$. So suppose that $\mathbf{v}'$ is another element in $F^{-1}(\mathbf{w})$. Then

$$F(\mathbf{v}') = F(\mathbf{v}) = \mathbf{w},$$

so

$$F(\mathbf{v}' - \mathbf{v}) = \mathbf{0}_W,$$

i.e.:

$$\mathbf{v}' - \mathbf{v} \in \text{Ker } F.$$

So any element $\mathbf{v}'$ of $F^{-1}(\mathbf{w})$ is written as $\mathbf{v}' = \mathbf{v} + (\mathbf{v}' - \mathbf{v}) = \mathbf{v} + \mathbf{z}$, with $\mathbf{z} \in \text{Ker } F$. Thus we have proved one inclusion.

Consider now $\mathbf{v} + \mathbf{z}$, with $\mathbf{z} \in \text{Ker } F$. Then

$$F(\mathbf{v} + \mathbf{z}) = F(\mathbf{v}) + F(\mathbf{z}) = \mathbf{w} + \mathbf{0}_W = \mathbf{w}.$$

This gives the other inclusion and we have shown the result. ■

6.2 LINEAR SYSTEMS

In Chapter 5, we defined the row rank of a matrix (see Definition 5.7.3). We now want to deepen the study of this notion and have a clearer view of the link between matrices, linear systems and transformations.

Given a matrix $A \in \text{M}_{m,n}(\mathbb{R})$, we can read its rows as vectors of $\mathbb{R}^n$ and its columns as vectors of $\mathbb{R}^m$. It is therefore natural to introduce the following definition.

Definition 6.2.1 We call *column rank* of a matrix $A \in \text{M}_{m,n}(\mathbb{R})$, the maximum number of linearly independent columns of A, i.e. the dimension of the subspace of $\mathbb{R}^m$ generated by the columns of A.

The following observation is already known and yet, given the its importance in the context that we are studying, we want to rexamine it.

Observation 6.2.2 If we write A as the matrix associated with the linear transformation $L_A : \mathbb{R}^n \longrightarrow \mathbb{R}^m$ with respect to the canonical bases, then the column rank of A is the dimension of the image of L_A. Indeed, the image is generated by the columns of the matrix A.

Although in general the row vectors and column vectors of a matrix $A \in \text{M}_{m,n}(\mathbb{R})$ are elements of different vector spaces, the row and column rank of A *always* coincide. This number is simply called *rank* of A, denoted by $\text{rk}(A)$.

Proposition 6.2.3 *If $A \in \text{M}_{m,n}(\mathbb{R})$, then the row rank of A is equal to the column rank of A.*

Proof. Let $L_A : \mathbb{R}^n \longrightarrow \mathbb{R}^m$ be the linear transformation associated with the matrix A with respect to the canonical bases. The kernel of L_A is the set of solutions of the homogeneous linear system associated with the matrix A and, by Proposition 5.7.4, has dimension $n - \mathrm{rr}(A)$, where $\mathrm{rr}(A)$ is the row rank of A. By the Rank Nullity Theorem 5.5.1, we also know that the dimension of $\mathrm{Ker}\, L_A$ is equal to $n - \dim(\mathrm{Im} L_A)$. It follows that $\mathrm{rr}(A) = \dim(\mathrm{Im} L_A)$, i.e. $\mathrm{rr}(A)$ is equal to the rank of columns A. ■

Let us see an example to clarify.

Example 6.2.4 Consider the matrix

$$A = \begin{pmatrix} 1 & 2 & 0 & -1 \\ 1 & 2 & 0 & -1 \end{pmatrix}.$$

The row rank of A is 1. Also the column rank is 1, since the vectors $(1,1)$, $(2,2)$, $(-1,-1)$ of $\mathbb{R}^2$ are linearly dependent. So $\mathrm{rk}(A) = 1$.

Observation 6.2.5 If $A \in \mathrm{M}_{m,n}(\mathbb{R})$ is a matrix in row echelon form, the definition of rank coincides with the Definition 1.3.4. In fact, by Proposition 4.3.3 the rows of a nonzero matrix A in row echelon form are linearly independent, hence the dimension of the subspace generated by the rows of A coincides with the number of nonzero rows of A. Computing the rank of a matrix in row echelon form is therefore immediate, while calculating the rank of a generic matrix requires more time.

Proposition 4.3.1 provides an effective method for calculating the rank of a matrix as the elementary operations on the rows of a matrix preserve the rank, since the subspace generated by the rows remains unchanged. Hence, to compute the rank of a matrix, A, we reduce A in row echelon form with the Gaussian algorithm and then we compute the rank of the reduced matrix, which simply amounts to counting the number of nonzero rows.

Example 6.2.6 We want to compute the rank of the matrix

$$A = \begin{pmatrix} 0 & 1 & 3 \\ 1 & -1 & 5 \\ -1 & 1 & 0 \end{pmatrix}.$$

We have:

$$\mathrm{rk}(A) = \mathrm{rk} \begin{pmatrix} 1 & -1 & 5 \\ 0 & 1 & 3 \\ 0 & 0 & 5 \end{pmatrix} = 3.$$

We now give a useful definition, already anticipated in the first chapter.

Definition 6.2.7 Let $A\mathbf{x} = \mathbf{b}$ a linear system. The linear system $A\mathbf{x} = \mathbf{0}$ is called the *homogeneous system* associated with the system $A\mathbf{x} = \mathbf{b}$.

Proposition 6.2.8 *Let $A\mathbf{x} = \mathbf{b}$ be a linear system of m equations in n unknowns that admits at least one solution. Then the set of solutions of the system is*

$$S = \{\mathbf{v} + \mathbf{z} \,|\, \mathbf{z} \in \operatorname{Ker} A\},$$

where $\mathbf{v}$ is a particular solution of the system and $\operatorname{Ker} A$ is the set of solutions of the associated homogeneous linear system $A\mathbf{x} = \mathbf{0}$.

Proof. This proposition is basically a rewriting of Proposition 6.1.4 with $F = L_A$. In fact, if we view the matrix A as the matrix associated to the linear transformation $L_A : \mathbb{R}^n \to \mathbb{R}^m$ with respect to the canonical bases, then determining the solutions of the linear system $A\mathbf{x} = \mathbf{b}$ is the same as determining the vectors $\mathbf{x} \in \mathbb{R}^n$ such that $L_A(\mathbf{x}) = \mathbf{b}$. In other words, it is the same as determining the preimage $L_A^{-1}(\mathbf{b})$ of the vector $\mathbf{b} \in \mathbb{R}^m$. Since by hypothesis the system admits solutions, $\mathbf{b}$ belongs to the image of L_A, i.e. $L_A(\mathbf{v}) = A\mathbf{v} = \mathbf{b}$ for a suitable vector $\mathbf{v}$ of $\mathbb{R}^n$. The set S of system solutions is $L_A^{-1}(\mathbf{b})$, and this, by Proposition 6.1.4, is exactly $\{\mathbf{v}+\mathbf{z} \,|\, \mathbf{z} \in \operatorname{Ker} L_A\}$. ■

The theorem that follows is the most important result in theory of linear systems.

Theorem 6.2.9 (Rouché-Capelli Theorem). *A linear system $A\mathbf{x} = \mathbf{b}$ of m equations in n unknowns admits solutions if and only if $\operatorname{rk}(A) = \operatorname{rk}(A|\mathbf{b})$. If this condition is satisfied, then the system has:*

1. *exactly one solution if and only if $\operatorname{rk}(A) = \operatorname{rk}(A|\mathbf{b}) = n$;*

2. *infinitely many solutions if and only if $\operatorname{rk}(A) = \operatorname{rk}(A|\mathbf{b}) < n$. In this case, the solutions of the system depend on $n - \operatorname{rk}(A)$ parameters.*

Proof. (1). We view the matrix A as the matrix associated with the linear transformation $L_A : \mathbb{R}^n \to \mathbb{R}^m$ with respect to the canonical bases. The solutions of the linear system $A\mathbf{x} = \mathbf{b}$ correspond to the vectors $\mathbf{x} \in \mathbb{R}^n$, such that $L_A(\mathbf{x}) = \mathbf{b}$. Hence, we need to determine the preimage $L_A^{-1}(\mathbf{b})$ of the vector $\mathbf{b} \in \mathbb{R}^m$. By definition, such preimage is not empty if and only if $\mathbf{b} \in \operatorname{Im} L_A$. In other words, the system has solutions if and only if $\mathbf{b} \in \operatorname{Im} L_A$. As $\operatorname{Im} L_A$ is generated by the column vectors of the matrix A, $\mathbf{b} \in \operatorname{Im} L_A$ if and only if the subspace generated by the column vectors of A coincides with the subspace generated by the column vectors of A and the column vector $\mathbf{b}$, that is, if and only if

$$\dim\langle\text{columns of } A\rangle = \dim\langle\text{columns of } (A|\mathbf{b})\rangle,$$

i.e. if and only if $\operatorname{rk}(A) = \operatorname{rk}(A|\mathbf{b})$.

(2). If $A\mathbf{x} = \mathbf{b}$ admits solutions then, by Proposition 6.2.8, all solutions are of the form

$$S = \{\mathbf{v} + \mathbf{z} \,|\, \mathbf{z} \in \operatorname{Ker} L_A\},$$

where $\mathbf{v}$ is a particular solution of the system. Then, we have only one of two cases:

1. $\dim(\operatorname{Ker} A) = 0$, i.e. $\operatorname{Ker} A = \{\mathbf{0}_{\mathbb{R}^n}\}$, so $S = \{\mathbf{v}\}$, i.e. the system admits only one solution;

2. $\dim(\operatorname{Ker} A) > 0$, thus $\operatorname{Ker} A$ contains infinitely many elements, being a real vector subspace of $\mathbb{R}^n$.

By the Rank Nullity Theorem, we have $\dim(\operatorname{Ker} A) + n = \dim \operatorname{Im} L_A = n - \operatorname{rk}(A)$, and therefore the solutions depend on $n - \operatorname{rk}(A)$ parameters. ■

Definition 6.2.10 Given a compatible linear system $A\mathbf{x} = \mathbf{b}$, i.e. a system that admits solution, the number $\operatorname{rk}(A) = \operatorname{rk}(A|\mathbf{b})$ is also called *the rank of the system*.

In Chapter 1, we learned to solve any linear system $A\mathbf{x} = \mathbf{b}$ in the following manner:

1. We write the complete matrix $(A|\mathbf{b})$ associated with the system;

2. We use the Gaussian algorithm to reduce $(A|\mathbf{b})$ to a row echelon matrix in the form $(A'|\mathbf{b}')$;

3. The starting linear system $A\mathbf{x} = \mathbf{b}$ is equivalent to the row echelon linear system $A'\mathbf{x} = \mathbf{b}'$;

4. The linear system $A'\mathbf{x} = \mathbf{b}'$ has solutions if and only if $\operatorname{rk}(A') = \operatorname{rk}(A'|\mathbf{b}')$. In this case, using subsequent substitutions, we obtain all the solutions of the system.

Rouché-Capelli theorem states that the linear system $A\mathbf{x} = \mathbf{b}$, which in general is not in row echelon form, admits solutions if and only if $\operatorname{rk}(A) = \operatorname{rk}(A|\mathbf{b})$. How do we reconcile Rouché-Capelli theorem with the method of resolution of linear systems just mentioned? It all works because, as shown in Proposition 4.3.1, the Gaussian algorithm preserves the rank of a matrix, hence $\operatorname{rk}(A) = \operatorname{rk}(A')$ and $\operatorname{rk}(A|\mathbf{b}) = \operatorname{rk}(A'|\mathbf{b}')$, therefore $\operatorname{rk}(A) = \operatorname{rk}(A|\mathbf{b})$ if and only if $\operatorname{rk}(A') = \operatorname{rk}(A'|\mathbf{b}')$.

Rouché-Capelli theorem gives all the information on linear systems, we have already seen in Chapter 1. In particular it states that:

A linear system with real coefficients which admits solutions has either one solution or infinitely many.

This is exactly the situation that we described in Chapter 1, reinterpreted in terms of linear transformations. In essence, a compatible linear system is a set of compatible conditions that are assigned on n real variables. These conditions then lower the number of degrees of freedom of the system: if the system rank is k, the set of the solutions no longer depends on n free variables, but on $n - k$. What counts is the rank of the system and not the number of equations, because the rank of the system quantifies independent conditions and eliminates those conditions that can be deduced from the others and so are redundant.

Here are some examples to illustrate the results shown above.

Example 6.2.11 Consider the linear transformation $L_A : \mathbb{R}^3 \to \mathbb{R}^3$ whose matrix with respect to the canonical basis (both in the domain and in the codomain) is:

$$A = \begin{pmatrix} 1 & 0 & 2 \\ 2 & 1 & 1 \\ 3 & 1 & 3 \end{pmatrix}.$$

We want to establish if the vector $\mathbf{b} = (3, 0, 3)$ belongs to $\mathrm{Im}(L_A)$ and, if so, to calculate the inverse image of $\mathbf{b}$ under L_A.

We observe that $\mathrm{Im}(L_A)$ is generated by the columns of the matrix A, therefore $\mathbf{b}$ belongs to $\mathrm{Im}(L_A)$ if and only if $\mathrm{rk}(A) = \mathrm{rk}(A|\mathbf{b})$. In addition, the inverse image of $\mathbf{b}$ under L_A consists of the vectors $(x, y, z) \in \mathbb{R}^3$ such that $L_A(x, y, z) = (3, 0, 3)$, i.e. the vectors (x, y, z) such that

$$A \begin{pmatrix} x \\ y \\ z \end{pmatrix} = \begin{pmatrix} 3 \\ 0 \\ 3 \end{pmatrix},$$

that is:

$$\begin{pmatrix} x + 2z \\ 2x + y + z \\ 3x + y + 3z \end{pmatrix} = \begin{pmatrix} 3 \\ 0 \\ 3 \end{pmatrix}.$$

In other words, computing the inverse image of $\mathbf{b}$ by L_A means solving the linear system

$$\begin{cases} x + 2z = 3 \\ 2x + y + z = 0 \\ 3x + y + 3z = 3. \end{cases}$$

To calculate the rank of matrix $(A|\mathbf{b})$ and compare it with the rank of A, we reduce the matrix $(A|\mathbf{b})$ and, simultaneously, the matrix A in row echelon form using the Gaussian algorithm and then we calculate the rank of the reduced matrices. We have:

$$(A|\mathbf{b}) \to \begin{pmatrix} 1 & 0 & 2 & 3 \\ 2 & 1 & 1 & 0 \\ 3 & 1 & 3 & 3 \end{pmatrix} \to \begin{pmatrix} 1 & 0 & 2 & 3 \\ 0 & 1 & -3 & -6 \\ 0 & 1 & -3 & -6 \end{pmatrix} \to$$

$$\to \begin{pmatrix} 1 & 0 & 2 & 3 \\ 0 & 1 & -3 & -6 \\ 0 & 0 & 0 & 0 \end{pmatrix}.$$

So $\mathrm{rk}(A) = \mathrm{rk}(A|\mathbf{b}) = 2$. This means that the vector $\mathbf{b}$ belongs to the image of L_A and $\dim(\mathrm{Ker}\, L_A) = 3 - 2 = 1$. The inverse image of $\mathbf{b}$ is given by the elements: $\mathbf{v} + \mathbf{z}$, where $\mathbf{v}$ is a particular solution of the system $A\mathbf{x} = \mathbf{b}$ and $\mathbf{z} \in \mathrm{Ker}\, A$, the kernel of the matrix A, i.e. the set of solutions of the homogeneous linear system $A\mathbf{x} = \mathbf{0}$. To compute $\mathbf{v}$, we observe that the starting system is equivalent to the system:

$$\begin{cases} x + 2z = 3 \\ y - 3z = -6. \end{cases}$$

Therefore, a particular solution can be calculated by substituting, for example, $z = 0$ in the equations found: if $z = 0$, we have $y = -6$, $x = 3$, then we can choose $\mathbf{v} = (3, -6, 0)$.

The homogeneous system $A\mathbf{x} = \mathbf{0}$ is equivalent to the homogeneous system:

$$\begin{cases} x + 2z = 0 \\ y - 3z = 0; \end{cases}$$

we can solve from the bottom with subsequent substitutions: $y = 3z$, $x = -2z$. The inverse image of $\mathbf{b}$ is thus $S = \{(3, -6, 0) + (-2z, 3z, z) \mid z \in \mathbb{R}\}$.

6.3   EXERCISES WITH SOLUTIONS

6.3.1 Compute the solutions of the following linear system in the unknowns x_1, x_2, x_3, x_4, x_5:

$$\begin{cases} x_1 - x_2 + 3x_3 + x_5 = 2 \\ 2x_1 + x_2 + 8x_3 - 4x_4 + 2x_5 = 3 \\ x_1 + 2x_2 + 5x_3 - 3x_4 + 4x_5 = 1. \end{cases}$$

Solution. The complete matrix associated with the system is:

$$(A|\mathbf{b}) = \left(\begin{array}{ccccc|c} 1 & -1 & 3 & 0 & 1 & 2 \\ 2 & 1 & 8 & -4 & 2 & 3 \\ 1 & 2 & 5 & -3 & 4 & 1 \end{array} \right),$$

which reduced to row echelon form becomes:

$$(A'|\mathbf{b}') = \left(\begin{array}{ccccc|c} 1 & -1 & 3 & 0 & 1 & 2 \\ 0 & 3 & 2 & -4 & 0 & -1 \\ 0 & 0 & 0 & 1 & 3 & 0 \end{array} \right).$$

We have $\mathrm{rk}(A') = \mathrm{rk}(A'|\mathbf{b}') = 3$ so the system is solvable, and the solutions depend on $5 - 3 = 2$ parameters.

The pivots are on the first, second and fourth column of A', so we can obtain the unknowns x_1, x_2 and x_4 in terms of x_3 and x_5.

The system associated with the row echelon form matrix $(A'|\mathbf{b}')$ is:

$$\begin{cases} x_1 - x_2 + 3x_3 + x_5 = 2 \\ 3x_2 + 2x_3 - 4x_4 = -1 \\ x_4 + 3x_5 = 0. \end{cases}$$

So we have $x_4 = -3x_5$, $x_2 = -\frac{1}{3} - 4x_5 - \frac{2}{3}x_3$, $x_1 = \frac{5}{3} - 5x_5 - \frac{11}{3}x_3$. The solutions are therefore:

$$\left\{ \left(-\frac{11}{3}x_3 - 5x_5 + \frac{5}{3}, -\frac{2}{3}x_3 - 4x_5 - \frac{1}{3}, x_3, -3x_5, x_5 \right) \mid x_3, x_5 \in \mathbb{R} \right\} =$$

$$\left\{ \left(\frac{5}{3}, -\frac{1}{3}, 0, 0, 0 \right) + \mathbf{z} \mid \mathbf{z} \in \left\langle \left(-\frac{11}{3}, -\frac{2}{3}, 1, 0, 0 \right), (-5, -4, 0, -3, 1) \right\rangle \right\}.$$

6.3.2 Determine the solutions of the following linear system in the unknown x_1, x_2, x_3:

$$\begin{cases} x_1 - x_2 + x_3 = 2 \\ 2x_1 - x_2 + 3x_3 = -1 \\ x_1 + 2x_3 = 1. \end{cases}$$

Solution. The complete matrix associated with the system is:

$$(A|\mathbf{b}) = \left(\begin{array}{ccc|c} 1 & -1 & 1 & 2 \\ 2 & -1 & 3 & -1 \\ 1 & 0 & 2 & 1 \end{array} \right),$$

which reduced to row echelon form becomes:

$$(A'|\mathbf{b}') = \left(\begin{array}{ccc|c} 1 & -1 & 1 & 2 \\ 0 & 1 & 1 & -5 \\ 0 & 0 & 0 & 4 \end{array} \right).$$

Thus, we have $\mathrm{rk}(A') = 2 \neq \mathrm{rk}(A'|\mathbf{b}') = 3$. Therefore, the system does not admit solutions by the Rouché-Capelli Theorem. We note that the linear system associated with the matrix $(A'|\mathbf{b}')$ is

$$\begin{cases} x_1 - x_2 + x_3 = 2 \\ x_2 + x_3 = -5 \\ 0 = 4, \end{cases}$$

which is clearly not compatible.

6.3.3 (a) Determine a linear system having

$$S = (1, 2, 1) + \langle (1, 1, 1) \rangle$$

as set of solutions.

(b) Determine, if possible, a linear system of three equations having

$$S = (1, 2, 1) + \langle (1, 1, 1) \rangle$$

as set of solutions.

(c) Determine, if possible, a linear system of rank 1 having

$$S = (1, 2, 1) + \langle (1, 1, 1) \rangle$$

as set of solutions.

Solution. Since S is a subset of $\mathbb{R}^3$, each linear system having S as set of solutions is a linear system in 3 unknowns. We indicate these unknowns with x, y, z. The set of solutions of a linear system of the form $A\mathbf{x} = \mathbf{b}$ is $S = \{\mathbf{v} + \mathbf{z} | \mathbf{z} \in +\ker A\}$, where $\mathbf{v}$ is a particular solution of the system. So if $A\mathbf{x} = \mathbf{b}$ is a linear system having S as set of solutions, $(1, 2, 1)$ is a system solution and $\ker A = \langle (1, 1, 1) \rangle$. In particular, we note that $(1, 2, 1) \notin \langle (1, 1, 1) \rangle$, so S is not a subspace of $\mathbb{R}^3$, so $\mathbf{b} \neq \mathbf{0}$. Also $\dim(\ker A) = 1 = 3 - \mathrm{rk}(A)$. So the system we seek has rank 2 and therefore must necessarily consist of at least 2 equations. This immediately allows us to answer the question (c): there is no linear system of rank 1 having S as a set of solutions.

To determine a linear system having S as a set of solutions and then answer question (a), we could then write a generic linear system consisting of two equations and impose that:

1. $(1, 2, 1)$ is system solution;

2. $(1, 1, 1)$ is the homogeneous system solution associated with $A\mathbf{x} = \mathbf{0}$.

This method, which certainly works, however, is not the most effective. We then choose a smarter approach.

What we want to do is to describe by equations the set of elements $(x, y, z) \in S$, that is the set of elements (x, y, z) of $\mathbb{R}^3$, such that :

$$(x, y, z) = (1, 2, 1) + z, \text{ with } z \in \langle (1, 1, 1) \rangle$$

or, equivalently,

$$(x, y, z) - (1, 2, 1) = z, \text{ with } z \in \langle (1, 1, 1) \rangle.$$

Note that the vector $(x, y, z) - (1, 2, 1) = (x - 1, y - 2, z - 1)$ belongs to the subspace $\langle (1, 1, 1) \rangle$ if and only if it is a multiple of $(1, 1, 1)$, i.e. if and only if

$$\mathrm{rk} \begin{pmatrix} 1 & 1 & 1 \\ x - 1 & y - 2 & z - 1 \end{pmatrix} = 1.$$

Using the Gaussian algorithm, we have that:

$$\mathrm{rk} \begin{pmatrix} 1 & 1 & 1 \\ x - 1 & y - 2 & z - 1 \end{pmatrix} = \mathrm{rk} \begin{pmatrix} 1 & 1 & 1 \\ 0 & y - 1 - x & z - x \end{pmatrix}.$$

The rank of this matrix is equal to 1 if and only if the second row of the matrix is null, that is if and only if

$$\begin{cases} -x + y - 1 = 0 \\ -x + z = 0. \end{cases}$$

We have therefore found a linear system having S as set of solutions. Naturally, every system equivalent to the one found has S as a set of solutions. In particular, to answer the question (b) we will have to determine a linear system of 3 equations equivalent to the one just written. Just add an equation that is a linear combination of the two equations found. For example:

$$\begin{cases} -x + y - 1 = 0 \\ -x + z = 0 \\ -2x + y + z - 1 = 0. \end{cases}$$

6.4   SUGGESTED EXERCISES

6.4.1 Let $F : \mathbb{R}^3 \to \mathbb{R}^3$ be the linear transformation defined by:

$$F(x, y, z) = (x - y, x + 2y, x + y + 3z).$$

Compute the inverse image of the vector $(1, 0, -2)$ under F.

6.4.2 Given the linear transformation $T_k : \mathbb{R}^4 \to \mathbb{R}^3$ defined by:

$$T_k(x_1, x_2, x_3, x_4) =$$

$$(x_1 - 5x_2 + kx_3 - kx_4, x_1 + kx_2 + kx_3 + 5x_4, 2x_1 - 10x_2 + (k+1)x_3 - 3kx_4),$$

determine for which values of k the vector $\mathbf{w}_k = (1, k, -1)$ belongs to $\mathrm{Im}(T_k)$. Set $k = 0$ and determine the preimage of $\mathbf{w}_0$ under T_0.

6.4.3 Given the linear transformation $T : \mathbb{R}^3 \to \mathbb{R}^3$ associated with the matrix

$$A = \begin{pmatrix} 3k & 3 & k+2 \\ 1 & k & k \\ 1 & 2 & 2 \end{pmatrix},$$

determine for which values of k the vector $(k, 3, 3)$ belongs to $\mathrm{Im}(T)$.
   Let $k = 2$; find all the vectors (x, y, z) such that $T(x, y, z) = (2, 3, 3)$.
   Let $k = 3$; find all vectors (x, y, z) such that $T(x, y, z) = (3, 3, 3)$.

6.4.4 Let $S = \{(1, 2, 0, 3) + \mathbf{z} \mid \mathbf{z} \in \langle (1, -1, 2, 1), (1, 5, -2, 5) \rangle\}$. Determine if S is a vector subspace of $\mathbb{R}^4$ and determine, if possible, a homogeneous linear system having S as set of solutions.

6.4.5 Construct, if possible a linear transformation $F : \mathbb{R}^3 \to \mathbb{R}^3$ such that $F^{-1}(1, 0, 0) = \{(1, 0, 0) + \mathbf{v} \mid \mathbf{v} \in \langle (1, 1, 1), (0, 1, -1) \rangle\}$. Establish whether such a transformation is unique.

6.4.6 Determine, if possible:

1. a linear equation having $S = (2, 1, 0, 1) + \langle (2, 1, 2, 2), (1, -1, 2, 1) \rangle$ as a set of solutions;

2. a linear system of two equations having $S = (2, 1, 0, 1) + \langle (2, 1, 2, 2), (1, -1, 2, 1) \rangle$ as a set of solutions;

3. a linear system of three equations having $S = (2, 1, 0, 1) + \langle (2, 1, 2, 2), (1, -1, 2, 1) \rangle$ as a set of solutions.

CHAPTER 7

Determinant and Inverse

In this chapter, we introduce two basic concepts: the *determinant* and the *inverse* of a square matrix. The importance of these two concepts will be summarized by Theorem 7.6.1, which contains essentially all we learnt about the linear maps from $\mathbb{R}^n$ to $\mathbb{R}^n$.

7.1 DEFINITION OF DETERMINANT

The concepts of linear algebra introduced so far are not sufficient to give a direct definition of the determinant. Hence, we shall introduce the determinant with an indirect definition and then we will compute it in some special cases, which will prove to be the most significant for us, and finally we will arrive at an algorithmic method to compute it in general.

Let A be a square matrix, i.e. an $n \times n$ matrix. We want to associate to it a real number, called *determinant* of A, which is calculated starting from the elements of the matrix A. The definition we give is apparently not a constructive one, however, we will see that, starting from simple rules, we can calculate the determinant of a matrix.

Before we begin, it is necessary to introduce the definition of the identity matrix.

Definition 7.1.1 The *identity matrix* or *identity matrix of order n*, is the $n \times n$ matrix having all the elements of the main diagonal equal to the number 1, while the remaining elements are equal to 0. Usually it is indicated with I_n, or with I, if there are no ambiguities.

For example, the identity matrix of order 3 is:

$$I = \begin{pmatrix} 1 & 0 & 0 \\ 0 & 1 & 0 \\ 0 & 0 & 1 \end{pmatrix}.$$

Definition 7.1.2 The *determinant* of a square matrix A of order n is a real number, denoted by $\det(A)$, with the following properties:

1. If the j-th row of A is the sum of two elements $\mathbf{u}$ and $\mathbf{v}$ of $\mathbb{R}^n$, then the

determinant of A is the sum of the determinants of the two matrices obtained by replacing the j-th row of A with $\mathbf{u}$ and $\mathbf{v}$, respectively.

2. If the j-th row of A is the product $\lambda \mathbf{u}$, where $\mathbf{u}$ is an element of $\mathbb{R}^n$ and λ is a scalar, then the determinant of A is the product of λ and the determinant of the matrix obtained by replacing the j-th row of A with $\mathbf{u}$.

3. If two rows of A are equal, then the determinant of A is zero.

4. If I is the identity matrix then $\det(I) = 1$.

Thanks to these properties, we can immediately calculate the determinant of some matrices. For example, by properties (2) and (4), if we consider the matrix

$$A = \begin{pmatrix} 2 & 0 \\ 0 & 3 \end{pmatrix},$$

then

$$\det(A) = 2\det\begin{pmatrix} 1 & 0 \\ 0 & 3 \end{pmatrix} = 2 \cdot 3 \det\begin{pmatrix} 1 & 0 \\ 0 & 1 \end{pmatrix} 2 \cdot 3 \det(I) = 6.$$

We will see later how to exploit these properties in a suitable manner to obtain the determinant of any matrix.

For the moment, we have defined the determinant as a function that has some properties, however this does not guarantee that such a function exists or, if it exists, that it is unique. The next proposition, which we do not prove, establishes such facts.

Proposition 7.1.3 *There is a function that satisfies the properties of Definition 7.1.2 and such function is unique.*

Thanks to the definition of determinant, we can immediately prove the following additional properties, which will be very useful.

Proposition 7.1.4 *Let A and B be two square matrices of order n.*

(a) If B is obtained from A by exchanging two rows, then:

$$\det(A) = -\det(B).$$

(b) If B is obtained from A by adding to a row any linear combination of the other rows, then:

$$\det(A) = \det(B).$$

(c) If A is an upper (or lower) triangular matrix, that is, the coefficients below (respectively above) the main diagonal are all equal to zero, then the determinant of A is the product of the elements that are located on its main diagonal.

Proof. (a) For a matrix A we use notation in compact form, $A = (R_1, R_2, \ldots, R_n)$, where $R_1, \ldots, R_n$ are the elements of $\mathbb{R}^n$ that make up the rows of A. Now consider the matrix $(R_1 + R_2, R_1 + R_2, R_3, \ldots, R_n)$. By property (3) of Definition 7.1.2 we have that:

$$\det(R_1 + R_2, R_1 + R_2, R_3, \ldots, R_n) = 0.$$

On the other hand, by property (1), we have:

$$\det(R_1 + R_2, \quad R_1 + R_2, R_3, \ldots, R_n) =$$

$$= \det(R_1 + R_2, R_1, R_3, \ldots R_n) +$$

$$+ \det(R_1 + R_2, R_2, R_3, \ldots, R_n)$$

$$= \det(R_1, R_1, R_3, \ldots R_n) + \det(R_1, R_2, R_3, \ldots, R_n) +$$

$$+ \det(R_2, R_1, R_3, \ldots, R_n) + \det(R_2, R_2, R_3, \ldots, R_n) =$$

$$= \det(R_1, R_2, R_3, \ldots, R_n) + \det(R_2, R_1, R_3, \ldots, R_n),$$

where at the last step we used again property (3). Then

$$\det(R_1, R_2, R_3, \ldots, R_n) + \det(R_2, R_1, R_3, \ldots, R_n) = 0,$$

from which (a) follows, relatively to the first two rows. It is clear that this can be repeated, in an identical way, for two generic rows.

(b) Let $A = (R_1, R_2, \ldots, R_n)$ and $B = (R_1 + \lambda_2 R_2 + \cdots + \lambda_n R_n, R_2, \ldots, R_n)$. Then, by properties (1) and (2) of Definition 7.1.2, we have:

$$\det(B) = \det(R_1, R_2, \ldots, R_n) + \lambda_2 \det(R_2, R_2, \ldots, R_n) + \cdots +$$

$$+ \lambda_n \det(R_n, R_2, \ldots, R_n) = \det(A),$$

since all the determinants, except the first, are equal to zero, as the corresponding matrices have two equal rows.

(c) Let A be a lower triangular matrix. Suppose for the moment that all the coefficients on the diagonal are different from zero. Using the Gaussian algorithm as described in the first chapter, we can add to the second row a multiple of the first so as to set to zero the coefficient in position $(2,1)$ and then proceed similarly by setting to zero all the coefficients in positions $(3,1), \ldots, (n,1)$. The determinant of the matrix obtained in this way does not change, by property (b). Then, we repeat the same procedure for the second row and then the third, up to the n-th. In this way, we obtain a diagonal matrix. By property (2) of Definition 7.1.2, we have:

$$\det \begin{pmatrix} d_1 & 0 & \ldots & 0 \\ 0 & d_2 & \ldots & 0 \\ \vdots & & & \vdots \\ 0 & \vdots & 0 & d_n \end{pmatrix} = d_1 d_2 \cdots d_n.$$

In the case when one or more coefficients on the diagonal are equal to zero, we cannot obtain a diagonal matrix, however, it is easy to see, by applying the Gaussian algorithm, that we obtain a matrix in which a row consists of all zeros and consequently the determinant is zero. We leave to the reader the details of this case. The reasoning for an upper triangular matrix is similar. ■

Given a square matrix A, using elementary row operations, we can always reduce A to a triangular form and then calculate the determinant using the properties seen above. One must pay attention to the fact that the elementary row operations may change the determinant: if we exchange two rows, we must remember that the determinant changes sign; if we multiply a row by a scalar, the determinant is multiplied by the same scalar; while finally, if we add to a row a linear combination of the others, the determinant does not change.

We see an explicit example, although the method that we describe is not the most efficient for the calculation of the determinant in general.

Example 7.1.5 We calculate the determinant of the matrix A:

$$A = \begin{pmatrix} 0 & 2 & 4 & 6 \\ 1 & 1 & 2 & 1 \\ 1 & 1 & 2 & -1 \\ 1 & 1 & 1 & 2 \end{pmatrix}.$$

We want to bring this matrix into triangular form, using the Gaussian algorithm, but we must take into account all the exchanges and all the multiplications by a scalar we make.

We exchange the first row with the second, in this case by Proposition 7.1.4 (a) the determinant changes sign and the matrix becomes:

$$A = \begin{pmatrix} 1 & 1 & 2 & 1 \\ 0 & 2 & 4 & 6 \\ 1 & 1 & 2 & -1 \\ 1 & 1 & 1 & 2 \end{pmatrix}.$$

Now we perform the following elementary operations, so as to set to zero all the coefficients in the first column:

- 3rd row → 3rd row - 1st row;

- 4th row → 4th row - 1st row.

In this way, the determinant does not change, by Proposition 7.1.4 (b), and we obtain:

$$\begin{pmatrix} 1 & 1 & 2 & 1 \\ 0 & 2 & 4 & 6 \\ 0 & 0 & 0 & -1 \\ 0 & 0 & -1 & 1 \end{pmatrix}.$$

Then, we exchange the second with the third row; by Proposition 7.1.4 (a) the determinant changes sign and the matrix becomes:

$$\begin{pmatrix} 1 & 1 & 2 & 1 \\ 0 & 2 & 4 & 6 \\ 0 & 0 & -1 & 1 \\ 0 & 0 & 0 & -1 \end{pmatrix}.$$

Now we can use property (c) of the previous proposition, which tells us that the determinant of a triangular matrix is the product of the coefficients on the diagonal.

$$\det \begin{pmatrix} 1 & 1 & 2 & 1 \\ 0 & 2 & 4 & 6 \\ 0 & 0 & -1 & 1 \\ 0 & 0 & 0 & -1 \end{pmatrix} = 1 \cdot 2 \cdot (-1) \cdot (-1) = 2.$$

To get the correct result, we must then multiply the determinant of A by (-1) as many times as the row swaps (in this case two), so:

$$\det(A) = 2 \cdot (-1) \cdot (-1) = 2.$$

7.2 CALCULATING THE DETERMINANT: CASES 2×2 AND 3×3

For 2×2 and 3×3 matrices, there are simple formulas for the calculation of the determinant.

We begin by examining the case of 2×2 matrices. Let A be the matrix:

$$A = \begin{pmatrix} a_{11} & a_{12} \\ a_{21} & a_{22} \end{pmatrix}.$$

We proceed with the algorithm we explained above, considering two cases.

Case 1. Suppose first that $a_{11} \neq 0$. In this case, we make the following elementary operation:

- 2nd row $\rightarrow$ 2nd row - $\frac{a_{21}}{a_{11}} \cdot$ 1st row.

In this way, by Proposition 7.1.4 (b), the determinant does not change and we obtain the triangular matrix:

$$\begin{pmatrix} a_{11} & a_{12} \\ 0 & a_{22} - \frac{a_{21}}{a_{11}} \cdot a_{12} \end{pmatrix}.$$

Now we simply take the product of the diagonal coefficients and we have that:

$$\det(A) = a_{11} \cdot \left(a_{22} - \frac{a_{21}}{a_{11}} \cdot a_{12} \right) = a_{11}a_{22} - a_{12}a_{21}.$$

Case 2. Suppose $a_{11} = 0$. We exchange the first and the second row; the determinant changes sign, and we get:

$$\begin{pmatrix} a_{21} & a_{22} \\ 0 & a_{12} \end{pmatrix}.$$

So, taking into account the change of sign,

$$\det(A) = -a_{21} \cdot a_{12} = a_{11}a_{22} - a_{12}a_{21},$$

being $a_{11} = 0$.

In both cases, then the following formula holds:

$$\det(A) = a_{11}a_{22} - a_{12}a_{21}.$$

So we can calculate the determinant of any matrix 2×2. For example:

$$\det \begin{pmatrix} 1 & 2 \\ 3 & 4 \end{pmatrix} = 1 \cdot 4 - 2 \cdot 3 = -2.$$

Consider now the case of 3×3 matrices. Let A be the matrix:

$$A = \begin{pmatrix} a_{11} & a_{12} & a_{13} \\ a_{21} & a_{22} & a_{23} \\ a_{31} & a_{32} & a_{33} \end{pmatrix}.$$

Proceeding in a similar manner as in the 2×2 case (obviously the Gaussian algorithm requires a greater number of steps), it is possible to show that the determinant is:

$$\det(A) = a_{11}a_{22}a_{33} + a_{12}a_{23}a_{31} + a_{13}a_{21}a_{32} - a_{13}a_{22}a_{31} - a_{12}a_{21}a_{33} - a_{11}a_{23}a_{32}.$$

Let us see a mnemonic aid to remember the formula. We rewrite the first two columns of A next to A to the right

$$\begin{pmatrix} a_{11} & a_{12} & a_{13} & a_{11} & a_{12} \\ a_{21} & a_{22} & a_{23} & a_{21} & a_{22} \\ a_{31} & a_{32} & a_{33} & a_{31} & a_{32} \end{pmatrix}.$$

To obtain the determinant of A it is necessary to take the sum of the products of the coefficients of the main diagonals $a_{11}a_{22}a_{33}$, $a_{12}a_{23}a_{31}$, $a_{13}a_{21}a_{32}$ and then subtract the sum of the products of the coefficients of the "opposite" diagonals: $a_{13}a_{22}a_{31}$, $a_{11}a_{23}a_{32}$, $a_{12}a_{21}a_{33}$.

Example 7.2.1 Let

$$A = \begin{pmatrix} 1 & 3 & 0 \\ 2 & 1 & -1 \\ 0 & -3 & 5 \end{pmatrix}.$$

Consider

$$\begin{pmatrix} 1 & 3 & 0 & 1 & 3 \\ 2 & 1 & -1 & 2 & 1 \\ 0 & -3 & 5 & 0 & -3 \end{pmatrix}.$$

We then have: $\det(A) = 5 + 0 - 0 - (0 + 3 + 30) = -28$.

7.3   CALCULATING THE DETERMINANT WITH A RECURSIVE METHOD

For matrices of order larger than two or three, but also as an alternative method, we give a method based on a *recursive* definition of determinant, completely equivalent to the one that we have already seen (though we will not prove such statement).

We begin by defining the concept of a *minor* of a matrix.

Definition 7.3.1 Let $A \in \mathrm{M}_{n \times n}(\mathbb{R})$ be a square matrix of order n (i.e. with n rows and n columns). We denote by A_{ij} the square submatrix of A obtained by deleting the i-th row and j-th column A. A_{ij} is called a *minor* of A of order $n - 1$.

Let us see some examples.

Example 7.3.2 If

$$A = \begin{pmatrix} -1 & 0 & 4 & 7 \\ 3 & -1 & 2 & 1 \\ 3 & 6 & -1 & 0 \\ -2 & -1 & 0 & -1 \end{pmatrix},$$

we have

$$A_{11} = \begin{pmatrix} -1 & 2 & 1 \\ 6 & -1 & 0 \\ -1 & 0 & -1 \end{pmatrix}, \quad A_{23} = \begin{pmatrix} -1 & 0 & 7 \\ 3 & 6 & 0 \\ -2 & -1 & -1 \end{pmatrix},$$

$$A_{42} = \begin{pmatrix} -1 & 4 & 7 \\ 3 & 2 & 1 \\ 3 & -1 & 0 \end{pmatrix}.$$

We now state a proposition that allows us to calculate the determinant of a matrix A by proceeding recursively on the order of A, that is, on the number of its rows (or columns). We will not give a proof of our statement.

Theorem 7.3.3 *Let A be a square matrix.*

- *If A has order 1, i.e. $A = (a_{11})$ has one row and one column, we set*

$$\det(A) = a_{11}.$$

- *Suppose now that we know how to compute the determinant of matrices order $n - 1$. Let*

$$\Gamma_{ij} = (-1)^{i+j} \det(A_{ij}),$$

then

$$\det(A) = a_{11}\Gamma_{11} + a_{12}\Gamma_{12} + \ldots + a_{1n}\Gamma_{1n} = \sum_{k=1}^{n} a_{1k}\Gamma_{1k}.$$

This is the method for the calculation of the determinant by *expanding* along the first row. Let us see how it works in practice, for 2×2 and 3×3 matrices: we find the results seen before.

In fact, we see at once that if $A = \begin{pmatrix} a & b \\ c & d \end{pmatrix}$ is a 2×2 matrix, we have

$$\det(A) = a\Gamma_{11} + b\Gamma_{12} = ad - bc.$$

For example: $\det\begin{pmatrix} 2 & 3 \\ 1 & 4 \end{pmatrix} = 2 \cdot 4 - 3 \cdot 1 = 8 - 3 = 5.$

Let us see now the case of 3×3 matrices. Let

$$A = \begin{pmatrix} a_{11} & a_{12} & a_{13} \\ a_{21} & a_{22} & a_{23} \\ a_{31} & a_{32} & a_{33} \end{pmatrix}.$$

We have, by definition:

$$\Gamma_{11} = (-1)^{1+1} \det A_{11} = \det\begin{pmatrix} a_{22} & a_{23} \\ a_{32} & a_{33} \end{pmatrix},$$

$$\Gamma_{12} = (-1)^{1+2} \det A_{12} = -\det\begin{pmatrix} a_{21} & a_{23} \\ a_{31} & a_{33} \end{pmatrix},$$

$$\Gamma_{13} = (-1)^{1+3} \det A_{13} = \det\begin{pmatrix} a_{21} & a_{22} \\ a_{31} & a_{32} \end{pmatrix}.$$

Hence

$$\det(A) = a_{11}\Gamma_{11} + a_{12}\Gamma_{12} + a_{13}\Gamma_{13} =$$

$$a_{11} \det\begin{pmatrix} a_{22} & a_{23} \\ a_{32} & a_{33} \end{pmatrix} - a_{12} \det\begin{pmatrix} a_{21} & a_{23} \\ a_{31} & a_{33} \end{pmatrix} + a_{13} \det\begin{pmatrix} a_{21} & a_{22} \\ a_{31} & a_{32} \end{pmatrix} =$$

$$= a_{11}(a_{22}a_{33} - a_{23}a_{32}) - a_{12}(a_{21}a_{33} - a_{23}a_{31}) + a_{13}(a_{21}a_{32} - a_{22}a_{31}) =$$

$$a_{11}a_{22}a_{33} + a_{12}a_{23}a_{31} + a_{13}a_{21}a_{32} - a_{11}a_{23}a_{32} - a_{12}a_{21}a_{33} - a_{13}a_{22}a_{31},$$

as we saw earlier.

It is possible to expand the determinant according according to the r-th row:

$$\det(A) = a_{r1}\Gamma_{r1} + a_{r2}\Gamma_{r2} + \ldots + a_{rn}\Gamma_{rn} = \sum_{k=1}^{n} a_{rk}\Gamma_{rk}.$$

The expansion of $\det(A)$ according to the s-th column is given by:

$$\det(A) = a_{1s}\Gamma_{1s} + a_{1s}\Gamma_{2s} + \ldots + a_{ns}\Gamma_{ns} = \sum_{k=1}^{n} a_{ks}\Gamma_{ks}.$$

It can be shown that, in general, expanding the determinant of a matrix $A \in M_n(\mathbb{R})$ according to any row or column, we always get the same number, so we can expand according to the row or column with the highest number of zeros.

Now let us see an example.

Example 7.3.4 Let

$$A = \begin{pmatrix} 1 & 3 & 0 & 0 \\ 2 & 0 & 0 & -1 \\ 0 & -3 & 5 & 0 \\ 1 & 0 & -1 & 0 \end{pmatrix}.$$

The expansion of $\det(A)$ according to the third row is:

$$\det(A) = 0\Gamma_{31} - 3\Gamma_{32} + 5\Gamma_{33} + 0\Gamma_{34} = -3\Gamma_{32} + 5\Gamma_{33}.$$

Now

$$\Gamma_{32} = (-1)^{3+2} \det \begin{pmatrix} 1 & 0 & 0 \\ 2 & 0 & -1 \\ 1 & -1 & 0 \end{pmatrix} = -(-1) = 1,$$

$$\Gamma_{33} = (-1)^{3+3} \det \begin{pmatrix} 1 & 3 & 0 \\ 2 & 0 & -1 \\ 1 & 0 & 0 \end{pmatrix} = +(-3) = -3,$$

then $\det(A) = -3 \cdot 1 + 5(-3) = -18$.

Let us now instead expand $\det(A)$ according to the second column:

$$\det(A) = 3\Gamma_{12} + 0\Gamma_{22} - 3\Gamma_{32} + 0\Gamma_{42} = 3\Gamma_{12} - 3\Gamma_{32}.$$

Now

$$\Gamma_{12} = (-1)^{1+2} \det \begin{pmatrix} 2 & 0 & -1 \\ 0 & 5 & 0 \\ 1 & -1 & 0 \end{pmatrix} = -(+5) = -5,$$

then $\det(A) = 3(-5) - 3 \cdot 1 = -18$.

This particular example shows that, if we expand the determinant according to the third row or the second column, the result does not change.

Next theorem is Binet theorem, and it is particularly important: it states that the determinant function has the multiplicative property, that is, the determinant of a product of matrices is the product of the determinants. Note that for the sum this is very far from being true. We will prove Binet theorem in the appendix.

Theorem 7.3.5 *Let A and B two square matrices $n \times n$, then*

$$\det(AB) = \det(A)\det(B).$$

7.4   INVERSE OF A MATRIX

In the set of real numbers, the inverse s of a number $t \in \mathbb{R} \setminus \{0\}$ is the number defined by the property: $st = ts = 1$ (s is denoted by $1/t$, as we all know). Since we can multiply square matrices, we can ask whether, given a square matrix A, there is a square matrix B such that $AB = BA = I$, where I is the identity matrix, which plays here the same role as the unit in real numbers. This matrix B is called the inverse of A. By Binet theorem, which we have just seen, it is clear that if $\det(A) = 0$, then it is not possible to find a matrix B with this property, because $\det(AB) = \det(A)\det(B) = \det(I) = 1$. We will see shortly that this condition is also sufficient.

We begin our discussion with the definition of the inverse of a square matrix and then move on to the various methods of calculation.

Definition 7.4.1 A square matrix A is called *invertible* if there is a matrix B (denoted by A^{-1}), such that: $AB = BA = I$.

We will compute the inverse of a matrix. A first direct method is given by the proof of the following theorem, which characterizes invertible matrices.

Theorem 7.4.2 *The matrix A is invertible if and only if its determinant is nonzero.*

Proof. We prove first that if A is invertible then its determinant is different from zero. By definition we have that $AA^{-1} = I$, then, by Binet theorem, $\det(A)\det(A^{-1}) = \det(AA^{-1}) = \det(I) = 1$, therefore $\det(A) \neq 0$.

Conversely, if the determinant of A is different from zero, we can construct the inverse of A in the following way. Let $\det(A_{ij})$ be the determinant of the matrix obtained from A by removing the i-th row and j-th column. We then have

$$(A^{-1})_{ij} = \frac{1}{\det(A)}(-1)^{i+j}\det(A_{ji}), \tag{7.1}$$

where $(A^{-1})_{ij}$ indicates the i, j entry of the matrix A^{-1}. We omit the proof of the fact that indeed this is the inverse of A. ◼

Observation 7.4.3 Assume we have two different square matrices A, B such that $AB = I$. Then, by Binet Theorem 7.3.5, $1 = \det(I) = \det(AB) = \det(A)\det(B)$, so $\det(A)$ and $\det(B)$ are both nonzero. By Theorem 7.4.2, both A and B are invertible, thus by multiplying (on the left) both sides of the equality $AB = I$ by A^{-1} we obtain that $B = A^{-1}$ is the inverse of A.

Thanks to this observation we get that, given two square matrices A and B, then $AB = I$ if and only if $BA = I$ and the inverse of a matrix is unique.

Now let us see the explicit formula for the inverse of a 2×2 matrix.

Let $A = \begin{pmatrix} a & b \\ c & d \end{pmatrix}$. If the determinant $\det(A) = ad - bc$ of A is not zero, the inverse of A can be calculated with formula (7.1) of the previous proposition. We have that:

$$(A^{-1})_{11} = \frac{1}{ad-bc}d, \qquad (A^{-1})_{12} = \frac{1}{ad-bc}(-b),$$

$$(A^{-1})_{21} = \frac{1}{ad-bc}(-c), \qquad (A^{-1})_{22} = \frac{1}{ad-bc}a,$$

then

$$A^{-1} = \begin{pmatrix} \frac{d}{ad-bc} & \frac{-b}{ad-bc} \\ \frac{-c}{ad-bc} & \frac{a}{ad-bc} \end{pmatrix}.$$

We leave as an exercise the easy verification that it is precisely the inverse of A, namely that:

$$\begin{pmatrix} \frac{d}{ad-bc} & \frac{-b}{ad-bc} \\ \frac{-c}{ad-bc} & \frac{a}{ad-bc} \end{pmatrix} \begin{pmatrix} a & b \\ c & d \end{pmatrix} = \begin{pmatrix} a & b \\ c & d \end{pmatrix} \begin{pmatrix} \frac{d}{ad-bc} & \frac{-b}{ad-bc} \\ \frac{-c}{ad-bc} & \frac{a}{ad-bc} \end{pmatrix} = \begin{pmatrix} 1 & 0 \\ 0 & 1 \end{pmatrix}.$$

Now we illustrate an alternative method for the calculation of the inverse of a matrix. It is difficult to say which of the two methods is faster, as it depends on the type of matrix.

7.5   CALCULATION OF THE INVERSE WITH THE GAUSSIAN ALGORITHM

Suppose we have an invertible matrix A, and we want to compute its inverse. Consider the matrix:

$$M = (A \,|\, I)$$

obtained by putting the identity matrix next to A. Then, through elementary operations on the rows of the matrix M, we can get the identity matrix on the left-hand side. We briefly indicate the procedure and then we will clarify it with examples.

We already know how to get a matrix C in row echelon form and, as the elementary row operations preserve the rank and the matrix A is invertible, the matrix C will have exactly n pivots. Dividing each row by a suitable number, we can assume that all the pivots of C are equal to 1. We then proceed from "bottom" up, using the last pivot and appropriate elementary operations on rows to obtain a new matrix where the last pivot is 1 and in its column there are all zeros except for the last pivot itself. Then we move to the pivot before the last and perform the same procedure, until in the end the matrix on the left hand side of M is the identity matrix. At this point, the matrix that appears to the right is the inverse matrix of A. We will not prove this procedure, but we will give examples.

Example 7.5.1 Consider the matrix

$$A = \begin{pmatrix} 1 & 3 \\ 1 & 4 \end{pmatrix}.$$

To compute the inverse, we must apply the Gaussian algorithm to the matrix:

$$\left(\begin{array}{cc|cc} 1 & 3 & 1 & 0 \\ 1 & 4 & 0 & 1 \end{array} \right).$$

We carry out the following elementary operation: 2nd row → 2nd row - 1st row, and we get:

$$\left(\begin{array}{cc|cc} 1 & 3 & 1 & 0 \\ 0 & 1 & -1 & 1 \end{array} \right).$$

Then we do: 1st row → 1st row - 3 · 2nd, obtaining:

$$\begin{pmatrix} 1 & 0 & | & 4 & -3 \\ 0 & 1 & | & -1 & 1 \end{pmatrix}.$$

We therefore have:

$$A^{-1} = \begin{pmatrix} 4 & -3 \\ -1 & 1 \end{pmatrix}.$$

We now look at another example, this time with a parameter.

Example 7.5.2 We want to determine for which values of k the matrix

$$A = \begin{pmatrix} 1 & k & 0 \\ 1 & 2k - 2 & 0 \\ 2k & 0 & k \end{pmatrix}$$

is invertible and for these values we want to compute the inverse.

First, we calculate the determinant of A, for example by expanding according to the third column:

$$\det(A) = 0 \cdot \Gamma_{13} + 0 \cdot \Gamma_{23} + k \cdot \Gamma_{33} = k(-1)^6 \det\begin{pmatrix} 1 & k \\ 1 & 2k - 2 \end{pmatrix} = k(k - 2).$$

A is invertible for $k \neq 0$ and $k \neq 2$. Consider now:

$$(A \,|\, I) = \begin{pmatrix} 1 & k & 0 & | & 1 & 0 & 0 \\ 1 & 2k - 2 & 0 & | & 0 & 1 & 0 \\ 2k & 0 & k & | & 0 & 0 & 1 \end{pmatrix}$$

and apply the Gaussian algorithm.

$$\begin{matrix} 2^a \text{ row} & \to & 2^a \text{ row} - 1^a \text{ row} \\ 3^a \text{ row} & \to & 3^a \text{ row} - 2k \cdot 1^a \text{ row} \end{matrix} \quad \to \begin{pmatrix} 1 & k & 0 & | & 1 & 0 & 0 \\ 0 & k - 2 & 0 & | & -1 & 1 & 0 \\ 0 & -2k^2 & k & | & -2k & 0 & 1 \end{pmatrix}$$

$$2^a \text{ row} \quad \to \quad \tfrac{1}{k-2} \cdot 2^a \text{ row} \; \to \begin{pmatrix} 1 & k & 0 & | & 1 & 0 & 0 \\ 0 & 1 & 0 & | & -\frac{1}{k-2} & \frac{1}{k-2} & 0 \\ 0 & -2k^2 & k & | & -2k & 0 & 1 \end{pmatrix}$$

$$3^a \text{ row} \quad \to \quad 3^a \text{ row} + 2k^2 \cdot 2^a \text{ row} \; \to \begin{pmatrix} 1 & k & 0 & | & 1 & 0 & 0 \\ 0 & 1 & 0 & | & -\frac{1}{k-2} & \frac{1}{k-2} & 0 \\ 0 & 0 & k & | & \frac{4k}{k-2} & \frac{2k^2}{k-2} & 1 \end{pmatrix}$$

$$\begin{matrix} 1^a \text{ row} & \to & 1^a \text{ row} - k \cdot 2^a \text{ row} \\ 3^a \text{ row} & \to & \tfrac{1}{k} \cdot 3^a \text{ row} \end{matrix} \quad \to \begin{pmatrix} 1 & 0 & 0 & | & \frac{2k-2}{k-2} & -\frac{k}{k-2} & 0 \\ 0 & 1 & 0 & | & -\frac{1}{k-2} & \frac{1}{k-2} & 0 \\ 0 & 0 & 1 & | & \frac{4}{k-2} & \frac{2k}{k-2} & \frac{1}{k} \end{pmatrix}$$

The inverse is therefore:

$$\begin{pmatrix} \frac{2k-2}{k-2} & -\frac{k}{k-2} & 0 \\ -\frac{1}{k-2} & \frac{1}{k-2} & 0 \\ \frac{4}{k-2} & \frac{2k}{k-2} & \frac{1}{k} \end{pmatrix}.$$

In the next section, we will relate the concept of determinant and inverse of a matrix with the properties of the linear transformation associated with it, once we have fixed the canonical bases in the domain and the codomain.

7.6   THE LINEAR MAPS FROM $\mathbb{R}^N$ TO $\mathbb{R}^N$

Now that we have introduced the concept of determinant and inverse of a matrix, we can give an important result that allows us to characterize invertible linear transformations from $\mathbb{R}^n$ to $\mathbb{R}^n$.

Theorem 7.6.1 *Let $F : \mathbb{R}^n \longrightarrow \mathbb{R}^n$ be a linear map, and let A be the matrix associated to F with respect to the canonical basis (in the domain and codomain). The following statements are equivalent.*

1. *F is an isomorphism.*

2. *F is injective.*

3. *F is surjective.*

4. *$\dim(\mathrm{Im}(F)) = n$.*

5. *$\mathrm{rk}(A) = n$.*

6. *The columns of A are linearly independent.*

7. *The rows of A are linearly independent.*

8. *The system $A\mathbf{x} = \mathbf{0}$ has a unique solution.*

9. *For every $\mathbf{b} \in \mathbb{R}^n$ the system $A\mathbf{x} = \mathbf{b}$ has a unique solution.*

10. *A is invertible.*

11. *The determinant of A is not zero.*

Proof. By Proposition 5.5.2, we immediately have the equivalence between (1), (2), (3). We now show that the statements (3) through (9) are equivalent, showing that each of them implies the next and then that (9) implies (2). We will show then, finally, that (1), (10), (11) are equivalent.

(3) implies (4), because if F is surjective, then $\mathrm{Im}F = \mathbb{R}^n$ has dimension n.

(4) implies (5), because $\mathrm{rk}(A) = \dim(\mathrm{Im}(F))$, by Observation 6.2.2.

(5) implies (6), by the definition of rank of a matrix (which is in particular is the column rank).

(6) implies (7), because the row rank of a matrix is equal to the column rank (Proposition 6.2.3).

(7) implies (8), because if the rows of A are linearly independent, when we apply the

Gaussian algorithm for solving the system $A\mathbf{x} = \mathbf{0}$, we find a row echelon matrix with exactly n pivots, so there is a unique solution.

We now show that (8) implies (9). If the system $A\mathbf{x} = \mathbf{0}$ has a unique solution, reducing the matrix A in row echelon form, we get a matrix A' with exactly n pivots. Then, reducing the matrix $A|\mathbf{b}$ in row echelon form, we get a matrix of the type $A'|\mathbf{b}'$, which also has exactly n pivots (those of A'). Then the system $A\mathbf{x} = \mathbf{b}$ admits a unique solution.

We show that (9) implies (2). By Proposition 5.4.8, it is enough to show that $\mathrm{Ker}\,(F) = \mathbf{0}$. But $\mathrm{Ker}\,(F)$ consists of the solutions of the homogeneous linear system $A \cdot \mathbf{x} = \mathbf{0}$, and thanks to (9), taking $\mathbf{x} = \mathbf{0}$, it has only a solution, which must be the zero solution, thus F is injective.

We have shown that the conditions (2) through (9) are equivalent

We show that (1) implies (10). Let G be the inverse of F, then $F \circ G = G \circ F = \mathrm{id}_{\mathbb{R}^n}$. Let B be the matrix associated with G with respect to the canonical basis. Then $AB = BA = I$, so B is the inverse of A.

(10) implies (1), because, if B is the inverse of A and $L_B : \mathbb{R}^n \longrightarrow \mathbb{R}^n$ is the linear map associated with it, then L_B is the inverse of F.

(10) is equivalent to (11) by Theorem 7.4.2.

■

7.7 EXERCISES WITH SOLUTIONS

7.7.1 Determine for which values of k the matrix

$$A = \begin{pmatrix} k - 3 & 2k \\ 2 & -1 \end{pmatrix}$$

is invertible, and for these values compute the inverse.
Solution. First we compute the determinant: $\det(A) = -(k - 3) - 4k = -5k + 3$, so A is invertible for $k \neq \frac{3}{5}$.

We use formula (7.1) for the inverse:
$(A^{-1})_{11} = \frac{1}{\det(A)}(-1)^{1+1}\det(A_{11}) = \frac{-1}{-5k+3}$,
$(A^{-1})_{12} = \frac{1}{\det(A)}(-1)^{1+2}\det(A_{21}) = -\frac{2}{-5k+3}$,
$(A^{-1})_{21} = \frac{1}{\det(A)}(-1)^{2+1}\det(A_{12}) = -\frac{2k}{-5k+3}$,
$(A^{-1})_{22} = \frac{1}{\det(A)}(-1)^{2+2}\det(A_{22}) = \frac{k-3}{-5k+3}$.
The inverse is therefore:

$$A^{-1} = \begin{pmatrix} \frac{1}{5k-3} & \frac{2}{5k-3} \\ \frac{2k}{5k-3} & \frac{k+1}{-5k+3} \end{pmatrix}.$$

7.7.2 Given the linear transformation $F : \mathbb{R}^3 \longrightarrow \mathbb{R}^3$ defined by: $F(\mathbf{e}_1) = 2\mathbf{e}_1 - \mathbf{e}_2$, $F(\mathbf{e}_2) = \mathbf{e}_2 + k\mathbf{e}_3$, $F(\mathbf{e}_3) = \mathbf{e}_1 - \mathbf{e}_2 + \mathbf{e}_3$, determine the values of k for which F is an isomorphism. Choosing an appropriate value of k compute F^{-1}.

Solution. The matrix associated with F with respect to the canonical bases of the domain and codomain is:

$$A = \begin{pmatrix} 2 & 0 & 1 \\ -1 & 1 & -1 \\ 0 & k & 1 \end{pmatrix}.$$

If we calculate the determinant of A with any of the methods that we have seen, for example expanding according to the first row, we get:

$$\det(A) = (-1)^{1+1} 2(1 + k) + 0 + (-1)^{1+3}(-k) = k + 2.$$

By Theorem 7.6.1 we know that F is an isomorphism if and only if the determinant of A is nonzero, and therefore F is isomorphism if and only if $k \neq 2$. We can therefore choose any value of k other than 2 to calculate F^{-1}. We choose $k = 0$, since this will simplify the calculations. The matrix associated with the inverse of F in the canonical bases for the domain and codomain is the inverse of the matrix A. We compute this inverse using formula 7.1:

$$A^{-1} = \begin{pmatrix} 2 & 0 & 1 \\ -1 & 1 & -1 \\ 0 & 0 & 1 \end{pmatrix}^{-1} = \begin{pmatrix} 1/2 & 0 & -1/2 \\ 1/2 & 1 & 1/2 \\ 0 & 0 & 1 \end{pmatrix}.$$

Although not necessary for this exercise, it is always a good idea to make sure that A^{-1} is actually the inverse of A. To that purpose, it is necessary to perform the rows by columns product of A and A^{-1} and verify that the result is the identity matrix:

$$AA^{-1} = \begin{pmatrix} 2 & 0 & 1 \\ -1 & 1 & -1 \\ 0 & 0 & 1 \end{pmatrix}\begin{pmatrix} 1/2 & 0 & -1/2 \\ 1/2 & 1 & 1/2 \\ 0 & 0 & 1 \end{pmatrix} = I.$$

Therefore the inverse of F is

$$F(x, y, z) = \left(\frac{1}{2}x - \frac{1}{2}z, \frac{1}{2}x + y + \frac{1}{2}z, z \right).$$

7.8 SUGGESTED EXERCISES

7.8.1 Let us consider the linear transformation $F : \mathbb{R}^3 \longrightarrow \mathbb{R}^3$ defined by: $F(x, y, z) = (x + 2z, y + z, z)$.
a) Determine if F is an isomorphism, motivating the answer.
b) If the answer to the question in point (a) is affirmative, calculate the inverse of F.

7.8.2 Consider the linear transformation $F : \mathbb{R}^3 \longrightarrow \mathbb{R}^3$ such that:
$F(x, y, z) = (x + y - 2z, -y + z, -z)$.
Determine a linear transformation $G : \mathbb{R}^3 \longrightarrow \mathbb{R}^3$ such that $F \circ G = id$.

7.8.3 Establish for which values of a the matrix

$$A = \begin{pmatrix} 1 & 2 \\ a & 3 \end{pmatrix}$$

is invertible.
Choose a value of a for which A is invertible and compute the inverse of A.

7.8.4 Establish for which values of a the matrix

$$A = \begin{pmatrix} 0 & 1 & 0 \\ 1 & 0 & 1 \\ 2 & a & 3 \end{pmatrix}$$

is invertible.

Choose one of the values for which it is invertible and compute the inverse.

7.8.5 Let $e_1 = (1,0,0)$, $e_2 = (0,1,0)$, $e_3 = (0,0,1)$ be the canonical basis of the real vector space $\mathbb{R}^3$ and let $a \in \mathbb{R}$. Let $T : \mathbb{R}^3 \longrightarrow \mathbb{R}^3$ be the linear transformation such that $T(e_1) = e_1 - ae_2$, $T(e_2) = e_2 + e_3$, $T(e_3) = ae_3$.
a) Find the values of a for which the map is invertible.
b) Choosing one of the values of a for which T is invertible, compute the inverse.

7.8.6 Calculate the inverse of the matrix A, if it exists:

$$A = \begin{pmatrix} -2 & 4 \\ -1 & 2 \end{pmatrix}.$$

7.8.7 a) Calculate the inverse of the matrix:

$$A = \begin{pmatrix} 1 & -1 & -1 \\ 2 & -1 & 0 \\ 0 & 0 & 1 \end{pmatrix}$$

with any method.
b) Let L_A be the linear transformation associated with A and let T be the linear transformation:

$$T : \quad \mathbb{R}^3 \quad \longrightarrow \quad \mathbb{R}^3$$
$$(x, y, z) \quad \longrightarrow \quad (-x + y - z, z, x - y).$$

Determine $T \circ L_A$.

7.9 APPENDIX

In this appendix, we want to give an alternative, but equivalent, definition of determinant of a square matrix $n \times n$. Instead of defining it indirectly through its properties, like we have done in the text, we will see a direct definition, through the concept of *permutation*, which is extremely important, even if does not appear so in our choice of exposition of the theory.

This appendix is not necessary to continue reading, but it represents a deepening of the concepts presented in this chapter. By the nature and depth of the topics of this appendix it will be impossible to give a complete treatment, and we refer the interested reader to the fundamental text of S. Lang, *Introduction to linear algebra* [5], for further details.

Definition 7.9.1 Let $\{1, \ldots, n\}$ be the set of the first n natural numbers. A *permutation* is a bijective function $\sigma : \{1, \ldots, n\} \longrightarrow \{1, \ldots, n\}$.

A permutation therefore associates a number between 1 and n to a number between 1 and n. For example, the function $\sigma : \{1, 2\} \longrightarrow \{1, 2\}$, defined by: $\sigma(1) = 2$, $\sigma(2) = 1$ is the permutation of $\{1, 2\}$ that *swaps* the number 1 with the number 2. A permutation that only exchanges two numbers i, j and leaves the others unchanged is called a *transposition*, and it is denoted with (i, j). For example, the transposition σ, described above, is written as $\sigma = (1, 2)$.

Another example of permutation is given by the cycle. A *cycle* is denoted with $(i_1, i_2, \ldots, i_s)$, where $\{i_1, i_2, \ldots, i_s\}$ is a subset of $\{1, \ldots, n\}$, and it is the permutation that sends each i_j to i_{j+1} for $j = 1, \ldots, n - 1$, and then sends i_n to i_1, leaving all the other integers unchanged. For example the cycle $(3, 1, 5)$ is the permutation of $\{1, \ldots, 5\}$ that sends 3 to 1, 1 to 5, 5 to 3 while 2 and 4 remain fixed.

We say that r cycles $s_1, \ldots, s_r$ are *disjoint* if each element of $\{1, \ldots, n\}$ is fixed by all cycles except at most one of them.

We denote the permutations also in this way:

$$\sigma = \begin{pmatrix} 1 & \cdots & n \\ \sigma(1) & \cdots & \sigma(n) \end{pmatrix}$$

and their set with S_n.

Let's see some examples.

Example 7.9.2   1. Consider the permutation

$$\sigma_1 = \begin{pmatrix} 1 & 2 & 3 & 4 \\ 1 & 4 & 3 & 2 \end{pmatrix}.$$

This permutation exchanges (or we also say *permutes*) the elements 2 and 4 and leaves 3 and 4 unchanged. It is also denoted for simplicity by $\sigma_1 = (2, 4)$ and, as we have seen, we call it a *transposition*.

2. Consider the permutation:

$$\sigma_2 = \begin{pmatrix} 1 & 2 & 3 & 4 \\ 2 & 3 & 1 & 4 \end{pmatrix}.$$

This permutation is the function that is obtained as a composition of the permutations $(2, 3)$ and $(1, 2)$, that is $\sigma_2 = (1, 2) \circ (2, 3)$. In fact, first we exchange 2 with 3 and then 1 with 2, obtaining the cycle $(1, 2, 3)$. Note that the fact that 4 does not appear in the cycle writing means that 4 is left unchanged by the permutation.

Let us now define the *parity* of a permutation σ.

Definition 7.9.3 Given a permutation σ, we consider the number of *inversions*, that is of pairs (i, j), such that $i < j$, but $\sigma(i) > \sigma(j)$. If such a number is even we will say that the permutation σ is *even*, or that its parity is $p(\sigma) = 1$, if it is odd, we say then that the permutation is *odd*, that is that its parity is $p(\sigma) = -1$. In other words, $p(\sigma) = (-1)^i$, where i is the number of inversions.

In the example considered above $p(\sigma_1) = -1$, while $p(\sigma_2) = 1$.

Each permutation can be written as a composition of transpositions, in fact one can show that every function from $\{1, \ldots, n\}$ to $\{1, \ldots, n\}$ can be realized making subsequent exchanges, i.e. transpositions. However, pay attention to the fact that in general there is not uniqueness of writing: the same permutation can appear in two different ways as a composition of transpositions. For example, the identity in S_3 is obtained as the composition $(1, 2) \circ (1, 2)$, but also as $(1, 3) \circ (1, 3)$.

The parity of a permutation σ can also be defined equivalently as $(-1)^m$, where m is the number of transpositions whose composition gives σ. Once we fix a permutation, it can be written as a product of transpositions in many different ways, but the number of transpositions appearing in each product is always of the same parity (i.e. always even or always odd), so the alternative definition of parity we have given is well posed. The proof of this fact and of the equivalence between the two proposed definitions of parity are not obvious at all, but it would take us too far from our goal, namely a direct definition of the determinant of a matrix.

We now state two fundamental results in the theory of permutations, omitting also in this case the proof for the reasons already described above.

Proposition 7.9.4 *Each permutation is written in a unique way as a composition of disjoint cycles.*

Proposition 7.9.5 *Parity has the following property:*

$$p(s_1 \circ s_2) = p(s_1)p(s_2),$$

for every $s_1, s_2 \in S_n$.

Now that we have introduced the concept of permutation, we can give an alternative definition of determinant.

Definition 7.9.6 Let $A = (a_{ij})$ be a square matrix $n \times n$. We define *determinant* of A the number:

$$\det(A) = \sum_{\sigma \in S_n} (-1)^{p(\sigma)} a_{1,\sigma(1)} \cdots a_{n,\sigma(n)},$$

where with $\sum_{\sigma \in S_n}$ we denote the fact that we are doing the sum of the elements $a_{1,\sigma(1)} \cdots a_{n,\sigma(n)}$ as σ varies among all permutations of S_n.

We verify that in the case of 2×2 and 3×3 matrices the new definition of determinant corresponds to that seen previously in 7.1.2.

Let A be a 2×2 matrix:

$$A = \begin{pmatrix} a_{11} & a_{12} \\ a_{21} & a_{22} \end{pmatrix}.$$

According to the new definition, its determinant is:

$$\det(A) = a_{11}a_{22} - a_{12}a_{21},$$

as S_2, the set of permutations of two elements, consists only of the identity and the transposition $(1, 2)$. We can immediately note that this expression coincides with the formula for the determinants of 2×2 matrices obtained in Section 7.2.

Let A be a 3×3 matrix:

$$A = \begin{pmatrix} a_{11} & a_{12} & a_{13} \\ a_{21} & a_{22} & a_{23} \\ a_{31} & a_{32} & a_{33} \end{pmatrix}.$$

The set of permutations of three elements is:

$$S_3 = \{\mathrm{id}, (1, 2), (2, 3), (1, 3), (3, 2, 1), (2, 3, 1)\},$$

with respective parities: $p(1, 2) = p(2, 3) = p(1, 3) = -1$, $p(\mathrm{id}) = p(3, 2, 1) = p(2, 3, 1) = 1$. The determinant of A is therefore given by:

$$\det(A) = a_{11}a_{22}a_{33} + a_{12}a_{23}a_{31} + a_{13}a_{21}a_{32} - a_{13}a_{22}a_{31} - a_{12}a_{21}a_{33} - a_{11}a_{23}a_{32}.$$

We introduce a notation that will be useful later.

Let $A = (a_{ij})_{\substack{i=1,\ldots,n \\ j=1,\ldots,n}}$ be a generic matrix. The ith row of A is given by:

$$(a_{i1}, a_{i2}, \ldots, a_{in}) = a_{i1}\mathbf{e}_1 + a_{i2}\mathbf{e}_2 + \cdots + a_{in}\mathbf{e}_n,$$

where $\mathbf{e}_i = (0, \ldots, 1, \ldots, 0)$ is the i-th basis vector of $\mathbb{R}^n$. Let us write the matrix A as the sequence of its rows:

$$A = \left(\sum_{k_1=1}^{n} a_{1k_1}\mathbf{e}_{k_1}, \sum_{k_2=1}^{n} a_{2k_2}\mathbf{e}_{k_2}, \ldots, \sum_{k_n=1}^{n} a_{nk_n}\mathbf{e}_{k_n} \right).$$

To clarify the new notation that we introduced, if $A = \begin{pmatrix} a_{11} & a_{12} \\ a_{21} & a_{22} \end{pmatrix}$, we write:

$$A = (a_{11}\mathbf{e}_1 + a_{12}\mathbf{e}_2, a_{21}\mathbf{e}_1 + a_{22}\mathbf{e}_2).$$

We now want to show that in the case of a 2×2 matrix the properties that define the determinant (see Definition 7.1.2) determine it in a unique way. Let $A = (a_{11}\mathbf{e}_1 + a_{12}\mathbf{e}_2, a_{21}\mathbf{e}_1 + a_{22}\mathbf{e}_2)$. Thanks to property (1) of Definition 7.1.2 we have:

$$\det(A) = \det(a_{11}\mathbf{e}_1 + a_{12}\mathbf{e}_2, a_{21}\mathbf{e}_1 + a_{22}\mathbf{e}_2) =$$

$$= \det(a_{11}\mathbf{e}_1, a_{21}\mathbf{e}_1 + a_{22}\mathbf{e}_2) + \det(a_{12}\mathbf{e}_2, a_{21}\mathbf{e}_1 + a_{22}\mathbf{e}_2) =$$

$$= \det(a_{11}\mathbf{e}_1, a_{21}\mathbf{e}_1) + \det(a_{11}\mathbf{e}_1, a_{22}\mathbf{e}_2) + \det(a_{12}\mathbf{e}_2, a_{21}\mathbf{e}_1) +$$

$$+ \det(a_{12}\mathbf{e}_2, a_{22}\mathbf{e}_2).$$

By property (2) of Definition 7.1.2, we can then write:

$$\det(A) = a_{11}a_{21}\det(\mathbf{e}_1, \mathbf{e}_1) + a_{11}a_{22}\det(\mathbf{e}_1, \mathbf{e}_2)+$$

$$+a_{12}a_{21}\det(\mathbf{e}_2, \mathbf{e}_1) + a_{12}a_{22}\det(\mathbf{e}_2, \mathbf{e}_2).$$

By Proposition 7.1.4, we have that: $\det(\mathbf{e}_1, \mathbf{e}_1) = \det(\mathbf{e}_2, \mathbf{e}_2) = 0$ (as they are matrices with two equal rows) and $\det(\mathbf{e}_1, \mathbf{e}_2) = -\det(\mathbf{e}_2, \mathbf{e}_1)$. Finally by property (3) of Definition 7.1.2, we have that $\det(\mathbf{e}_1, \mathbf{e}_2) = 1$. Therefore

$$\det(A) = (a_{11}a_{22} - a_{12}a_{21})\det(\mathbf{e}_1, \mathbf{e}_2) = a_{11}a_{22} - a_{12}a_{21}.$$

This shows that the function defined in 7.1.2, must necessarily be expressed by the formula in 7.9.6, and therefore this function is unique.

The procedure we have described for $2{\times}2$ matrices can be replicated identically in the case of $n \times n$ matrices, allowing us to get the equivalence of the two definitions of determinant. Let us look at this in more detail in the proof of the following theorem, which is the most significant result of this appendix.

Theorem 7.9.7 *The function defined in 7.1.2 exists and is unique, and it is expressed by the formula of the Definition 7.9.6. The two Definitions 7.9.6 and 7.1.2 of determinant are therefore equivalent.*

Proof. Let

$$A = \left(\sum_{k_1=1}^{n} a_{1k_1}\mathbf{e}_{k_1}, \sum_{k_2=1}^{n} a_{2k_2}\mathbf{e}_{k_2}, \ldots, \sum_{k_n=1}^{n} a_{nk_n}\mathbf{e}_{k_n} \right)$$

be a $n \times n$ matrix written according to the notation introduced above (i.e. as a sequence of its rows). Let $\det(A)$ be the number defined in 7.1.2. In a similar way to what we saw for the case of 2×2 matrices, by properties (1) and (2) of Definition 7.1.2 we can write immediately

$$\det(A) = \sum_{1 \leq k_1 \ldots k_n \leq n} a_{1,k_1} \ldots a_{n,k_n} \det(\mathbf{e}_{k_1}, \mathbf{e}_{k_2}, \ldots, \mathbf{e}_{k_n}).$$

Note that the matrix $(\mathbf{e}_{k_1}, \mathbf{e}_{k_2}, \ldots, \mathbf{e}_{k_n})$ (written as a sequence of rows, according to our convention) has, in the i-th row, 1 in the k_i position and zero elsewhere. It is therefore clear that if there are some values repeated among $k_1, \ldots, k_n$ the matrix $(\mathbf{e}_{k_1}, \mathbf{e}_{k_2}, \ldots, \mathbf{e}_{k_n})$ has two equal rows and therefore by property (3) of Definition 7.1.2 its determinant is zero. So $\det(\mathbf{e}_{k_1}, \mathbf{e}_{k_2}, \ldots, \mathbf{e}_{k_n}) \neq 0$ only if $k_1, \ldots, k_n$ are all distinct, that is, the function s defined by $s(i) = k_i$ is a permutation of $\{1, \ldots, n\}$. At this point we have

$$\det(\mathbf{e}_{k_1}, \mathbf{e}_{k_2}, \ldots, \mathbf{e}_{k_n}) = \begin{cases} 1 & \text{if } p(s) = 1 \\ -1 & \text{if } p(s) = -1 \end{cases}.$$

because we can reorder the rows of $(\mathbf{e}_{k_1}, \mathbf{e}_{k_2}, \ldots, \mathbf{e}_{k_n})$ to get the identity matrix (which has determinant 1), and to do this we made a number of exchanges corresponding to the parity of s. We therefore obtained that the determinant of A, as defined in 7.1.2, must be expressed by the formula:

$$\det(A) = \sum_{s \in S_n} (-1)^{p(s)} a_{1,s(1)} \cdots a_{n,s(n)}.$$

This proves the equivalence between the two definitions, but also how the number defined by the four properties in 7.1.2 exists and is unique, that is how these properties determine it uniquely. ■

We have an important corollary.

Corollary 7.9.8 *The determinant is also given by the formula:*

$$det(A) = \sum_{\sigma \in S_n} (-1)^{p(\sigma)} a_{\sigma(1),1} \cdots a_{\sigma(n),n}$$

(note that, with respect to Definition 7.9.6, the permutations are carried out on the row and not on the column indexes). In particular we have that

$$\det(A) = \det(A^T).$$

Proof. By the previous theorem we have:

$$\det(A) = \sum_{\sigma \in S_n} (-1)^{p(\sigma)} a_{1,\sigma(1)} \cdots a_{n,\sigma(n)}.$$

We now observe that

$$a_{\sigma(r),r} = a_{s,\sigma^{-1}(s)},$$

where $s = \sigma(r)$, since $\sigma^{-1}(s) = \sigma^{-1}(\sigma(r)) = r$. Also, since s is a bijection of $\{1, \ldots, n\}$ into itself, we have that

$$a_{1,\sigma(1)} \cdots a_{n,\sigma(n)} = a_{\sigma^{-1}(1),1} \cdots a_{\sigma^{-1}(n),n}.$$

Therefore:

$$\det(A) = \sum_{\sigma \in S_n} (-1)^{p(\sigma)} a_{1,\sigma(1)} \cdots a_{n,\sigma(n)}$$

$$= \sum_{\sigma \in S_n} (-1)^{p(\sigma)} a_{\sigma^{-1}(1),1} \cdots a_{\sigma^{-1}(n),n} =$$

$$= \sum_{\tau \in S_n} (-1)^{p(\tau)} a_{\tau(1),1} \cdots a_{\tau(n),n},$$

because, if σ varies among all the permutations in S_n, also $\tau = \sigma^{-1}$ varies among all permutations in S_n and moreover $p(\tau) = p(\sigma)$. ■

Thanks to this new definition and to the previous theorem, we can prove what we have just stated in the text concerning the determinant calculation procedures.

We now want to get a proof of Theorem 7.3.3, which provides us with a valid tool for calculating the determinant. Before going to the proof of the formula that appears in Theorem 7.3.3, also known as the formula for *Laplace expansion* of the determinant, we need a technical lemma.

Lemma 7.9.9 *If B is a square matrix of the type:*

$$B = \begin{pmatrix} 1 & 0 & 0 & \ldots & 0 \\ b_{21} & b_{22} & b_{23} & \ldots & b_{2n} \\ \vdots & \vdots & \vdots & \ldots & \vdots \\ b_{n1} & b_{n2} & b_{n3} & \ldots & b_{nn} \end{pmatrix},$$

then

$$\det(B) = \det\begin{pmatrix} b_{22} & b_{23} & \ldots & b_{2n} \\ \vdots & \vdots & \ldots & \vdots \\ b_{n2} & b_{n3} & \ldots & b_{nn} \end{pmatrix}.$$

Proof. We have $\det(B) = \sum_{\sigma \in S_n} (-1)^{p(\sigma)} b_{1,\sigma(1)} \ldots b_{n,\sigma(n)}$. Since $b_{1,j} = 0$ for every $j \neq 1$, the permutations $\sigma \in S_n$ such that $\sigma(1) \neq 1$ give no contribution to the sum, thus we have that

$$\det(B) = \sum_{\sigma \in S_n} (-1)^{p(\sigma)} b_{1,1} b_{2,\sigma(2)} \ldots b_{n,\sigma(n)}.$$

Now the set of permutations σ of $\{1, \ldots, n\}$ that fix 1 can be seen as the set of all permutations τ of the set $\{2, \ldots, n\}$. Considering that $b_{1,1} = 1$ we have:

$$\det(B) = \sum_{\tau \in S_{n-1}} (-1)^{p(\tau)} b_{2,\tau(2)} \ldots b_{n,\tau(n)},$$

and this is exactly the determinant of the matrix:

$$\begin{pmatrix} b_{22} & b_{23} & \ldots & b_{2n} \\ \vdots & \vdots & \ldots & \vdots \\ b_{n2} & b_{n3} & \ldots & b_{nn} \end{pmatrix}.$$

■

Theorem 7.9.10 (Laplace theorem). *Let A be a $n \times n$ matrix and $i \in \{1, \ldots, n\}$ a fixed index. Then we have that:*

$$\det(A) = a_{i,1}\Gamma_{i,1} + \ldots a_{i,n}\Gamma_{i,n}, \qquad \text{and}$$

$$\det(A) = a_{1,i}\Gamma_{1,i} + \ldots a_{n,i}\Gamma_{n,i},$$

where Γ_{kl} denotes determinant of the matrix obtained from A by deleting the k-th row and the l-th column, multiplied by $(-1)^{k+l}$.

Proof. We look at the proof of the first of these properties, the second is quite similar. We also suppose $i = 1$. The general case is only more complicated to write, but does not offer any additional conceptual difficulty. We write A as $A = (\sum_{j=1}^{n} a_{1j}\mathbf{e}_j, A_2, \ldots, A_n)$. By properties (1) and (2) of Definition 7.1.2, we have that:

$$\det(A) = \sum_{j=1}^{n} a_{1j} \det(\mathbf{e}_j, A_2, \ldots, A_n).$$

It is therefore sufficient to calculate the determinant of matrices of the type:

$$M_j = \det(\mathbf{e}_j, A_2, \ldots, A_n) =$$

$$= \begin{pmatrix} 0 & 0 & \cdots & 0 & 1 & 0 & \cdots & 0 \\ a_{21} & a_{22} & \cdots & a_{2(j-1)} & a_{2j} & a_{2(j+1)} & \cdots & a_{2n} \\ \vdots & \vdots & \ddots & \vdots & \vdots & \vdots & \ddots & \vdots \\ a_{n1} & a_{n2} & \cdots & a_{n(j-1)} & a_{nj} & a_{n(j+1)} & \cdots & a_{2n} \end{pmatrix}.$$

We now observe that, as a consequence of property (1) of Proposition 7.1.4 and of Corollary 7.9.8, if two columns of a matrix are exchanged, the determinant changes sign, therefore

$$\det(M_j) =$$

$$= (-1)^{j-1} \det \begin{pmatrix} 1 & 0 & \cdots & 0 & 0 & 0 & \cdots & 0 \\ a_{2j} & a_{21} & a_{22} & \cdots & a_{2(j-1)} & a_{2(j+1)} & \cdots & a_{2n} \\ \vdots & \vdots & \vdots & \cdots & \vdots & & & \\ a_{nj} & a_{n1} & a_{n2} & \cdots & a_{n(j-1)} & a_{n(j+1)} & \cdots & a_{nn} \end{pmatrix}.$$

Observing that $(-1)^{j-1} = (-1)^{1+j}$ and using Lemma 7.9.9 we now have that:

$$\det(M_j) = (-1)^{1+j} \det \begin{pmatrix} a_{21} & a_{22} & \cdots & a_{2(j-1)} & a_{2(j+1)} & \cdots & a_{2n} \\ \vdots & \vdots & \cdots & \vdots & \vdots & \cdots & \vdots \\ a_{n1} & a_{n2} & \cdots & a_{n(j-1)} & a_{n(j+1)} & \cdots & a_{nn} \end{pmatrix},$$

and this is precisely the matrix obtained from A by deleting the first row and the j-th column.

Then we get:

$$\det(A) = \sum_{j=1}^{n} a_{1j}(-1)^{1+j}$$

$$\det \begin{pmatrix} a_{21} & a_{22} & \cdots & a_{2(j-1)} & a_{2(j+1)} & \cdots & a_{2n} \\ \vdots & \vdots & \cdots & \vdots & \vdots & \cdots & \vdots \\ a_{n1} & a_{n2} & \cdots & a_{n(j-1)} & a_{n(j+1)} & \cdots & a_{nn} \end{pmatrix},$$

that is, the desired formula. ∎

Let us now prove Binet Theorem 7.3.5; as the two definitions of determinant are equivalent we can use the one that best fits what we want to do.

Theorem 7.9.11 (Binet Theorem). *Let* $A = (a_{ij})_{\substack{i=1,\ldots,n \\ j=1,\ldots,n}}$ *and* $B = (b_{ij})_{\substack{i=1,\ldots,n \\ j=1,\ldots,n}}$ *be two square matrices of order n. Then:*

$$\det(AB) = \det(A)\det(B).$$

Proof. We know that AB is the matrix whose coefficient of place i, j is $\sum_{k=1}^{n} a_{ik}b_{kj}$, then apply Definition 7.9.6:

$$
\det(AB) = \sum_{\sigma \in S_n} (-1)^{p(\sigma)} \left(\sum_{k_1=1}^{n} a_{1k_1} b_{k_1,\sigma(1)}\right) \cdots \left(\sum_{k_n=1}^{n} a_{nk_n} b_{k_n,\sigma(n)}\right) =
$$

$$
= \sum_{1 \le k_1,\ldots,k_n \le n} \sum_{\sigma \in S_n} (-1)^{p(\sigma)} a_{1,k_1} b_{k_1,\sigma(1)} \cdots a_{n,k_n} b_{k_n,\sigma(n)} =
$$

$$
= \left(\sum_{1 \le k_1,\ldots,k_n \le n} a_{1,k_1} \cdots a_{n,k_n}\right)
$$

$$
\left(\sum_{\sigma \in S_n} (-1)^{p(\sigma)} b_{k_1,\sigma(1)} \cdots b_{k_n,\sigma(n)}\right)
$$

rearranging the terms. Let us fix $k_1, \ldots, k_n$ and consider the number:

$$
\sum_{\sigma \in S_n} (-1)^{p(\sigma)} b_{k_1,\sigma(1)} \cdots b_{k_n,\sigma(n)}.
$$

This is the determinant of the matrix having as rows the rows $k_1, \ldots, k_n$ of the matrix B, i.e.:

$$
\sum_{\sigma \in S_n} (-1)^{p(\sigma)} b_{k_1,\sigma(1)} \cdots b_{k_n,\sigma(n)} = \det\left(\sum_{i_1} b_{k_1,i_1} \mathbf{e}_{i_1}, \ldots, \sum_{i_n} b_{k_1,i_n} \mathbf{e}_{i_n}\right),
$$

using the notation introduced previously. By property (3) of Definition 7.1.2, we have that if $k_1, \ldots, k_n$ are not all distinct, that determinant is zero. So from now on in the expression $\det(AB)$ we sum up only the terms where $k_1, \ldots, k_n$ are distinct. In this case, we note that the matrix $(\sum_{i_1} b_{k_1,i_1} \mathbf{e}_{i_1}, \ldots, \sum_{i_n} b_{k_1,i_n} \mathbf{e}_{i_n})$ is obtained starting from B with a number of row exchanges corresponding to the parity of the permutation τ defined by $\tau(i) = k_i$. Therefore:

$$
\det\left(\sum_{i_1} b_{k_1,i_1} \mathbf{e}_{i_1}, \ldots, \sum_{i_n} b_{k_1,i_n} \mathbf{e}_{i_n}\right) = (-1)^{p(\tau)} \det(B).
$$

Since $(k_1, \ldots, k_n) = (\tau(1), \ldots, \tau(n))$, considering all n-tuples with $k_1, \ldots, k_n$ distinct is equivalent to having τ varying among all permutations of S_n.

We have therefore obtained:

$$
\det(AB) = \sum_{1 \le k_1,\ldots,k_n \le n} (-1)^{p(\tau)} a_{1,k_1} \cdots a_{n,k_n} \det(B) =
$$

$$
= \sum_{\tau \in S_n} (-1)^{p(\tau)} a_{1,\tau(1)} \cdots a_{n,\tau(n)} \det(B) = \det(A) \det(B).
$$

∎

If A is a square matrix we denote with $A_1, \ldots, A_n$ the rows of A and with $\tilde{\Gamma}_i$ the column vector $(\Gamma_{i1}, \ldots, \Gamma_{kn})^T$, where the numbers Γ_{ij} are defined as in Laplace theorem, and they are called the *algebraic complements* of the A matrix. The first formula of Laplace theorem can then be written as:

$$
A_i \tilde{\Gamma}_i = \det(A), \quad \text{for each } i = 1, \ldots, n, \tag{7.2}
$$

where the product is the usual product rows by columns.

It makes sense to ask if the product $A_i \tilde{\Gamma}_j$ has any meaning even when $i \ne j$. The answer is given by following proposition.

Proposition 7.9.12 *Let the notation be as above. Then:*

$$A_i \widetilde{\Gamma}_j = 0, \quad \text{for each } i, j = 1, \ldots, n, \quad i \neq j. \tag{7.3}$$

Proof. To prove (7.3) we consider the matrix $A' = (a'_{ij})_{\substack{i=1,\ldots,n \\ j=1,\ldots,n}}$ obtained from A by replacing the j-th row with the i-th row. Then we have $a'_{rs} = a_{rs}$ for every $r, s = 1, \ldots, n$, with $r \neq j$, and $a'_{js} = a_{is}$ for every $s = 1, \ldots, n$.

Let Γ'_{rs} be the algebraic complement of A'. Since the j-th row of A' does not appear in the calculation of Γ'_{js}, we have $\Gamma'_{js} = \Gamma_{js}$ for every $s = 1, \ldots, n$. We now compute the determinant of A' using Laplace theorem and expanding according to the j-th row:

$$\det(A') = a'_{j,1}\Gamma'_{j,1} + \cdots + a'_{j,n}\Gamma'_{j,n} = a_{i,1}\Gamma_{j,1} + \cdots + a_{i,n}\Gamma_{j,n}.$$

On the other hand, the matrix A' has two equal rows (the i-th and the j-th), then by property (3) of Definition 7.1.2 the determinant of A' is equal to zero. This proves precisely formula (7.3), that is, we have:

$$a_{i,1}\Gamma_{j,1} + \cdots + a_{i,n}\Gamma_{j,n} = 0 \quad \text{per ogni } i, j = 1, \ldots, n, \quad i \neq j.$$

■

Let us now prove formula (7.1) for the computation of the inverse matrix of a square matrix A with non zero determinant. Let $\det(A_{ij})$ be the determinant of the matrix obtained from A by removing the i-th row and the j-th column and consider the matrix B whose elements are defined by:

$$(B)_{ij} = \frac{1}{\det(A)}(-1)^{i+j}\det(A_{ji}). \tag{7.4}$$

Note that the j-th column of the matrix B is $\frac{1}{\det(A)}\widetilde{\Gamma}_j$. We now compute the product, rows by columns, AB. The element of place i, j is $A_i B_j = \frac{1}{\det(A)}A_i\widetilde{\Gamma}_j$ and by formulas (7.2) and (7.3) this is $\frac{1}{\det(A)}\det(A) = 1$ if $i = j$ and 0 if $i \neq j$.

So $AB = I$. Similarly, starting from Laplace theorem and expanding the determinant according to the columns, we will have that $BA = I$, hence B is the inverse of A.

Finally, we prove the correctness of the method described in Section 7.5 to compute the inverse of an invertible matrix using the Gaussian algorithm.

Let $A = (a_{ij})_{\substack{i=1,\ldots,n \\ j=1,\ldots,n}}$ be an invertible square matrix and let $B = (b_{ij})_{\substack{i=1,\ldots,n \\ j=1,\ldots,n}}$ be its inverse. We have $AB = I$, where I is the identity matrix of order n. Let $A_1, \ldots, A_n$ be the row vectors of A and $\widetilde{B}_1, \ldots, \widetilde{B}_n$ the column vectors of B. The coefficients $b_{11}, \ldots, b_{n1}$ of $\widetilde{B}_1$ satisfy the relations: $A_1\widetilde{B}_1 = (AB)_{11} = 1$, $A_2\widetilde{B}_1 = (AB)_{21} = 0, \ldots, A_n\widetilde{B}_1 = (AB)_{n1} = 0$ (where the products are rows by columns), i.e. they are a solution of the linear system associated with the matrix:

$$\left(\begin{array}{cccc|c} a_{11} & a_{12} & \ldots & a_{1n} & 1 \\ a_{21} & a_{22} & \ldots & a_{2n} & 0 \\ \vdots & \vdots & & \vdots & \vdots \\ a_{n1} & a_{n2} & \ldots & a_{nn} & 0 \end{array} \right). \tag{7.5}$$

Since A is invertible, by Theorem 7.6.1, the system admits a unique solution, and then solving this system we determine uniquely the elements of the column $\widetilde{B}_1$. As described in Section 7.5, through elementary operations on the rows, it is possible to obtain the identity matrix on the left, that is

$$\left(\begin{array}{cccc|c} 1 & 0 & \cdots & 0 & c_{11} \\ 0 & 1 & \cdots & 0 & c_{21} \\ \vdots & \vdots & & \vdots & \vdots \\ 0 & 0 & \cdots & 1 & c_{n1} \end{array}\right). \tag{7.6}$$

Since the two systems associated with the matrices in (7.5) and (7.6) have the same solutions, it must be $b_{j1} = c_{j1}$ for each $j = 1, \ldots, n$, i.e. in (7.6) the column on the left is precisely $\widetilde{B}_1$.

We proceed in the same way for the generic column $\widetilde{B}_i$, whose coefficients are the solutions of the linear system:

$$\left(\begin{array}{cccc|c} a_{11} & a_{12} & \cdots & a_{1n} & 0 \\ \vdots & \vdots & & \vdots & \vdots \\ a_{i1} & a_{i2} & \cdots & a_{in} & 1 \\ \vdots & \vdots & & \vdots & \vdots \\ a_{n1} & a_{n2} & \cdots & a_{nn} & 0 \end{array}\right). \tag{7.7}$$

To solve this system, we can perform on the rows of A *exactly the same elementary operations* that we did in the case of $\widetilde{B}_1$, and thus we obtain:

$$\left(\begin{array}{cccc|c} 1 & 0 & \cdots & 0 & c_{1i} \\ 0 & 1 & \cdots & 0 & c_{2i} \\ \vdots & \vdots & & \vdots & \vdots \\ 0 & 0 & \cdots & 1 & c_{ni} \end{array}\right) \tag{7.8}$$

and again, since the two systems associated with the matrices in (7.7) and (7.8) have the same solutions, it must be $b_{ji} = c_{ji}$ for each $j = 1, \ldots, n$, i.e. in (7.8) the column on the left is just $\widetilde{B}_i$. Since we have to solve n linear systems that all have the same matrix of coefficients, we can solve them at the same time by considering the matrix

$$A|I.$$

For all we said, after performing the elementary operations on the rows needed to reduce A to the identity matrix, we obtain a matrix of the type:

$$\left(\begin{array}{cccc|cccc} 1 & 0 & \cdots & 0 & c_{11} & c_{12} & \cdots & c_{1n} \\ 0 & 1 & \cdots & 0 & c_{21} & c_{22} & \cdots & c_{2n} \\ \vdots & \vdots & & \vdots & \vdots & & & \\ 0 & 0 & \cdots & 1 & c_{n1} & c_{n2} & \cdots & c_{nn} \end{array}\right),$$

where $c_{ji} = b_{ji}$ for every $i, j = 1, \ldots, n$, i.e. the matrix on the left is precisely the inverse B of A.

Let us now summarize the key properties of the determinant of a matrix.

Properties of the determinant

The determinant has the following properties:

- The determinant of a matrix A is zero if A has a zero row or a zero column.

- The determinant of a matrix A is zero if and only if the matrix A has one row (or a column), which is a linear combination of the others.

- If the matrix A' is obtained from the matrix A by exchanging two rows (or two columns) the determinant of A' is the opposite of the determinant of A.

- If the matrix A' is obtained from the matrix A by multiplying a row (or a column) by a scalar λ, the determinant of A' is the product of λ by the determinant of A.

- If the matrix A' is obtained from the matrix A adding to a row (or a column) a linear combination of the others the determinant of A' is equal to the determinant of A.

- The determinant of the identity matrix is 1.

- The determinant of a diagonal matrix is the product of the elements on the diagonal.

- $\det(A) = \det(A^T)$, for each matrix A.

- $\det(AB) = \det(BA) = \det(A)\det(B)$ for each pair of square matrices A and B of the same order (but caution, in general $AB \neq BA$!).

- $\det(A^{-1}) = \det(A)^{-1}$, for each invertible square matrix A.

CHAPTER 8

Change of Basis

In this chapter, we want to address one of the most technical topics of this theory, i.e. the change of basis within a vector space. We will also understand how to change the matrix associated with a linear transformation, if we change the bases in the domain and codomain.

8.1 LINEAR TRANSFORMATIONS AND MATRICES

As we have seen in Chapter 5, if we fix the canonical bases in $\mathbb{R}^n$ and $\mathbb{R}^m$, we can identify linear transformations $F : \mathbb{R}^n \longrightarrow \mathbb{R}^m$ and $m \times n$ matrices $\mathrm{M}_{m,n}(\mathbb{R})$:

$$\{\text{linear transformations}\, \mathbb{R}^n \longrightarrow \mathbb{R}^m\} \quad \leftrightarrow \quad \mathrm{M}_{m,n}(\mathbb{R})$$

$$F : \mathbb{R}^n \longrightarrow \mathbb{R}^m \qquad \mapsto \quad (F(\mathbf{e}_1), \ldots, F(\mathbf{e}_n)).$$

The matrix $(F(\mathbf{e}_1), \ldots, F(\mathbf{e}_n))$, associated with the linear transformation $F : \mathbb{R}^n \longrightarrow \mathbb{R}^m$, in such one to one correspondence, has as columns the images of the vectors of the canonical basis of $\mathbb{R}^n$. Let us see an example.

Consider the linear transformation $F : \mathbb{R}^2 \longrightarrow \mathbb{R}^2$, $F(\mathbf{e}_1) = \mathbf{e}_1 - \mathbf{e}_2$, $F(\mathbf{e}_2) = 3\mathbf{e}_2$. This transformation is associated, with respect to the canonical basis in the domain and codomain, to the matrix:

$$A = \begin{pmatrix} 1 & 0 \\ -1 & 3 \end{pmatrix}.$$

We know that the choice of the canonical basis to represent the vectors in $\mathbb{R}^n$ is arbitrary, while being extremely convenient. For example, we have seen that a vector expressed with respect to two different-ordered bases has obviously different coordinates. Up to now, using a basis, other than the canonical one, to represent vectors seemed unnecessary. However, as we shall see in the next chapter, it provides us the key to understanding the concepts of eigenvalues and eigenvectors, which are of fundamental importance not only in linear algebra but also in its applications.

We now want to generalize the correspondence between matrices and linear transformations described above. Let us start with some observations (see 5).

Let V and W be two vector spaces of finite dimension and let $\mathcal{B} = \{\mathbf{v}_1, \ldots, \mathbf{v}_n\}$ and $\mathcal{B}' = \{\mathbf{w}_1, \ldots, \mathbf{w}_m\}$ be two ordered bases of V and W, respectively. If $F : V \to W$ is a linear transformation, by Theorem 5.1.7, we know that F is uniquely determined by $F(\mathbf{v}_1), \ldots, F(\mathbf{v}_n)$.

Let:
$$\begin{aligned}
F(\mathbf{v}_1) &= a_{11}\mathbf{w}_1 + a_{21}\mathbf{w}_2 + \cdots + a_{m1}\mathbf{w}_m, \\
F(\mathbf{v}_2) &= a_{12}\mathbf{w}_1 + a_{22}\mathbf{w}_2 + \cdots + a_{m2}\mathbf{w}_m, \\
&\vdots \\
F(\mathbf{v}_n) &= a_{1n}\mathbf{w}_1 + a_{2n}\mathbf{w}_2 + \cdots + a_{mn}\mathbf{w}_m,
\end{aligned} \tag{8.1}$$

and let $\mathbf{v} = x_1\mathbf{v}_1 + \cdots + x_n\mathbf{v}_n$ be a generic vector in V. Similarly to what was done in Chapter 5, we want determine $F(\mathbf{v})$.

We have:
$$\begin{aligned}
F(\mathbf{v}) &= F(x_1\mathbf{v}_1 + x_2\mathbf{v}_2 \cdots + x_n\mathbf{v}_n) = \\[6pt]
&= x_1 F(\mathbf{v}_1) + x_2 F(\mathbf{v}_2) + \cdots + x_n F(\mathbf{v}_n) = \\[6pt]
&= x_1(a_{11}\mathbf{w}_1 + a_{21}\mathbf{w}_2 + \cdots + a_{m1}\mathbf{w}_m) + \\[6pt]
&\quad + x_2(a_{12}\mathbf{w}_1 + a_{22}\mathbf{w}_2 + \cdots + a_{m2}\mathbf{w}_m) + \ldots \\[6pt]
&\quad + x_n(a_{1n}\mathbf{w}_1 + a_{2n}\mathbf{w}_2 + \cdots + a_{mn}\mathbf{w}_m) = \\[6pt]
&= (a_{11}x_1 + a_{12}x_2 + \ldots + a_{1n}x_n)\mathbf{w}_1 + \\[6pt]
&\quad + (a_{21}x_1 + a_{22}x_2 + \ldots + a_{2n}x_n)\mathbf{w}_2 + \ldots \\[6pt]
&\quad + (a_{m1}x_1 + a_{m2}x_2 + \ldots + a_{mn}x_n)\mathbf{w}_m.
\end{aligned}$$

So, if the coordinates of $\mathbf{v}$ with respect to the base $\mathcal{B}$ are $(\mathbf{v})_\mathcal{B} = (x_1, \ldots, x_n)$, we have that the coordinates of $F(\mathbf{v})$ with respect to the base $\mathcal{B}'$ are:
$$(F(\mathbf{v}))_{\mathcal{B}'} = \begin{pmatrix} a_{11}x_1 + a_{12}x_2 + \ldots + a_{1n}x_n \\ a_{21}x_1 + a_{22}x_2 + \ldots + a_{2n}x_n \\ \vdots \\ a_{m1}x_1 + a_{m2}x_2 + \ldots + a_{mn}x_n \end{pmatrix} = A \cdot (\mathbf{v})_\mathcal{B},$$

where A is the matrix defined in (8.1), which has as columns the coordinates of the vectors $F(\mathbf{v}_1), \ldots, F(\mathbf{v}_n)$ with respect to base $\mathcal{B}' = \{\mathbf{w}_1, \ldots, \mathbf{w}_m\}$, and $A \cdot (\mathbf{v})_\mathcal{B}$ denotes the product rows by columns of the matrix A and the vector $(\mathbf{v})_\mathcal{B}$, which represents the coordinates of $\mathbf{v}$ with respect to the basis $\mathcal{B}$.

We are therefore able to associate a $m \times n$ matrix A to F once we fix *arbitrary ordered bases* $\mathcal{B}$ and $\mathcal{B}'$ in the domain and codomain. If $V = \mathbb{R}^n$, $W = \mathbb{R}^m$ and we fix as $\mathcal{B}$ and $\mathcal{B}'$ the canonical bases of the domain and the codomain, we have that the matrix A is precisely the matrix associated to F defined in the previous chapters and recalled the beginning of this chapter. This matrix has as columns $F(\mathbf{e}_1), \ldots, F(\mathbf{e}_n)$, namely the coordinates, with respect to the canonical basis, of the the images of the vectors of the canonical basis (in $\mathbb{R}^n$, if we do not specify otherwise, we always consider the coordinates of vectors with respect to the canonical basis).

We now want to formalize what we observed in the following definition.

Let us make an important distinction. While until now we interchangeably used rows or columns to indicate the coordinates, in this chapter we require more accuracy, since the coordinates of a vector with respect to a given basis will form a column vector, which must then be multiplied (rows by columns) by a matrix. Henceforth, we shall denote the coordinates of a vector with respect to a given basis via a *column vector*.

Definition 8.1.1 Let $F : V \longrightarrow W$ be a linear transformation, where V and W are vector spaces, and let $\mathcal{B} = \{\mathbf{v}_1, \ldots, \mathbf{v}_n\}$, $\mathcal{B}' = \{\mathbf{w}_1, \ldots, \mathbf{w}_m\}$ be ordered bases of the domain and the codomain, respectively.

Let

$$(\mathbf{v})_{\mathcal{B}} = \begin{pmatrix} x_1 \\ \vdots \\ x_n \end{pmatrix}, \qquad (F(\mathbf{v}))_{\mathcal{B}'} = \begin{pmatrix} y_1 \\ \vdots \\ y_m \end{pmatrix}$$

be the coordinates of a vector $\mathbf{v}$ of V and the coordinates of its image under F, respectively.

The matrix $A_{\mathcal{B},\mathcal{B}'}$ associated to F with respect to the ordered bases $\mathcal{B}$ and $\mathcal{B}'$ is, by definition, the $m \times n$ matrix such that:

$$(F(\mathbf{v}))_{\mathcal{B}'} = \begin{pmatrix} y_1 \\ \vdots \\ y_m \end{pmatrix} = A_{\mathcal{B},\mathcal{B}'} \begin{pmatrix} x_1 \\ \vdots \\ x_n \end{pmatrix}.$$

From previous observations, we have that the i-th column of $A_{\mathcal{B},\mathcal{B}'}$ is given by the coordinates of $F(\mathbf{v}_i)$ with respect to the basis $\mathcal{B}'$.

In case of a linear transformation $F : V \longrightarrow V$, where we choose the same basis $\mathcal{B}$ in the domain and codomain, the matrix associated to F with respect to the basis $\mathcal{B}$ will be denoted just with $A_{\mathcal{B}}$.

Let us return to the example shown above. We choose $\mathcal{B} = \{\mathbf{v}_1 = \mathbf{e}_1 - \mathbf{e}_2, \mathbf{v}_2 = 3\mathbf{e}_2\}$ as a basis of $\mathbb{R}^2$, and we ask the following questions:

1. What are the coordinates of the vectors $\mathbf{v}_1$ and $\mathbf{v}_2$ with respect to the basis $\mathcal{B}$?

2. How can we write the matrix associated with F with respect to the canonical basis in the domain and the basis $\mathcal{B}$ in the codomain?

The answer to the first question is obvious. The vectors $\mathbf{v}_1$ and $\mathbf{v}_2$ have, respectively, coordinates $(1, 0)^T$ and $(0, 1)^T$ (T denotes the transpose, namely the fact that the coordinates represent a column vector). Indeed, $\mathbf{v}_1 = 1 \cdot \mathbf{v}_1 + 0 \cdot \mathbf{v}_2$, $\mathbf{v}_2 = 0 \cdot \mathbf{v}_1 + 1 \cdot \mathbf{v}_2$.

The answer to the second question is also quite simple. Indeed, we have already seen that changing the basis that we choose to represent vectors within a vector space, does not change vectors, but only how we write them, i.e. their coordinates. In the above example, the vectors $\mathbf{v}_1$ and $\mathbf{v}_2$ are the same, what changes passing from the canonical basis to the basis $\mathcal{B}$ are just their coordinates, which change from $(\mathbf{v}_1)_\mathcal{C} = (1, -1)^T$, $(\mathbf{v}_2)_\mathcal{C} = (0, 3)^T$ to $(\mathbf{v}_1)_\mathcal{B} = (1, 0)^T$, $(\mathbf{v}_2)_\mathcal{B} = (0, 1)^T$.

The same reasoning is valid for linear transformations, where the concept of coordinates is replaced by the concept of matrix associated with the transformation, with respect to two given ordered bases in the domain and codomain. Let us see a concrete example.

Consider the linear transformation $F : \mathbb{R}^2 \longrightarrow \mathbb{R}^2$, such that $F(\mathbf{e}_1) = \mathbf{v}_1$, $F(\mathbf{e}_2) = \mathbf{v}_2$. Let us now represent F using the coordinates with respect to the canonical basis $\mathcal{C} = \{\mathbf{e}_1, \mathbf{e}_2\}$ in the domain and the coordinates with respect to the basis $\mathcal{B} = \{\mathbf{v}_1, \mathbf{v}_2\}$ in the codomain. We have that $(F(\mathbf{e}_1))_\mathcal{B} = (1, 0)^T$, $(F(\mathbf{e}_2))_\mathcal{B} = (0, 1)^T$, where we use an index to remind us that we use coordinates with respect to a certain basis, which is not necessarily the canonical one. So we have that the matrix associated to F with respect to the bases $\mathcal{C}$ of the domain and $\mathcal{B}$ of the codomain is

$$A_{\mathcal{C}, \mathcal{B}} = \begin{pmatrix} 1 & 0 \\ 0 & 1 \end{pmatrix}.$$

In fact, taking the product rows by columns, we see that:

$$A_{\mathcal{C}, \mathcal{B}} \begin{pmatrix} 1 \\ 0 \end{pmatrix} = \begin{pmatrix} 1 \\ 0 \end{pmatrix} = (F(\mathbf{e}_1))_\mathcal{B}$$

$$A_{\mathcal{C}, \mathcal{B}} \begin{pmatrix} 0 \\ 1 \end{pmatrix} = \begin{pmatrix} 0 \\ 1 \end{pmatrix} = (F(\mathbf{e}_2))_\mathcal{B}.$$

Therefore, the matrix associated to F with respect to the bases $\mathcal{C}$ in the domain and $\mathcal{B}$ in the codomain is just $A_{\mathcal{C}, \mathcal{B}}$, that is, the identity matrix.

We can easily generalize what we have just said.

Proposition 8.1.2 *Let $F : V \longrightarrow V$ be a linear transformation such that $F(\mathbf{v}_1) = \mathbf{w}_1, \ldots, F(\mathbf{v}_n) = \mathbf{w}_n$, where $\mathcal{B} = \{\mathbf{v}_1, \ldots, \mathbf{v}_n\}$, $\mathcal{B}' = \{\mathbf{w}_1, \ldots, \mathbf{w}_n\}$ are two ordered bases of V. Then the matrix associated to F, with respect to the base $\mathcal{B}$ in the domain and the base $\mathcal{B}'$ in the codomain is the identity matrix.*

The proof is an easy exercise and follows the previous reasoning.

8.2   THE IDENTITY MAP

From now on, our focus will be exclusively concentrated on examining the case when $V = \mathbb{R}^n$ for concreteness, even if everything we say can easily be generalized to the case of a generic vector space, provided that the dimension is finite.[1] Our choice to

---

[1] A vector space V of finite dimension can always be identified with $\mathbb{R}^n$ *provided that we fix a basis.* We want to point out, however, that *this is not the case* in this chapter: everything we will say from now on can be generalized to a vector space V of dimension n, without identifying it with $\mathbb{R}^n$.

avoid to treat the change of basis in this generality is dictated only by the hope to increase the clarity, but does not involve conceptual issues.

We now ask a question in some way related to the previous ones.

What is the matrix associated to the identity id $: \mathbb{R}^n \longrightarrow \mathbb{R}^n$ *with respect to the bases* $\mathcal{B} = (\mathbf{v}_1 \ldots \mathbf{v}_n)$, $\mathcal{B}' = (\mathbf{w}_1 \ldots \mathbf{w}_n)$, *respectively, in the domain and in the codomain?*

Certainly we know that the identity matrix is associated to the identity map if we fix the canonical bases in the domain and codomain, however we already know from the previous example that changing the basis can radically change the appearance of the matrix associated to the same linear transformation.

We now look at a simple example to help our understanding.

Example 8.2.1 Consider the identity map id $: \mathbb{R}^2 \longrightarrow \mathbb{R}^2$ and fix the basis $\mathcal{B} = \{\mathbf{v}_1, \mathbf{v}_2\}$ in the domain and the canonical basis $\mathcal{C}$ in the codomain, with $\mathbf{v}_1 = 2\mathbf{e}_1$, $\mathbf{v}_2 = \mathbf{e}_1 + \mathbf{e}_2$.

The identity map always behaves in the same way even if we change the way we represent it: id still sends a vector to itself. Let us see what happens:

$$\mathrm{id}(\mathbf{v}_1) = \mathbf{v}_1 = 2\mathbf{e}_1, \qquad \mathrm{id}(\mathbf{v}_2) = \mathbf{v}_2 = \mathbf{e}_1 + \mathbf{e}_2.$$

We write the coordinates of the vectors with respect to the canonical basis:

$$(\mathrm{id}(\mathbf{v}_1))_\mathcal{C} = \begin{pmatrix} 2 \\ 0 \end{pmatrix}, \qquad (\mathrm{id}(\mathbf{v}_2))_\mathcal{C} = \begin{pmatrix} 1 \\ 1 \end{pmatrix}.$$

Therefore, the matrix associated to the identity with respect to the basis $\mathcal{B}$ in the domain and the canonical basis $\mathcal{C}$ in the codomain is:

$$I_{\mathcal{B},\mathcal{C}} = \begin{pmatrix} 2 & 1 \\ 0 & 1 \end{pmatrix}.$$

Now we wonder what happens if we want to represent the identity using the canonical basis $\mathcal{C}$ in the domain and the basis $\mathcal{B}$ in the codomain. The identity always associates to each vector itself; the problem is to understand what are the right coordinates.

$$\mathrm{id}(\mathbf{e}_1) = \mathbf{e}_1 = (1/2)\mathbf{v}_1, \qquad \mathrm{id}(\mathbf{e}_2) = \mathbf{e}_2 = -(1/2)\mathbf{v}_1 + \mathbf{v}_2.$$

So:

$$(\mathrm{id}(\mathbf{e}_1))_\mathcal{B} = \begin{pmatrix} 1/2 \\ 0 \end{pmatrix}, \qquad (\mathrm{id}(\mathbf{e}_2))_\mathcal{B} = \begin{pmatrix} -1/2 \\ 1 \end{pmatrix}.$$

Therefore, the matrix associated to the identity with respect to the canonical basis $\mathcal{C}$ of the domain and the basis $\mathcal{B}$ of the codomain is:

$$I_{\mathcal{C},\mathcal{B}} = \begin{pmatrix} 1/2 & -1/2 \\ 0 & 1 \end{pmatrix}.$$

In this very simple example, it was possible to calculate easily the coordinates of $\mathbf{e}_1$

and e_2 with respect to the basis $\mathcal{B} = \{v_1, v_2\}$; in general this is not always so easy. However, in this example, note that $I_{\mathcal{C},\mathcal{B}} = I_{\mathcal{B},\mathcal{C}}^{-1}$. Therefore, the coordinates of the vectors e_1 and e_2 with respect to the basis $\mathcal{B}$ can be read from the columns of the matrix $I_{\mathcal{B},\mathcal{C}}^{-1}$. Remember that the matrix $I_{\mathcal{B},\mathcal{C}}$ can be easily calculated and it has as columns the coordinates of the vectors v_1, v_2 with respect to the canonical basis.

This is true in general: we get the coordinates of a vector v with respect to a basis $\mathcal{B}$ through the multiplication $(v)_\mathcal{B} = I_{\mathcal{B},\mathcal{C}}^{-1}(v)_\mathcal{C}$, but we still need some work to prove it. Before this, we look at another special case.

Proposition 8.2.2 *Let $\mathcal{B}$ be a basis of $\mathbb{R}^n$; then the matrix associated to the identity id : $\mathbb{R}^n \to \mathbb{R}^n$ if we fix the basis $\mathcal{B}$ in the domain and the codomain is the identity matrix.*

Proof. The proof is immediate; it is a consequence of Proposition 8.1.2. We can see it, however, also directly. If $\mathcal{B} = \{v_1, \ldots, v_n\}$, then the i-th column of such matrix is given by the coordinates of $\mathrm{id}(v_i) = v_i$ with respect to the basis $\mathcal{B}$, i.e. it is the vector that has zero everywhere and 1 in the i-th position. ■

Observation 8.2.3 In general, if we have the composition of linear transformations $V \xrightarrow{F} W \xrightarrow{G} Z$, and we denote by A_F the matrix associated to F and with A_G the matrix associated to G (with respect to any three fixed bases for V, W and Z), then the composition $G \circ F$ is associated with the matrix $A_G \cdot A_F$, where the product is intended as the usual product rows by columns. The proof of this fact is quite similar to the proof Corollary 5.3.4.

We can now formalize the equality $I_{\mathcal{C},\mathcal{B}} = I_{\mathcal{B},\mathcal{C}}^{-1}$ that we have seen in the particular case of Example 8.2.1 along with other essential facts regarding the calculation of the coordinates of a vector with respect to a given basis.

Theorem 8.2.4 *Let id : $\mathbb{R}^n \longrightarrow \mathbb{R}^n$ be the identity transformation, which associates to each vector, the vector itself. Let $\mathcal{B} = \{v_1, \ldots, v_n\}$ be an odered basis for $\mathbb{R}^n$. Then we have:*

- *the matrix associated to id with respect to the basis $\mathcal{B}$ in the domain and the canonical basis $\mathcal{C}$ in codomain is:*

$$I_{\mathcal{B},\mathcal{C}} = ((v_1)_\mathcal{C}, \ldots, (v_n)_\mathcal{C}),$$

where $(v_i)_\mathcal{C}$ are the coordinates of the vector v_i with respect to the canonical basis;

- *the matrix associated to id with respect to the canonical basis $\mathcal{C}$ in the domain and $\mathcal{B}$ in the codomain is $I_{\mathcal{B},\mathcal{C}}^{-1}$.*

Proof. The first point is clear; substantially it is what we saw in Example 8.2.1. Indeed $(\mathrm{id}(v_1))_\mathcal{C} = (v_1)_\mathcal{C}$ are precisely the coordinates of the vector v_1 in the canonical basis. The same is true for $(\mathrm{id}(v_2))_\mathcal{C}, \ldots, (\mathrm{id}(v_n))_\mathcal{C}$.

To show the second point, we show that $I_{C,B}$ is the inverse of $I_{B,C}$, that is, $I_{B,C}I_{C,B} = I_{C,B}I_{B,C} = I$, where I is the identity matrix.

Consider the composition of the identity with itself, with respect to different bases, as indicated in the following diagram:

$$\mathbb{R}^n \xrightarrow{\text{id}} \mathbb{R}^n \xrightarrow{\text{id}} \mathbb{R}^n$$
$$\mathcal{B} \qquad\quad \mathcal{C} \qquad\quad \mathcal{B}\ .$$

The composite function $\text{id} \circ \text{id} = \text{id}$ is still the identity, and if we consider the matrices associated with it, by Observation 8.2.3 we get: $I_{C,B}I_{B,C} = I_B$. Now, by Proposition 8.2.2 we have that $I_B = I$, so that $I_{C,B}I_{B,C} = I$. Similarly, considering the diagram:

$$\mathbb{R}^n \xrightarrow{\text{id}} \mathbb{R}^n \xrightarrow{\text{id}} \mathbb{R}^n$$
$$\mathcal{C} \qquad\quad \mathcal{B} \qquad\quad \mathcal{C}\ ,$$

we get that $I_{B,C} = I_{C,B}^{-1}$. This concludes the proof. ■

This theorem also answers another question, which was asked previously, that is, how we can write the coordinates of a vector $\mathbf{v}$ with respect to a given basis $\mathcal{B}$.

So far we have responded with a very explicit calculation in each case, but now we can state a corollary that contains the answer in general.

Corollary 8.2.5 *Let $\mathcal{C}$ be a basis of $\mathbb{R}^n$, and let $\mathbf{v}$ be a vector of $\mathbb{R}^n$. Then the coordinates of $\mathbf{v}$ with respect to the basis $\mathcal{B}$ are given by:*

$$(\mathbf{v})_{\mathcal{B}} = I_{B,C}^{-1}(\mathbf{v})_{\mathcal{C}}.$$

Proof. We have that $\mathbf{v} = \text{id}(\mathbf{v})$, so we choose the canonical basis $\mathcal{C}$ in the domain and the basis $\mathcal{B}$ in the codomain. Now by Theorem 8.2.4, we know that, with respect to these bases, the identity map is represented by the matrix $I_{C,B} = I_{B,C}^{-1}$. ■

Let us see an example to illustrate these concepts.

Example 8.2.6 Consider the basis $\mathcal{B} = \{\mathbf{v}_1, \mathbf{v}_2\}$ of $\mathbb{R}^2$, where $\mathbf{v}_1 = -\mathbf{e}_1 + 2\mathbf{e}_2$, $\mathbf{v}_2 = -\mathbf{e}_2$. We want find the coordinates of the vector $\mathbf{v} = -\mathbf{e}_1 + 3\mathbf{e}_2$ with respect to the basis $\mathcal{B}$. We know that $(\mathbf{v})_{\mathcal{C}} = (-1, 3)$ and $I_{B,C} = \begin{pmatrix} -1 & 0 \\ 2 & -1 \end{pmatrix}$. With an easy calculation, we obtain that $I_{B,C}^{-1} = \begin{pmatrix} -1 & 0 \\ -2 & -1 \end{pmatrix}$, then the coordinates are:

$$(\mathbf{v})_{\mathcal{B}} = I_{B,C}^{-1} \cdot (\mathbf{v})_{\mathcal{C}} = \begin{pmatrix} -1 & 0 \\ -2 & -1 \end{pmatrix}\begin{pmatrix} -1 \\ 3 \end{pmatrix} = \begin{pmatrix} 1 \\ -1 \end{pmatrix}.$$

Indeed:

$$\mathbf{v} = 1 \cdot \mathbf{v}_1 - 1 \cdot \mathbf{v}_2 = -\mathbf{e}_1 + 2\mathbf{e}_2 + \mathbf{e}_2 = -\mathbf{e}_1 + 3\mathbf{e}_2.$$

8.3 CHANGE OF BASIS FOR LINEAR TRANSFORMATIONS

In this section, we want to deal with a generalization of the problem studied in the previous section.

Let $F : \mathbb{R}^n \longrightarrow \mathbb{R}^m$ be a linear transformation, which is associated to the matrix $A_{\mathcal{C},\mathcal{C}'}$ with respect to the canonical bases $\mathcal{C}$ and $\mathcal{C}'$ in the domain and codomain, respectively. We want to find the matrix associated to F, if we fix the bases $\mathcal{B}$ for $\mathbb{R}^n$ and $\mathcal{B}'$ for $\mathbb{R}^m$.

Both the transformation F and the vector space vectors behave independently from the basis that we arbitrarily choose to represent them. The example of the identity is particularly useful, because the identity is the transformation that associates to each vector the vector itself, but if we change the basis in which we represent it, the associated matrix undergoes drastic changes.

Consider the following linear transformations, where we place next to each vector space the basis we choose to represent the vectors; in the upper row, we choose the canonical bases, at the bottom, we choose arbitrarily the two bases $\mathcal{B}$ and $\mathcal{B}'$.

$$
\begin{array}{ccccc}
\mathcal{C} & \mathbb{R}^n & \xrightarrow{\ F\ } & \mathbb{R}^m & \mathcal{C}' \\[2mm]
& \text{id}\uparrow & & \downarrow\text{id} & \\[2mm]
\mathcal{B} & \mathbb{R}^n & \xrightarrow{\ F\ } & \mathbb{R}^m & \mathcal{B}'
\end{array}
$$

This is what is called a *commutative diagram*, because the path we choose in the diagram does not influence the result:

$$F = \text{id} \circ F \circ \text{id}.$$

The equality we wrote appears as a tautology and not particularly interesting, however, when we associate with each transformation its matrix with respect to the fixed bases in the domain and codomain, this same equality will provide a complete answer to the question we set at the beginning of this section, and that is perhaps the most technical point of our linear algebra notes.

Therefore, we associate to each linear transformation the corresponding matrix on the same diagram using coordinates, that is, using matrices to represent linear transformations. As we know from the previous section, we get:

$$
\begin{array}{ccccc}
 & \mathbb{R}^n & \xrightarrow{\ F\ } & \mathbb{R}^m & \\
I_{\mathcal{B},\mathcal{C}}(\mathbf{v})_{\mathcal{B}} & (\mathbf{v})_{\mathcal{C}} & \mapsto & A_{\mathcal{C},\mathcal{C}'}(\mathbf{v})_{\mathcal{C}} & (\mathbf{v})_{\mathcal{C}'} \\[3mm]
\uparrow & \text{id}\uparrow & & \downarrow\text{id} & \downarrow \\[3mm]
(\mathbf{v})_{\mathcal{B}} & \mathbb{R}^n & \xrightarrow{\ F\ } & \mathbb{R}^m & I_{\mathcal{C}',\mathcal{B}'}(\mathbf{v})_{\mathcal{C}'} \\[3mm]
& (\mathbf{v})_{\mathcal{B}} & \mapsto & A_{\mathcal{B},\mathcal{B}'}(\mathbf{v})_{\mathcal{B}} &
\end{array}
$$

We want to use this diagram to determine $A_{\mathcal{B},\mathcal{B}'}$, that is, the matrix associated to F with respect to the bases $\mathcal{B}$ in the domain and $\mathcal{B}'$ in the codomain.

Thanks to Observation 8.2.3, we have that the equality:

$$F = \mathrm{id} \circ F \circ \mathrm{id}$$

corresponds to:

$$A_{\mathcal{B},\mathcal{B}'} = I_{\mathcal{C}',\mathcal{B}'} A_{\mathcal{C},\mathcal{C}'} I_{\mathcal{B},\mathcal{C}} = I_{\mathcal{B}',\mathcal{C}'}^{-1} A_{\mathcal{C},\mathcal{C}'} I_{\mathcal{B},\mathcal{C}},$$

where, for the last equality, we have used Theorem 8.2.4.

Thus, we have proved the following theorem.

Theorem 8.3.1 *Let $F : \mathbb{R}^n \longrightarrow \mathbb{R}^m$ be a linear transformation and let $A_{\mathcal{C},\mathcal{C}'}$ be the matrix associated with F with respect to the canonical bases $\mathcal{C}$ and $\mathcal{C}'$ in the domain and codomain, respectively. Then the matrix associated to F, with respect to the bases $\mathcal{B}$ in the domain and $\mathcal{B}'$ in the codomain, is given by:*

$$A_{\mathcal{B},\mathcal{B}'} = I_{\mathcal{B}',\mathcal{C}'}^{-1} A_{\mathcal{C},\mathcal{C}'} I_{\mathcal{B},\mathcal{C}},$$

where:

- *$A_{\mathcal{C},\mathcal{C}'}$ has, as columns, the coordinates of the vectors which are the images of the vectors of the canonical basis, i.e. $F(\mathbf{e}_1), \ldots, F(\mathbf{e}_n)$, expressed in terms of the canonical basis of $\mathbb{R}^m$;*

- *$I_{\mathcal{B}',\mathcal{C}'}$ has, as columns, the coordinates of the vectors of the basis $\mathcal{B}'$ expressed with respect to the canonical basis of $\mathbb{R}^m$;*

- *$I_{\mathcal{B},\mathcal{C}}$ has, as columns, the coordinates of the vectors of the basis $\mathcal{B}$, expressed in terms of the canonical basis of $\mathbb{R}^n$.*

Let us see an example to clarify how we can explicitly calculate $A_{\mathcal{B},\mathcal{B}'}$.

Example 8.3.2 Consider the linear transformation $F : \mathbb{R}^2 \longrightarrow \mathbb{R}^3$ associated to the matrix $A_{\mathcal{C},\mathcal{C}'}$ with respect to the canonical bases, where

$$A_{\mathcal{C},\mathcal{C}'} = \begin{pmatrix} 1 & 2 \\ 0 & 1 \\ -1 & 0 \end{pmatrix}.$$

We want to find the matrix $A_{\mathcal{B},\mathcal{B}'}$ associated with F with respect to the bases: $\mathcal{B} = \{\mathbf{e}_1 + \mathbf{e}_2, -\mathbf{e}_1 - 2\mathbf{e}_2\}$ in the domain and $\mathcal{B}' = \{2\mathbf{e}_3, \mathbf{e}_1 + \mathbf{e}_3, \mathbf{e}_1 + \mathbf{e}_2\}$ in the codomain.

We find the matrices $I_{\mathcal{B},\mathcal{C}}$ and $I_{\mathcal{B}',\mathcal{C}'}$:

$$I_{\mathcal{B},\mathcal{C}} = \begin{pmatrix} 1 & -1 \\ 1 & -2 \end{pmatrix}, \qquad I_{\mathcal{B}',\mathcal{C}'} = \begin{pmatrix} 0 & 1 & 1 \\ 0 & 0 & 1 \\ 2 & 1 & 0 \end{pmatrix}.$$

We calculate $I_{\mathcal{B}',\mathcal{C}'}^{-1}$ with any method:

$$I_{\mathcal{B}',\mathcal{C}'}^{-1} = \begin{pmatrix} -1/2 & 1/2 & 1/2 \\ 1 & -1 & 0 \\ 0 & 1 & 0 \end{pmatrix}.$$

The matrix $A_{\mathcal{B},\mathcal{B}'}$ is given by:

$$A_{\mathcal{B},\mathcal{B}'} = I_{\mathcal{B}',\mathcal{C}'}^{-1} A_{\mathcal{C},\mathcal{C}'} I_{\mathcal{B},\mathcal{C}} = \begin{pmatrix} -1/2 & 1/2 & 1/2 \\ 1 & -1 & 0 \\ 0 & 1 & 0 \end{pmatrix} \begin{pmatrix} 1 & 2 \\ 0 & 1 \\ -1 & 0 \end{pmatrix} \begin{pmatrix} 1 & -1 \\ 1 & -2 \end{pmatrix} =$$

$$= \begin{pmatrix} -3/2 & 2 \\ 2 & -3 \\ 1 & -2 \end{pmatrix}.$$

We conclude with a general observation.

Observation 8.3.3 Theorem 5.2.2 establishes a one to one correspondence between linear transformations from $\mathbb{R}^n$ to $\mathbb{R}^m$ and matrices in $M_{m,n}(\mathbb{R})$, once we have fixed the canonical bases in $\mathbb{R}^n$ and $\mathbb{R}^m$. However, the attentive reader will easily get convinced that the choice of the canonical bases is totally arbitrary and therefore such one to one correspondence holds also when we choose arbitrary bases in $\mathbb{R}^n$ and $\mathbb{R}^m$, respectively. Obviously the matrix associated to the same linear transformation will be in general different, depending on the bases that we decide to fix. We can calculate this matrix in a direct way, simply by knowing the coordinates of the vectors of the two bases with respect to the canonical bases, thanks to Theorem 8.3.1.

Given two vector spaces V and W, with $\dim(V) = n$ and $\dim(W) = m$, we can also determine in a similar way a bijective correspondence between linear transformation from V to W and matrices in $M_{m,n}(\mathbb{R})$. We invite the reader to go through the calculations we made before Theorem 5.2.2, replacing $\mathbb{R}^n$ and $\mathbb{R}^m$ with the vector spaces V and W, respectively, and replacing the canonical bases with two arbitrary bases V and W, respectively.

8.4   EXERCISES WITH SOLUTIONS

8.4.1 Let $F : M_2(\mathbb{R}) \longrightarrow \mathbb{R}$ be the linear transformation that associates to each matrix its trace, that is:

$$F \begin{pmatrix} a_{11} & a_{12} \\ a_{21} & a_{22} \end{pmatrix} = a_{11} + a_{22}.$$

Determine the matrix associated with F with respect to the canonical bases of $M_2(\mathbb{R})$ and $\mathbb{R}$.

Solution. We observe that $M_2(\mathbb{R})$ has dimension 4 and $\mathbb{R}$ has dimension 1, so A is a 4×1 matrix. The first "column" of A is $F(a_{11}) = F \begin{pmatrix} 1 & 0 \\ 0 & 0 \end{pmatrix} = 1 + 0 = 1$, the second column is $F(\mathbf{e}_{12}) = F \begin{pmatrix} 0 & 1 \\ 0 & 0 \end{pmatrix} = 0 + 0 = 0$, the third is $F(\mathbf{e}_{21}) = F \begin{pmatrix} 0 & 0 \\ 1 & 0 \end{pmatrix} = 0$ and

the fourth is $F(\mathbf{e}_{22}) = F \begin{pmatrix} 0 & 0 \\ 0 & 1 \end{pmatrix} = 1$, thus $A = (1, 0, 0, 1)$.

8.4.2 Let $F : \mathbb{R}^3 \longrightarrow \mathbb{R}^2$ be the linear transformation defined by: $F(\mathbf{e}_1) = 2\mathbf{e}_1 - \mathbf{e}_2$, $F(\mathbf{e}_2) = \mathbf{e}_1$, $F(\mathbf{e}_3) = \mathbf{e}_1 + \mathbf{e}_2$. Let $\mathcal{B} = \{2\mathbf{e}_1 - \mathbf{e}_2, \mathbf{e}_1 - \mathbf{e}_2\}$ be a basis of $\mathbb{R}^2$. Determine the matrix $A_{\mathcal{C},\mathcal{B}}$ associated with F with respect to the canonical basis in the domain and the basis $\mathcal{B}$ in the codomain.

Solution. The matrix associated with F with respect to the canonical bases of the domain and codomain is:

$$A_{\mathcal{C},\mathcal{C}'} = \begin{pmatrix} 2 & 1 & 1 \\ -1 & 0 & 1 \end{pmatrix}.$$

We want to change basis in the codomain, therefore we consider the following composition of functions:

$$\mathbb{R}^3 \xrightarrow{F} \mathbb{R}^2 \xrightarrow{id} \mathbb{R}^2$$
$$\quad \mathcal{C} \qquad \mathcal{C}' \qquad \mathcal{B} \quad .$$

The composition is $id \circ F = F$, and the matrix associated to it is $A_{\mathcal{C},\mathcal{B}} = I_{\mathcal{C}',\mathcal{B}} A_{\mathcal{C},\mathcal{C}'}$. In addition, $I_{\mathcal{C}',\mathcal{B}} = I_{\mathcal{B},\mathcal{C}'}^{-1}$, where $I_{\mathcal{B},\mathcal{C}}$ is the matrix that has as columns the coordinates of the vectors of $\mathcal{B}$ with respect to the canonical basis $\mathcal{C}'$, thus: $I_{\mathcal{B},\mathcal{C}'} = \begin{pmatrix} 2 & 1 \\ -1 & -1 \end{pmatrix}$.

We have that $I_{\mathcal{B},\mathcal{C}'}^{-1} = \begin{pmatrix} 1 & 1 \\ -1 & -2 \end{pmatrix}$, therefore

$$A_{\mathcal{C},\mathcal{B}} = I_{\mathcal{C}',\mathcal{B}} A_{\mathcal{C},\mathcal{C}'} = \begin{pmatrix} 1 & 1 \\ -1 & -2 \end{pmatrix} \begin{pmatrix} 2 & 1 & 1 \\ -1 & 0 & 1 \end{pmatrix} = \begin{pmatrix} 1 & 1 & 2 \\ 0 & -1 & -3 \end{pmatrix}.$$

8.4.3 Let $\mathcal{B} = \{(2, -1), (1, 1)\}$ be a basis of $\mathbb{R}^2$ and let $G : \mathbb{R}^2 \longrightarrow \mathbb{R}^3$ be the linear transformation defined by: $G(2, -1) = (1, -1, 0)$, $G(1, 1) = (2, 1, -2)$. Determine the matrix $A_{\mathcal{C},\mathcal{C}'}$ associated with G with respect to the canonical bases of the domain and of the codomain.

Solution. The matrix associated with G with respect to the basis $\mathcal{B}$ of the domain and the canonical basis $\mathcal{C}'$ of the codomain is:

$$A_{\mathcal{B},\mathcal{C}'} = \begin{pmatrix} 1 & 2 \\ -1 & 1 \\ 0 & -2 \end{pmatrix}.$$

Since we have changed the basis in the domain, we consider the following composition of functions:

$$\mathbb{R}^2 \xrightarrow{id} \mathbb{R}^2 \xrightarrow{F} \mathbb{R}^3$$
$$\quad \mathcal{B} \qquad \mathcal{C} \qquad \mathcal{C}' \quad .$$

The composition is $G \circ id = G$, and the matrix associated with it is $A_{\mathcal{B},\mathcal{C}'} = A_{\mathcal{C},\mathcal{C}'} I_{\mathcal{B},\mathcal{C}}$, therefore, multiplying to the right-both members by $I_{\mathcal{B},\mathcal{C}}^{-1}$ we get: $A_{\mathcal{C},\mathcal{C}'} = A_{\mathcal{B},\mathcal{C}'} I_{\mathcal{B},\mathcal{C}}^{-1}$.

We have $I_{\mathcal{B},\mathcal{C}} = \begin{pmatrix} 2 & 1 \\ -1 & 1 \end{pmatrix}$ and $I_{\mathcal{B},\mathcal{C}'}^{-1} = \begin{pmatrix} \frac{1}{3} & -\frac{1}{3} \\ \frac{1}{3} & \frac{2}{3} \\ \frac{1}{3} & \frac{2}{3} \end{pmatrix}$, so

$$A_{\mathcal{C},\mathcal{C}'} = \begin{pmatrix} 1 & 2 \\ -1 & 1 \\ 0 & -2 \end{pmatrix} \begin{pmatrix} \frac{1}{3} & -\frac{1}{3} \\ \frac{1}{3} & \frac{2}{3} \\ \frac{1}{3} & \frac{2}{3} \end{pmatrix} = \begin{pmatrix} 1 & 1 \\ 0 & 1 \\ -\frac{2}{3} & -\frac{4}{3} \end{pmatrix}.$$

8.5 SUGGESTED EXERCISES

8.5.1 Write the coordinates of the vector $\mathbf{v} = (1, 2, 3)$ with respect to the basis $\mathcal{B} = \{\mathbf{e}_1 - \mathbf{e}_2, \mathbf{e}_1 + \mathbf{e}_2 - \mathbf{e}_3, -\mathbf{e}_1 + \mathbf{e}_3\}$ of $\mathbb{R}^3$.

8.5.2 Consider the linear transformation $F : \mathbb{R}^2 \longrightarrow \mathbb{R}^3$ defined by: $F(\mathbf{e}_1) = -\mathbf{e}_2 + \mathbf{e}_3$, $F(\mathbf{e}_2) = 3\mathbf{e}_1 - 2\mathbf{e}_3$.
a) Determine the matrix associated with F with respect to the canonical basis and the matrix $A_{\mathcal{B},\mathcal{B}'}$ associated to F with respect to the bases $\mathcal{B} = \{(1, -2), (-2, 1)\}$ of $\mathbb{R}^2$ and $\mathcal{B}' = \{(2, 0, 5), (0, -1, 0), (1, 1, 3)\}$ of $\mathbb{R}^3$.
b) Determine the coordinates of the vector $\mathbf{v} = (1, -1, 1)$ with respect to the basis $\mathcal{B}'$ of $\mathbb{R}^3$.

8.5.3 Consider the linear transformation $F : \mathbb{R}^4 \longrightarrow \mathbb{R}^2$ defined by: $F(\mathbf{e}_1) = 2\mathbf{e}_2$, $F(\mathbf{e}_2) = \mathbf{e}_1 - 3\mathbf{e}_2$, $F(\mathbf{e}_3) = -2\mathbf{e}_1$, $F(\mathbf{e}_4) = \mathbf{e}_1 + \mathbf{e}_2$.
a) Determine the matrix associated with F with respect to the canonical basis, and the matrix $A_{\mathcal{B},\mathcal{C}}$ associated to F with respect to the bases $\mathcal{B} = \{\mathbf{e}_2, \mathbf{e}_4, \mathbf{e}_1, \mathbf{e}_3\}$ of $\mathbb{R}^4$ and the canonical basis $\mathcal{C}$ of $\mathbb{R}^2$.
b) Determine the coordinates of the vector $\mathbf{v} = (6, -1, 3, -2)$ with respect to the basis $\mathcal{B}$ of $\mathbb{R}^4$.

8.5.4 Consider the linear transformation $F : \mathbb{R}^3 \longrightarrow \mathbb{R}^3$ defined by:

$$F(x, y, z) = (8x - 9y, 6x - 7y, 2x - 3y - z).$$

a) Determine the matrix associated to F with respect to the canonical basis and the matrix $A_{\mathcal{B}}$ associated to F with respect to the basis $\mathcal{B} = \{-\mathbf{e}_1 + \mathbf{e}_2, -2\mathbf{e}_1 + 3\mathbf{e}_2, -\mathbf{e}_3\}$.
b) Say if F is an isomorphism and give a motivation for your answer.
c) If the answer in point (b) is affirmative, compute the inverse of F.

8.5.5 a) Given the linear transformation $T : \mathbb{R}^3 \to \mathbb{R}^2$ defined by:
$T(x, y, z) = (3kx + y - 2kz, 3x + ky - 2z)$, determine for which values of k we have that T is surjective, motivating the procedure followed. Set $k = 0$ and determine $\ker T$.
b) Let $\mathcal{B} = \{4\mathbf{e}_1 - 2\mathbf{e}_2, -\mathbf{e}_1 + \mathbf{e}_2\}$ be another $\mathbb{R}^2$ basis. Set $k = 0$. Determine the matrix $A_{\mathcal{C},\mathcal{B}}$ associated with T with respect to the canonical basis $\mathcal{C}$ of $\mathbb{R}^3$ in the domain and at the basis $\mathcal{B}$ in the codomain.

8.5.6 a) Let $F : \mathbb{R}^3 \to \mathbb{R}^3$ be the linear transformation defined by:
$F(x, y, z) = (x - 4y - 2z, -x + ky + kz, kx - 4ky + z)$.
Determine for which values of k we have that F is surjective.

b) Set $k = 0$ and determine, if possible, a linear transformation $G : \mathbb{R}^3 \to \mathbb{R}^3$ such that $G \circ F$ is the identity.

c) Let $\mathcal{B} = \{e_1 + e_2, -e_1 + e_3, 2e_2\}$ be another basis of $\mathbb{R}^3$. Set $k = 0$. Determine the matrix $A_{\mathcal{C},\mathcal{B}}$ associated with F with respect to the basis $\mathcal{B}$ in the domain and the canonical basis $\mathcal{C}$ of $\mathbb{R}^3$ in the codomain.

8.5.7 Consider the linear transformation $D : \mathbb{R}_3[x] \longrightarrow \mathbb{R}_3[x]$ that associates its derivative to each polynomial. Determine the matrix associated with D with respect to the basis $\{x^3, x^2, x, 1\}$ of $\mathbb{R}_3[x]$.

8.5.8 Determine a linear transformation $F : \mathbb{R}^3 \to \mathbb{R}^3$ such that $\ker F = \langle e_1 + e_2, e_2 - e_3 \rangle$ and $\mathrm{Im}F = \langle 2e_1 + e_3 \rangle$. (Suggestion: it is advisable to choose appropriate bases in the domain and codomain).

Determine the matrix associated with F with respect to the canonical basis.

CHAPTER 9

Eigenvalues and Eigenvectors

In this chapter, we want to address one of the most important questions of linear algebra, namely the problem of diagonalizing a linear transformation together with the concepts of eigenvalue and eigenvector.

9.1 DIAGONALIZABILITY

The idea behind the problem of diagonalizability is very simple: given a linear transformation $F : \mathbb{R}^n \longrightarrow \mathbb{R}^n$, we ask if there is a basis, both for the domain and the codomain, such that the matrix associated to F, with respect to this basis, has the simplest possible form, namely the diagonal one. Let us see an example.

***Example* 9.1.1** Let $\phi : \mathbb{R}^2 \longrightarrow \mathbb{R}^2$ be the linear transformation defined by: $\phi(\mathbf{e}_1) = \mathbf{e}_2$, $\phi(\mathbf{e}_2) = \mathbf{e}_1$. With respect to the canonical basis in domain and codomain ϕ is represented by the matrix:

$$A = \begin{pmatrix} 0 & 1 \\ 1 & 0 \end{pmatrix}.$$

We observe that:

$$\phi(\mathbf{e}_1 + \mathbf{e}_2) = \mathbf{e}_1 + \mathbf{e}_2, \qquad \phi(\mathbf{e}_1 - \mathbf{e}_2) = -(\mathbf{e}_1 - \mathbf{e}_2).$$

So, if we choose the basis $\mathcal{B} = \{\mathbf{v}_1, \mathbf{v}_2\}$, with $\mathbf{v}_1 = \mathbf{e}_1 + \mathbf{e}_2$ and $\mathbf{v}_2 = \mathbf{e}_1 - \mathbf{e}_2$, we have that $\phi(\mathbf{v}_1) = \mathbf{v}_1$ and $\phi(\mathbf{v}_2) = -\mathbf{v}_2$. Therefore, the matrix associated with ϕ, with respect to the basis $\mathcal{B}$ (in domain and codomain) is a diagonal matrix:

$$A_\mathcal{B} = \begin{pmatrix} 1 & 0 \\ 0 & -1 \end{pmatrix}.$$

We can see this right away without calculations, however, to convince ourselves, we can just use Definition 8.1.1 or the formula for changing the basis in Chapter 8.

In this case, it is very simple to see what happens geometrically. The transformation ϕ is the reflection of the plane with respect to the line $x = y$. Indeed, $\phi(\mathbf{e}_1) = \mathbf{e}_2$,

$\phi(\mathbf{e}_2) = \mathbf{e}_1$. We see, geometrically, that the vector $\mathbf{v}_1$, lying on the straight line $y = x$, is fixed by the transformation, while the vector $\mathbf{v}_2$, which is perpendicular to the line $y = x$, is sent to $-\mathbf{v}_2$. Based on these observations, we can conclude without any calculation that, with respect to the basis $\mathcal{B} = \{\mathbf{v}_1, \mathbf{v}_2\}$, the matrix associated with ϕ is in the specified diagonal form.

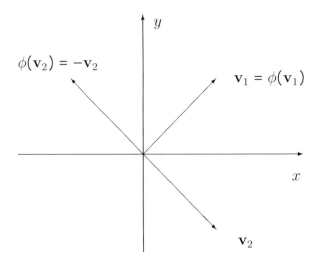

Definition 9.1.2 A linear transformation $T : \mathbb{R}^n \longrightarrow \mathbb{R}^n$ is said to be *diagonalizable*, if there is a basis $\mathcal{B}$ for $\mathbb{R}^n$, such that the matrix $A_\mathcal{B}$ associated with T with respect to $\mathcal{B}$ (in domain and codomain) is a diagonal matrix.

In the above example, the transformation ϕ is diagonalizable and $\mathcal{B} = \{\mathbf{e}_1 + \mathbf{e}_2, \mathbf{e}_1 - \mathbf{e}_2\}$ is the basis with respect to which matrix associated with ϕ is diagonal.

Just as we gave the definition of diagonalizable linear map, we can also give the definition of diagonalizable matrix: this is essentially a matrix associated with a diagonalizable linear transformation. We now see the precise definition.

Definition 9.1.3 A square matrix A is called *diagonalizable*, if there is a invertible matrix P, such that $P^{-1}AP$ is diagonal.

Proposition 9.1.4 *Let $T : \mathbb{R}^n \longrightarrow \mathbb{R}^n$ be a linear transformation associated with the matrix A with respect to the canonical basis (in domain and codomain). Then T is diagonalizable if and only if A is diagonalizable. Also, if T and A are diagonalizable, then the coordinates of the vectors forming the basis $\mathcal{B}$ are the columns of the matrix P such that $P^{-1}AP$ is diagonal.*

Proof. The statement is straightforward, if we remember how to make basis changes from the previous chapter. Suppose that T is diagonalizable. Then there is a basis $\mathcal{B}$ with respect to which the matrix $A_\mathcal{B}$ associated with T is diagonal. If $P = I_{\mathcal{B},\mathcal{C}}$, we have that the formula in Theorem 8.3.1 becomes:

$$A_\mathcal{B} = P^{-1}AP,$$

So, by definition, A is diagonalizable. Conversely, if A is diagonalizable, it means that there is a matrix P, such that $P^{-1}AP$ is diagonal. Consider P as the basis change matrix $I_{\mathcal{B},\mathcal{C}}$, where $\mathcal{B}$ consists of the column vectors of P. The matrix associated with T with respect to the basis $\mathcal{B}$ is just $I_{\mathcal{B},\mathcal{C}}^{-1}AI_{\mathcal{B},\mathcal{C}} = P^{-1}AP$, and it is diagonal. Therefore, T is diagonalizable. ∎

Observation 9.1.5 At this point, it is clear that a linear map is diagonalizable if and only if the matrix associated with it with respect to any basis is diagonalizable. We leave the proof of this statement as an exercise.

Two questions arise spontaneously now:

1) In general, is a linear transformation $T : \mathbb{R}^n \longrightarrow \mathbb{R}^n$ always diagonalizable? Equivalently, given a square matrix A, do we always have an invertible matrix P, such that $P^{-1}AP$ is diagonal?

2) Is there a procedure for "diagonalizing" a linear transformation or a matrix? That is, is there a procedure to determine the basis $\mathcal{B}$ or equivalently the matrix P?

The answer to the first question in general is no, while for the second question the answer is yes; we will see the answer in detail in the next section. Geometrically, we see that there are many matrices that are not diagonalizable. Let us see an example.

Example 9.1.6 Consider the linear transformation $\psi : \mathbb{R}^2 \longrightarrow \mathbb{R}^2$ defined by: $\psi(\mathbf{e}_1) = -\mathbf{e}_2$, $\psi(\mathbf{e}_2) = \mathbf{e}_1$. The matrix associated with ψ with respect to the canonical basis is given by:

$$A = \begin{pmatrix} 0 & 1 \\ -1 & 0 \end{pmatrix}.$$

Although very similar to the previous example, there is an important difference. Geometrically, ψ corresponds to a clockwise rotation of 90^0 around the origin of the plane. In order for A, or equivalently for ψ, to be diagonalizable there must be two linearly independent vectors $\mathbf{v}_1, \mathbf{v}_2$, such that $\psi(\mathbf{v}_1) = \lambda_1\mathbf{v}_1$, $\psi(\mathbf{v}_2) = \lambda_2\mathbf{v}_2$ for $\lambda_1, \lambda_2 \in \mathbb{R}$, i.e. two vectors that *preserve* their direction.

In fact, the matrix of ψ with respect to the basis $\mathcal{B} = (\mathbf{v}_1, \mathbf{v}_2)$ would be:

$$A_{\mathcal{B}} = \begin{pmatrix} \lambda_1 & 0 \\ 0 & \lambda_2 \end{pmatrix}.$$

But it is easy to see that a rotation is not fixing any direction, hence $\mathbf{v}_1$ and $\mathbf{v}_2$ cannot exist, that is, A is not diagonalizable.

Note that the argument would be different, if we allowed the scalars to take complex values. In fact, in this case there would be two vectors namely $\mathbf{v}_1 = (1, -i)$ and $\mathbf{v}_2 = (i, -1)$, such that $\psi(\mathbf{v}_1) = -i\mathbf{v}_1$, $\psi(\mathbf{v}_2) = i\mathbf{v}_2$.

This example suggests a third question:

3) If we allow scalars to take complex values, then, can we always diagonalize a given matrix (or linear transformation)?

The answer is no, but we can always bring a matrix to a form which is almost diagonal; this is called the Jordan form, and we will not discuss it, because it would take us too far.

9.2 EIGENVALUES AND EIGENVECTORS

As we saw in the previous section, regarding the diagonalizability of a linear transformation, the vectors whose direction is not changed by the linear transformation play a fundamental role. In fact, these vectors form a basis $\mathcal{B}$, with respect to which the linear transformation is associated with a diagonal matrix. Equivalently, the coordinates of these vectors, with respect to the canonical basis, form the columns of the matrix P, such that $P^{-1}AP$ is diagonal (where A is the matrix associated with the given transformation with respect to the canonical basis).

These vectors are called *eigenvectors*. Let us look at the precise definition.

Definition 9.2.1 Given a linear transformation $T : V \longrightarrow V$, we say that a nonzero vector $\mathbf{v} \in V$ is an *eigenvector* of T of *eigenvalue* λ, if $T(\mathbf{v}) = \lambda\mathbf{v}$, with $\lambda \in \mathbb{R}$.

It should be noted that an eigenvalue λ can be zero, while an eigenvector, by definition must be different from the zero vector.

Given a matrix A, we call *eigenvalues* and *eigenvectors of A* the eigenvalues and eigenvectors of the linear transformation L_A associated to A with respect to the canonical basis (in domain and codomain).

It is important to understand that the fact that a vector is an eigenvector and that a scalar is an eigenvalue of a linear transformation T *does not depend on basis chosen to represent T!*

In Example 9.1.1, the vectors $\mathbf{v}_1 = \mathbf{e}_1 + \mathbf{e}_2$ and $\mathbf{v}_2 = \mathbf{e}_1 - \mathbf{e}_2$ are eigenvectors of the transformation ϕ of eigenvalues 1 and -1, respectively. If we take the basis consisting of the eigenvectors $\mathcal{B} = (\mathbf{v}_1, \mathbf{v}_2)$, we have that $\mathbf{v}_1 = (1, 1)_C = (1, 0)_{\mathcal{B}}$, $\mathbf{v}_2 = (1, -1)_C = (0, 1)_{\mathcal{B}}$. If we change the coordinates, that is, if we change the way to write these vectors, the vectors $\mathbf{v}_1$ and $\mathbf{v}_2$ do not change and also the eigenvalues associated with them do not change. In fact, in the example, we have seen that the matrix associated with ϕ is

$$A_{\mathcal{B}} = \begin{pmatrix} 1 & 0 \\ 0 & -1 \end{pmatrix}.$$

So the eigenvalues are -1, 1. We summarize our discussion in the following observation.

Observation 9.2.2 A scalar λ is an eigenvalue of a transformation T if and only if it is an eigenvalue of the matrix $A_{\mathcal{B}}$ associated with T with respect to any basis $\mathcal{B}$.

We want to characterize the notion of diagonalizability for a linear transformation.

Proposition 9.2.3 *A linear transformation $T : V \longrightarrow V$ is diagonalizable if and only if there is a basis $\mathcal{B}$ of V consisting of eigenvectors of T.*

The proof of this statement is particularly instructive as it contains basic facts for understanding eigenvalues and eigenvectors.

Proof. If we have a basis $\mathcal{B} = \{\mathbf{v}_1, \ldots, \mathbf{v}_n\}$ of eigenvectors, then the matrix associated with T with respect to the basis $\mathcal{B}$ is diagonal. Indeed:

$$T(\mathbf{v}_1) = \lambda_1 \mathbf{v}_1 = \lambda_1 \mathbf{v}_1 + 0\mathbf{v}_2 + \cdots + 0\mathbf{v}_n,$$

$$\ldots$$

$$T(\mathbf{v}_n) = \lambda_n \mathbf{v}_n = 0\mathbf{v}_1 + 0\mathbf{v}_2 + \cdots + \lambda_n \mathbf{v}_n,$$

so the matrix associated with T respect to the basis $\mathcal{B}$ is:

$$A_\mathcal{B} = \begin{pmatrix} \lambda_1 & \cdots & 0 \\ \vdots & & \vdots \\ 0 & \cdots & \lambda_n \end{pmatrix}.$$

So the transformation T is diagonalizable by definition.

Conversely, if T is diagonalizable, it means that there is a basis $\mathcal{B}$ with respect to which the matrix associated with T is diagonal, i.e.

$$A_\mathcal{B} = \begin{pmatrix} \lambda_1 & \cdots & 0 \\ \vdots & & \vdots \\ 0 & \cdots & \lambda_n \end{pmatrix}.$$

But then, this matrix has precisely the eigenvalues on the diagonal because

$$(T(\mathbf{v}_1))_\mathcal{B} = A(\mathbf{v}_1)_\mathcal{B} = \begin{pmatrix} \lambda_1 & \cdots & 0 \\ \vdots & & \vdots \\ 0 & \cdots & \lambda_n \end{pmatrix} \begin{pmatrix} 1 \\ \vdots \\ 0 \end{pmatrix} = \begin{pmatrix} \lambda_1 \\ \vdots \\ 0 \end{pmatrix}$$

$$\vdots$$

$$(T(\mathbf{v}_n))_\mathcal{B} = A(\mathbf{v}_n)_\mathcal{B} = \begin{pmatrix} \lambda_1 & \cdots & 0 \\ \vdots & & \vdots \\ 0 & \cdots & \lambda_n \end{pmatrix} \begin{pmatrix} 0 \\ \vdots \\ 1 \end{pmatrix} = \begin{pmatrix} 0 \\ \vdots \\ \lambda_n \end{pmatrix}.$$

Therefore $T(\mathbf{v}_1) = \lambda_1 \mathbf{v}_1, \ldots, T(\mathbf{v}_n) = \lambda_n \mathbf{v}_n$ and the basis $\mathcal{B}$ consists of eigenvectors.

■

We now want to give a concrete method for calculating eigenvalues and eigenvectors of a given matrix or linear transformation.

Definition 9.2.4 Given a square matrix A, we define *characteristic polynomial* p_A of A the following polynomial in x:

$$p_A(x) = \det(A - xI),$$

where det denotes the determinant and I the identity matrix.

The fact that $p_A(x)$ is actually a polynomial in x, for example, follows from the recursive calculation of the determinant expanded according to a row (or a column) of $A - xI$.

In the following, if A is a matrix, for the sake of brevity we will denote by Ker A the kernel of the linear transformation L_A associated with A with respect to the canonical basis (in the domain and codomain). Equivalently, Ker A is the set of solutions of the homogeneous linear system $A\mathbf{x} = \mathbf{0}$.

Theorem 9.2.5 *A scalar λ is an eigenvalue of A if and only if it is a zero of the characteristic polynomial, i.e. $\det(A - \lambda I) = 0$.*

Proof. If λ is an eigenvalue of A, then there exists $\mathbf{v} \neq \mathbf{0}$ such that $A\mathbf{v} = \lambda\mathbf{v}$, that is, $A\mathbf{v} - \lambda\mathbf{v} = \mathbf{0}$, i.e. $(A - \lambda I)\mathbf{v} = \mathbf{0}$ and so $\mathbf{v} \in$ Ker $(A - \lambda I)$. Thus by Theorem 7.6.1 the determinant of the matrix $A - \lambda I$ is zero.

Conversely, if $\det(A - \lambda I) = 0$, by Theorem 7.6.1 there is a nonzero vector $\mathbf{v} \in$ Ker $(A - \lambda I)$. Then $(A - \lambda I)\mathbf{v} = \mathbf{0}$, so $A\mathbf{v} = \lambda\mathbf{v}$ and $\mathbf{v}$ is an eigenvector of eigenvalue λ. ∎

Definition 9.2.6 Let A and B be two $n \times n$ matrices. A and B are said *similar*, if there exists an invertible matrix $n \times n$ P such that:

$$B = P^{-1}AP.$$

Observation 9.2.7   • If A and B are similar then A and B represent the *same* linear transformation with respect to different bases. This immediately follows from our discussion on basis change.

   • A is diagonalizable if and only if it is similar to a diagonal matrix.

Theorem 9.2.8 *Similar matrices have the same characteristic polynomial.*

(Warning, the reverse is not true!).

Proof. Let A and B be similar matrices, with $B = P^{-1}AP$. By Binet Theorem 7.3.5 and the properties of associativity and distributivity of matrix product, we have that

$$p_B(x) = \det(P^{-1}AP - xI) = \det(P^{-1}AP - P^{-1}PxI) =$$

$$\det(P^{-1}AP - P^{-1}xIP) = \det(P^{-1}(A - xI)P) =$$

$$\det P^{-1} \det(A - xI) \det P = \det(A - xI) = p_A(x).$$

(In this proof, we have also used the fact that if P is a matrix and x is a scalar, then $P(xI) = (xI)P$). ∎

Observation 9.2.9 We observe that in Theorem 9.2.8 the reverse implication is false. Namely, the two matrices

$$A = \begin{pmatrix} 3 & 0 \\ 0 & 3 \end{pmatrix} \qquad B = \begin{pmatrix} 3 & 0 \\ 1 & 3 \end{pmatrix}$$

have the same characteristic polynomial $p_A(x) = (3-x)^2 = p_B(x)$, but using the techniques we will learn at the end of this chapter it is easy to deduce that A and B are not similar, because B is not diagonalizable.

Observation 9.2.10 From the proof of Theorem 9.2.5, we have that a vector $\mathbf{v} \neq \mathbf{0}$ is an eigenvector of a linear transformation T with associated eigenvalue λ if and only if it belongs to $\mathrm{Ker}\,(A - \lambda I)$, where A is the matrix associated with T with respect to the canonical basis.

Definition 9.2.11 If λ is an eigenvalue of $T : \mathbb{R}^n \longrightarrow \mathbb{R}^n$ then $V_\lambda = \{\mathbf{v} \in \mathbb{R}^n \mid T(\mathbf{v}) = \lambda\mathbf{v}\}$ is called the *eigenspace* associated to the eigenvalue λ.

Observation 9.2.12 We observe that the elements of V_λ are exactly the eigenvectors of eigenvalue λ and the zero vector, so $V_\lambda = \mathrm{Ker}\,(A - \lambda I)$, where A is the matrix associated with T with respect to the canonical basis. Obviously V_λ is a vector subspace, being the kernel of a linear transformation.

We now apply what we learned theoretically to compute eigenvalues and eigenvectors of an arbitrary matrix.

Example 9.2.13 We want to find eigenvalues and eigenvectors of the matrix

$$A = \begin{pmatrix} 5 & -4 \\ 3 & -2 \end{pmatrix}$$

and determine whether the matrix A is diagonalizable.

- The characteristic polynomial of A is:

$$p_A(x) = \det \begin{pmatrix} 5 - x & -4 \\ 3 & -2 - x \end{pmatrix} = x^2 - 3x + 2.$$

  The associated equation has solutions: 1 and 2, so there are two eigenvalues for A, $\lambda = 1$ and $\lambda = 2$.

- We compute the eigenspace corresponding to the eigenvalue $\lambda = 1$. It consists of all vectors $(x, y) \in \mathbb{R}^2$, such that:

$$A \begin{pmatrix} x \\ y \end{pmatrix} = \begin{pmatrix} x \\ y \end{pmatrix},$$

  that is, the solutions of the system

$$\begin{cases} 5x - 4y = x \\ 3x - 2y = y\,. \end{cases}$$

This is an homogeneous linear system, and the matrix of coefficients is:

$$\begin{pmatrix} 5-1 & -4 \\ 3 & -2-1 \end{pmatrix},$$

which is the matrix we used to compute the characteristic polynomial where we replaced the eigenvalue 1 in place of x.

We therefore have

$$V_1 = \mathrm{Ker}\,(A-I) = \mathrm{Ker}\,\begin{pmatrix} 4 & -4 \\ 3 & -3 \end{pmatrix} = \langle (1,1) \rangle.$$

- To compute the eigenspace relative to the eigenvalue 2, we proceed in a similar way, and we obtain:

$$V_2 = \mathrm{Ker}\,(A-2I) = \mathrm{Ker}\,\begin{pmatrix} 3 & -4 \\ 3 & -4 \end{pmatrix} = \langle (4/3,1) \rangle.$$

- The vectors $(1,1), (4/3,1)$ are linearly independent, so we have that A is diagonalizable, and it is similar to the diagonal matrix:

$$D = \begin{pmatrix} 1 & 0 \\ 0 & 2 \end{pmatrix} = P^{-1}AP,$$

$$\text{where } P = \begin{pmatrix} 1 & 4/3 \\ 1 & 1 \end{pmatrix}.$$

P is the matrix of the change of basis, which allows us to pass from the canonical basis to the basis formed by the eigenvectors of A. By the theory on the basis change (see Chapter 8), the columns of P consist of the coordinates of the eigenvectors with respect to the canonical basis.

The following theorem will also be very useful.

Theorem 9.2.14 *Let T be a linear transformation, and let $\mathbf{v}_1, \ldots, \mathbf{v}_n$ be eigenvectors of T associated to the eigenvalues $\lambda_1, \ldots, \lambda_n$, respectively. Assume that $\lambda_1, \ldots, \lambda_n$ are all distinct. Then $\mathbf{v}_1, \ldots, \mathbf{v}_n$ are linearly independent.*

Proof. The proof consists of n steps, all similar.
  - Step 1. We have that $\mathbf{v}_1$ is a linearly independent vector, for $\mathbf{v}_1 \neq \mathbf{0}$.
  - Step 2. We show that $\mathbf{v}_1, \mathbf{v}_2$ are linearly independent. Let $\beta_1, \beta_2 \in \mathbb{R}$ such that:

$$\beta_1 \mathbf{v}_1 + \beta_2 \mathbf{v}_2 = \mathbf{0}. \tag{9.1}$$

Applying T to both members and taking into account that $T(\mathbf{v}_1) = \lambda_1 \mathbf{v}_1$, $T(\mathbf{v}_2) = \lambda_2 \mathbf{v}_2$, we obtain:

$$\beta_1 \lambda_1 \mathbf{v}_1 + \beta_2 \lambda_2 \mathbf{v}_2 = \mathbf{0}. \tag{9.2}$$

Now we subtract from the second equality the first equality multiplied by λ_2, and we get:

$$\beta_1(\lambda_1 - \lambda_2)\mathbf{v}_1 = \mathbf{0}.$$

As $\mathbf{v}_1 \neq \mathbf{0}$, we have that $\beta_1(\lambda_1 - \lambda_2) = 0$, and so $\beta_1 = 0$, being $\lambda_1 \neq \lambda_2$. By replacing $\beta_1 = 0$ in (9.1), we get that $\beta_2\mathbf{v}_2 = \mathbf{0}$. But $\mathbf{v}_2$ is an eigenvector, thus $\mathbf{v}_2 \neq \mathbf{0}$, and it follows that $\beta_2 = 0$, as we wanted.

After $k - 1$ steps, we have shown that the vectors $\mathbf{v}_1, \ldots, \mathbf{v}_{k-1}$ are linearly independent.

- Step k. We show that $\mathbf{v}_1, \ldots, \mathbf{v}_{k-1}, \mathbf{v}_k$ are linearly independent. Let $\beta_1, \ldots, \beta_{k-1}, \beta_k \in \mathbb{R}$, such that:

$$\beta_1\mathbf{v}_1 + \cdots + \beta_{k-1}\mathbf{v}_{k-1} + \beta_k\mathbf{v}_k = \mathbf{0}. \tag{9.3}$$

Applying T to both sides and taking into account that $T(\mathbf{v}_i) = \lambda_i\mathbf{v}_i$ we obtain:

$$\beta_1\lambda_1\mathbf{v}_1 + \cdots + \beta_{k-1}\lambda_{k-1}\mathbf{v}_{k-1} + \beta_k\lambda_k\mathbf{v}_k = \mathbf{0}. \tag{9.4}$$

Now we subtract from equality (9.4) equality (9.3) multiplied by λ_k and we get:

$$\beta_1(\lambda_1 - \lambda_k)\mathbf{v}_1 + \cdots + \beta_k(\lambda_k - \lambda_k)\mathbf{v}_k = \mathbf{0}.$$

As $\mathbf{v}_1, \ldots, \mathbf{v}_{k-1}$ are linearly independent and $\lambda_i - \lambda_k \neq 0$ for every $i = 1, \ldots, k - 1$ (remember that the eigenvalues are all distinct), it follows that $\beta_i = 0$ for each $i = 1, \ldots, k - 1$. Now equality (9.3) becomes:

$$\beta_k\mathbf{v}_k = \mathbf{0},$$

therefore also $\beta_k = 0$, being $\mathbf{v}_k \neq \mathbf{0}$. So the β_i are all zero, and $\mathbf{v}_1, \ldots, \mathbf{v}_k$ are linearly independent, as we wanted.

After n steps we get what wanted, namely that $\mathbf{v}_1, \ldots, \mathbf{v}_n$ are linearly independent. ■

An immediate consequence of the previous theorem is the following corollary.

Corollary 9.2.15 *A matrix $A \in \mathrm{M}_n(\mathbb{R})$ with n distinct eigenvalues is diagonalizable.*

Proof. The n eigenvectors associated with the n distinct eigenvalues of A are linearly independent by Theorem 9.2.14, and thus form a basis of $\mathbb{R}^n$. ■

Let us see a concrete example of what we have just seen.

Example 9.2.16 We want to find eigenvalues and eigenvectors of the matrix

$$A = \begin{pmatrix} -1 & 2 & 0 \\ 1 & 1 & 0 \\ -1 & 1 & 4 \end{pmatrix}$$

and determine if it is diagonalizable.

Let us see schematically how to proceed.

- We calculate the characteristic polynomial of A:

$$p_A(x) = \det \begin{pmatrix} -1-x & 2 & 0 \\ 1 & 1-x & 0 \\ -1 & 1 & 4-x \end{pmatrix} = (x^2 - 3)(4-x).$$

- We find the roots of the characteristic polynomial, namely the solutions of $(x^2 - 3)(4-x) = 0$:

$$x_1 = \sqrt{3}, \qquad x_2 = -\sqrt{3}, \qquad x_3 = 4.$$

These roots are the eigenvalues of A. Since A has three distinct eigenvalues, by the previous theorem there is a basis of $\mathbb{R}^3$ consisting of eigenvectors of A. So we can immediately answer one of the questions: the matrix A is diagonalizable.

- We calculate the eigenspace V_4 corresponding to the eigenvalue 4, which consists of the vectors $(x, y, z) \in \mathbb{R}^3$, such that:

$$A \begin{pmatrix} x \\ y \\ z \end{pmatrix} = 4 \begin{pmatrix} x \\ y \\ z \end{pmatrix},$$

that is, the solutions of the system

$$\begin{cases} -x + 2y = 4x \\ x + y = 4y \\ -x + y + 4z = 4z. \end{cases}$$

We note that the matrix associated with this linear system is:

$$\begin{pmatrix} -5 & 2 & 0 \\ 1 & -3 & 0 \\ -1 & 1 & 0 \end{pmatrix},$$

which is the matrix $A - 4I$.

By reducing the matrix with a Gauss algorithm, we obtain:

$$\begin{pmatrix} -1 & 1 & 0 \\ 0 & -2 & 0 \\ 0 & 0 & 0 \end{pmatrix}.$$

So the solutions of the system depend on $3-2 = 1$ parameters. We can determine x and y, while z takes an arbitrary value s. We get that $V_4 = \{(0, 0, s) \,|\, s \in \mathbb{R}\} = \langle (0, 0, 1) \rangle$, and it has dimension 1.

- We calculate the eigenspace $V_{\sqrt{3}}$ corresponding eigenvalue $\sqrt{3}$, which consists of the vectors $(x, y, z) \in \mathbb{R}^3$ such that:

$$A \begin{pmatrix} x \\ y \\ z \end{pmatrix} = \sqrt{3} \begin{pmatrix} x \\ y \\ z \end{pmatrix}.$$

So, similarly to what we did before, we need to find the solutions of the homogeneous system associated with the matrix $A - \sqrt{3}I$:

$$A = \begin{pmatrix} -1 - \sqrt{3} & 2 & 0 \\ 1 & 1 - \sqrt{3} & 0 \\ -1 & 1 & 4 - \sqrt{3} \end{pmatrix}.$$

By reducing the matrix with the Gauss method we obtain

$$\begin{pmatrix} 1 & 1 - \sqrt{3} & 0 \\ 0 & 2 - \sqrt{3} & 4 - \sqrt{3} \\ 0 & 0 & 0 \end{pmatrix},$$

so $V_{\sqrt{3}} = \{(-1 - 3\sqrt{3})z, -(5 + 2\sqrt{3})z, z) \mid z \in \mathbb{R}\} = \langle(-1 - 3\sqrt{3}, -5 - 2\sqrt{3}, 1)\rangle$ and it has dimension 1.

- We calculate the eigenspace corresponding to the eigenvalue $-\sqrt{3}$, which consists of the vectors $(x, y, z) \in \mathbb{R}^3$, such that:

$$A \begin{pmatrix} x \\ y \\ z \end{pmatrix} = -\sqrt{3} \begin{pmatrix} x \\ y \\ z \end{pmatrix},$$

that is, the solutions of the homogeneous linear system associated with the matrix $A + \sqrt{3}I$:

$$A = \begin{pmatrix} -1 + \sqrt{3} & 2 & 0 \\ 1 & 1 + \sqrt{3} & 0 \\ -1 & 1 & 4 + \sqrt{3} \end{pmatrix}.$$

By reducing the matrix with the Gauss method we obtain

$$\begin{pmatrix} 1 & 1 + \sqrt{3} & 0 \\ 0 & 2 + \sqrt{3} & 4 + \sqrt{3} \\ 0 & 0 & 0 \end{pmatrix},$$

so $V_{-\sqrt{3}} = \{((3\sqrt{3} - 1)z, (2\sqrt{3} - 5)z, z) \mid z \in \mathbb{R}\} = \langle(3\sqrt{3} - 1, 2\sqrt{3} - 5, 1)\rangle$, and it has dimension 1. The matrix

$$D = \begin{pmatrix} 4 & 0 & 0 \\ 0 & \sqrt{3} & 0 \\ 0 & 0 & -\sqrt{3} \end{pmatrix}$$

is similar to A, and we have that $D = P^{-1}AP$, where

$$P = \begin{pmatrix} 0 & -1 - 3\sqrt{3} & 3\sqrt{3} - 1 \\ 0 & -5 - 2\sqrt{3} & 2\sqrt{3} - 5 \\ 1 & 1 & 1 \end{pmatrix}.$$

We now return to the general theory. We know that the eigenvalues of a matrix A are the roots of its characteristic polynomial. The fact that $p_A(\lambda) = 0$ is equivalent to saying that $x - \lambda$ divides $p_A(x)$, i.e. we can write $p_A(x) = (x - \lambda)f(x)$, where $f(x)$ is a polynomial in x.

We now want to be more precise.

Definition 9.2.17 Let A be a square matrix. A scalar λ is called an eigenvalue of *algebraic multiplicity* m if $p_A(x) = (x - \lambda)^m f(x)$, where $f(x)$ is a polynomial, such that $f(\lambda) \neq 0$. In practice $(x - \lambda)^m$ is the maximum power of $x - \lambda$ that divides $p_A(x)$.

If λ is an eigenvalue of A, the dimension of $\mathrm{Ker}\,(A - \lambda I)$ is called the *geometric multiplicity* of λ.

Observation 9.2.18 If λ is an eigenvalue of A, by definition there is at least one nonzero eigenvector, so that $\mathrm{Ker}\,(A - \lambda I)$ has dimension greater or equal to one. This tells us that the geometric multiplicity of an eigenvalue is always greater than or equal to one.

Proposition 9.2.19 *If A is a square matrix and λ is an eigenvalue of A, then the geometric multiplicity of λ is less than or equal to its algebraic multiplicity.*

Proof. Suppose that A is a square matrix of order n and let $T : \mathbb{R}^n \longrightarrow \mathbb{R}^n$ be the linear transformation associated with A with respect to the canonical basis. The geometric multiplicity s of λ is the dimension of $\mathrm{Ker}\,(A - \lambda I) = \mathrm{Ker}\,(T - \lambda \mathrm{id}) = V_\lambda$. Let $\{\mathbf{v}_1, \ldots, \mathbf{v}_s\}$ be a basis of $\mathrm{Ker}\,(T - \lambda \mathrm{id})$ and let us complete it to a basis $\mathcal{B} = \{\mathbf{v}_1, \ldots, \mathbf{v}_s, \ldots, \mathbf{v}_n\}$ of $\mathbb{R}^n$. (Note that we can always do it by the Completion Theorem 4.2.1). Since $T(\mathbf{v}_i) = \lambda \mathbf{v}_i$ for $i = 1, \ldots, s$, the matrix $A_\mathcal{B}$ associated to T with respect to the basis $\mathcal{B}$ is of the type:

$$A_\mathcal{B} = \begin{pmatrix} \lambda & 0 & \cdots & 0 & b_{1\,s+1} & \cdots & b_{1\,n} \\ 0 & \lambda & \cdots & 0 & b_{2\,s+1} & \cdots & b_{2\,n} \\ \vdots & \vdots & \ddots & \vdots & \vdots & \ddots & \vdots \\ 0 & 0 & \cdots & \lambda & b_{s\,s+1} & \cdots & b_{s\,n} \\ 0 & 0 & \cdots & 0 & b_{s+1\,s+1} & \cdots & b_{s+1\,n} \\ \vdots & \vdots & \ddots & \vdots & \vdots & \ddots & \vdots \\ 0 & 0 & \cdots & 0 & b_{n\,s+1} & \cdots & b_{n\,n} \end{pmatrix}.$$

We observe now that, by Theorem 9.2.8, we have $p_A(x) = p_{A_\mathcal{B}}(x)$, because the two matrices A and $A_\mathcal{B}$ are similar. Compute $\det(A - xI) = \det(A_\mathcal{B} - xI)$ developing according to the first column, and then again according to the first column of the only minor of order $n - 1$ with nonzero determinant appearing in the formula, and so on. We get:

$$\det(A_{\mathcal{B}} - xI) =$$

$$\det \begin{pmatrix} \lambda - x & 0 & \cdots & 0 & b_{1\,s+1} & \cdots & b_{1\,n} \\ 0 & \lambda - x & \cdots & 0 & b_{2\,s+1} & \cdots & b_{2\,n} \\ \vdots & \vdots & \ddots & \vdots & \vdots & \ddots & \vdots \\ 0 & 0 & \cdots & \lambda - x & b_{s\,s+1} & \cdots & b_{s\,n} \\ 0 & 0 & \cdots & 0 & b_{s+1\,s+1} - x & \cdots & b_{s+1\,n} \\ \vdots & \vdots & \ddots & \vdots & \vdots & \ddots & \vdots \\ 0 & 0 & \cdots & 0 & b_{n\,s+1} & \cdots & b_{n\,n} - x \end{pmatrix} =$$

$$(\lambda - x)\det \begin{pmatrix} \lambda - x & \cdots & 0 & b_{2\,s+1} & \cdots & b_{2\,n} \\ \vdots & \ddots & \vdots & \vdots & \ddots & \vdots \\ 0 & \cdots & \lambda - x & b_{s\,s+1} & \cdots & b_{s\,n} \\ 0 & \cdots & 0 & b_{s+1\,s+1} - x & \cdots & b_{s+1\,n} \\ \vdots & \ddots & \vdots & \vdots & \ddots & \vdots \\ 0 & \cdots & 0 & b_{n\,s+1} & \cdots & b_{n\,n} - x \end{pmatrix} =$$

$$= (\lambda - x)^{s-1}\det \begin{pmatrix} \lambda - x & b_{s\,s+1} & \cdots & b_{s\,n} \\ 0 & b_{s+1\,s+1} - x & \cdots & b_{s+1\,n} \\ \vdots & \vdots & \ddots & \vdots \\ 0 & b_{n\,s+1} & \cdots & b_{n\,n} - x \end{pmatrix} =$$

$$= (\lambda - x)^{s}\det \begin{pmatrix} b_{s+1\,s+1} - x & \cdots & b_{s+1\,n} \\ \vdots & \ddots & \vdots \\ b_{n\,s+1} & \cdots & b_{n\,n} - x \end{pmatrix}.$$

Let $B = \begin{pmatrix} b_{s+1\,s+1} & \cdots & b_{s+1\,n} \\ \vdots & \ddots & \vdots \\ b_{n\,s+1} & \cdots & b_{n\,n} \end{pmatrix}$; we get that $p_A(x) = p_{A_{\mathcal{B}}}(x) = (x - \lambda)^{s}p_B(x)$, that is, $(x - \lambda)^{s}$ divides $p_A(x)$. As the algebraic multiplicity m of λ is the maximum power of $x - \lambda$ that divides $p_A(x)$, we have that $s \leq m$, i.e. the geometric multiplicity of λ is less than or equal to its algebraic multiplicity. ■

Observation 9.2.20 Combining the previous proposition and Observation 9.2.18 we obtain that if λ is an eigenvalue of a matrix, and λ has algebraic multiplicity 1, then the geometric multiplicity of λ is exactly 1. We will see that this will be of great help in the exercises.

The following proposition allows us to have a clear strategy for figuring out if a certain $n \times n$ matrix is diagonalizable.

Proposition 9.2.21 *Let A be a square matrix of order n, and let $\lambda_1, \ldots, \lambda_r$ be its distinct eigenvalues, with geometric multiplicity, respectively, $n_1, \ldots, n_r$. Then A is diagonalizable if and only if $n_1 + \cdots + n_r = n$.*

Proof. Let T be the linear application associated with A with respect to the canonical basis. For each eigenvalue λ_i, we consider a basis $\mathcal{B}_i = \{\mathbf{v}_{i1}, \ldots, \mathbf{v}_{in_i}\}$ of the eigenspace V_{λ_i}. We then show that the union $\mathcal{B} = \{\mathbf{v}_{11}, \ldots, \mathbf{v}_{1n_1}, \ldots, \mathbf{v}_{r1}, \ldots, \mathbf{v}_{rn_r}\}$ of such bases is a basis for $\mathbb{R}^m$. Once shown this, by Proposition 9.2.3, we have that T is diagonalizable, because $\mathcal{B}$ is a basis of eigenvectors.

Since $n_1 + \cdots + n_r = n$ the set $\mathcal{B}$ contains n vectors, so, by Proposition 4.2.6, to prove that $\mathcal{B}$ is a basis of $\mathbb{R}^n$, it is enough to prove that the vectors of $\mathcal{B}$ are linearly independent. Suppose

$$\lambda_{11}\mathbf{v}_{11} + \cdots + \lambda_{1n_1}\mathbf{v}_{1n_1} + \cdots + \lambda_{r1}\mathbf{v}_{r1} + \cdots + \lambda_{rn_r}\mathbf{v}_{rn_r} = \mathbf{0}. \tag{9.5}$$

Observe that $\mathbf{w}_1 = \lambda_{11}\mathbf{v}_{11} + \cdots + \lambda_{1n_1}\mathbf{v}_{1n_1}, \ldots, \mathbf{w}_r = \lambda_{r1}\mathbf{v}_{r1} + \cdots + \lambda_{rn_r}\mathbf{v}_{rn_r}$ belong to $V_{\lambda_1}, \ldots, V_{\lambda_r}$, respectively. Observe that we can read equality (9.5) as:

$$1 \cdot \mathbf{w}_1 + 1 \cdot \mathbf{w}_2 + \cdots + 1 \cdot \mathbf{w}_r = \mathbf{0}. \tag{9.6}$$

If some of the $\mathbf{w}_i$ were not zero, by equality (9.6) such $\mathbf{w}_i$ would be linearly dependent, but this contradicts the fact that eigenvectors with distinct eigenvalues are linearly independent (Theorem 9.2.14). So $\mathbf{w}_i = \mathbf{0}$ for each $i = 1, \ldots, r$, i.e. $\lambda_{i1}\mathbf{v}_{i1} + \cdots + \lambda_{in_i}\mathbf{v}_{in_i} = 0$, for each $i = 1, \ldots, r$. Let us now exploit the fact that the vectors $\mathbf{v}_{i1}, \ldots, \mathbf{v}_{in_i}$ are linearly independent, because they form a basis of V_{λ_i}. We get: $\lambda_{i1} = \cdots = \lambda_{in_i} = 0$. Thus, in the linear combination in the first member of equality (9.5), the coefficients must be all zero, and this shows that the vectors of $\mathcal{B}$ are linearly independent, so $\mathcal{B}$ is a basis of eigenvectors.

To see the reverse implication, let us assume that A is diagonalizable and let $\mathcal{B} = \{\mathbf{v}_1, \ldots, \mathbf{v}_n\}$ be a basis of $\mathbb{R}^n$ consisting of eigenvectors of T. The matrix D associated to T with respect to the basis $\mathcal{B}$ is a digonal matrix which is similar to A, and its characteristic polynomial is the product of n linear terms of the type $x - \lambda_i$, where λ_i is an eigenvalue of A. By Proposition 9.2.8, similar matrices have the same characeristic polyniomial, thus $p_A(x) = p_D(x) = (x - \lambda_1)^{m_1} \cdots (x - \lambda_r)^{m_r}$, where m_i is the algebraic multiplicity of the eigenvalue λ_i, but it also is the number of distinct eigenvectors of eigenvalue λ_i in $\mathcal{B}$, thus $m_i \leq n_i = \dim(V_{\lambda_i})$ for all $i = 1, \ldots, r$. Moreover, $m_1 + \cdots + m_r = n$. As $n_i \leq m_i$ by Proposition 9.2.19, it follows that $n_i = m_i$ for all $i = 1, \ldots, r$. Therefore $n_1 + \cdots + n_r = n$, as we wanted to prove. ■

We can now describe how to proceed to determine whether a square matrix of order n is diagonalizable.

- We calculate the roots of the characteristic polynomial. If we get n distinct roots, then we have n distinct eigenvalues corresponding to n linearly independent eigenvectors, and thus A is diagonalizable.

- If the eigenvalues are not distinct, for each eigenvalue λ we calculate its geometric multiplicity. If the sum of all geometric multiplicities is n, this will allow us to find n linearly independent eigenvectors and therefore a basis of V; consequently A is diagonalizable. If the sum of all geometric multiplicities is less than n, then A is not diagonalizable.

We conclude with an observation, which relates the calculation of the kernel of a linear transformation and the fact that it has a zero eigenvalue.

Observation 9.2.22 Suppose that the linear transformation $T : \mathbb{R}^n \to \mathbb{R}^n$ has an eigenvalue equal to zero. What can we say about T? The eigenspace V_0 has a dimension of at least 1 by Observation 9.2.18. Also, if A is the matrix associated to T with respect to the canonical basis, V_0 consists of the solutions of the linear system associated with the matrix $A - 0I = A$, that is V_0 is the kernel of T. So if the linear transformation T has an eigenvalue equal to zero, we can say that $V_0 = \operatorname{Ker} T$ and T is not injective.

9.3 EXERCISES WITH SOLUTIONS

9.3.1 Establish if the linear transformation $L : \mathbb{R}^2 \to \mathbb{R}^2$ defined by $L(x, y) = (x - y, x + 3y)$ is diagonalizable. Furthermore, if possible, determine a diagonal matrix D that is similar to A and a matrix B that is not similar to A.

Solution. We first write the matrix associated with L with respect to the canonical basis:

$$A = \begin{pmatrix} 1 & -1 \\ 1 & 3 \end{pmatrix}.$$

Now we compute the characteristic polynomial of this matrix:

$$p_A(x) = \det \begin{pmatrix} 1 - x & -1 \\ 1 & 3 - x \end{pmatrix} = x^2 - 4x + 4 = (x - 2)^2.$$

The characteristic polynomial has the unique root $x = 2$, with algebraic multiplicity 2. So L has only one eigenvalue.

To find the eigenvectors of eigenvalue 2, we solve the homogeneous linear system associated with the matrix $A - 2I$:

$$\begin{pmatrix} -1 & -1 \\ 1 & 1 \end{pmatrix}.$$

We get that $V_2 = \langle (-1, 1) \rangle$ has dimension 1. So it is not possible to find a basis of $\mathbb{R}^2$ consisting of eigenvectors and so the matrix A is not diagonalizable.
Since A is not diagonalizable, no diagonal matrix D can be found that is similar to A. In particular, if we take for example $B = \begin{pmatrix} -1 & 0 \\ 0 & 5 \end{pmatrix}$, certainly B is not similar to A.

9.3.2 Let $L : \mathbb{R}^3 \longrightarrow \mathbb{R}^3$ be the linear transformation defined by $L(\mathbf{e}_1) = 8\mathbf{e}_1 + 3\mathbf{e}_2$, $L(\mathbf{e}_2) = -18\mathbf{e}_1 - 7\mathbf{e}_2$, $L(\mathbf{e}_3) = 9\mathbf{e}_1 + 3\mathbf{e}_2 - \mathbf{e}_3$. Determine if A is diagonalizable and, if so, determine a matrix D similar to A, where A is the matrix associated with L with respect to the canonical basis. Furthermore, if possible, determine two distinct matrices P_1 and P_2 such that $P_1^{-1}AP_1 = P_2^{-1}AP_2 = D$.

Solution. The A matrix associated with L with respect to the canonical basis is:

$$A = \begin{pmatrix} 8 & -18 & 9 \\ 3 & -7 & 3 \\ 0 & 0 & -1 \end{pmatrix}.$$

Let us compute the characteristic polynomial of A:

$$p_A(x) = \det \begin{pmatrix} 8-x & -18 & 9 \\ 3 & -7-x & 3 \\ 0 & 0 & -1-x \end{pmatrix} = -(1+x)(x^2-x-2).$$

We now find the roots of the characteristic polynomial, i.e. the solutions of $(1+x)(x^2-x-2) = 0$, that is $(1+x)^2(x-2) = 0$:

$$x_1 = -1, \qquad x_2 = 2.$$

Such roots are the eigenvalues of A. We notice that the eigenvalue -1 has algebraic multiplicity 2 and the eigenvalue 2 has algebraic multiplicity 1.

By Observation 9.2.20, we know that the eigenspace V_2 has dimension 1, while by Proposition 9.2.19 we can only say that the dimension of V_{-1} is at most 2. If V_{-1} has dimension 2, then, by Proposition 9.2.21, we have that A is diagonalizable; if instead V_{-1} has a dimension smaller than 2, then it is not possible to find a basis of $\mathbb{R}^3$ consisting of eigenvectors and therefore A is not diagonalizable.

So we compute the eigenspace corresponding to the eigenvalue $x = -1$; it consists of the solutions of the homogeneous linear system associated with the matrix $A + I$:

$$\begin{pmatrix} 8+1 & -18 & 9 \\ 3 & -7+1 & 3 \\ 0 & 0 & -1+1 \end{pmatrix}.$$

Reducing the matrix with the Gaussian algorithm we obtain

$$\begin{pmatrix} 1 & -2 & 1 \\ 0 & 0 & 0 \\ 0 & 0 & 0 \end{pmatrix},$$

so the system solutions depend on $3 - 1 = 2$ parameters. We can determine x while y, z can assume arbitrary values s, t. We obtain that $V_{-1} = \{(2s - t, s, t) \mid s \in \mathbb{R}\} = \langle(2,1,0),(-1,0,1)\rangle$. Hence it has dimension 2. So A is diagonalizable.

The required matrices P_i are matrices whose columns must consist of the coordinates (with respect to the canonical basis) of a basis of eigenvectors, therefore it is necessary to determine also the eigenspace V_2.

We have to solve the homogeneous linear system associated with the matrix $A-2I$:

$$\begin{pmatrix} 8-2 & -18 & 9 \\ 3 & -7-2 & 3 \\ 0 & 0 & -1-2 \end{pmatrix}.$$

Reducing the matrix with the Gaussian algorithm we obtain

$$\begin{pmatrix} 1 & -3 & 1 \\ 0 & 0 & 3 \\ 0 & 0 & 0 \end{pmatrix},$$

so the system solutions depend on $3 - 2 = 1$ parameter. We get x, z while y has an arbitrary value s. We obtain that $V_2 = \{(3s, s, 0) \mid s \in \mathbb{R}\} = \langle \{(3, 1, 0)\} \rangle$ Hence, it has 1 dimension, as we expected.

Now a basis of eigenvectors is for example $\mathcal{B}_1 = \{(2, 1, 0), (-1, 0, 1), (3, 1, 0)\}$ and the matrix associated to L with respect to this ordered basis has on the diagonal the eigenvalues corresponding to the basis vectors, i.e.:

$$D = \begin{pmatrix} -1 & 0 & 0 \\ 0 & -1 & 0 \\ 0 & 0 & 2 \end{pmatrix}.$$

A matrix P_1 such that $P_1^{-1} A P_1 = D$ is, for example, the basis change matrix $I_{\mathcal{B}_1, \mathcal{C}}$:

$$P_1 = \begin{pmatrix} 2 & -1 & 3 \\ 1 & 0 & 1 \\ 0 & 1 & 0 \end{pmatrix}.$$

If we want another matrix P_2 such that $P_2^{-1} A P_2 = D$, we must choose another ordered basis of $\mathbb{R}^3$ constisting of eigenvectors, making sure that the first two columns consist always of the coordinates of eigenvectors with eigenvalue -1, while in third columns we have the coordinates of an eigenvector with eigenvalue 2, for example $\mathcal{B}_1 = \{(-3, 0, 3), (2, 1, 0), (-3, -1, 0\}$. In this case, the basis change matrix $P_2 = I_{\mathcal{B}_2, \mathcal{C}}$ is:

$$P_2 = \begin{pmatrix} -3 & 2 & -3 \\ 0 & 1 & -1 \\ 3 & 0 & 0 \end{pmatrix}.$$

9.3.3 Establish if the matrix

$$A = \begin{pmatrix} -8 & 18 & 2 \\ -3 & 7 & 1 \\ 0 & 0 & 1 \end{pmatrix}$$

is diagonalizable.

Solution. We determine the characteristic polynomial of A:

$$p_A(x) = \det \begin{pmatrix} -8 - x & 18 & 2 \\ -3 & 7 - x & 1 \\ 0 & 0 & 1 - x \end{pmatrix} = (1 - x)(x^2 + x - 2).$$

We find the roots of the characteristic polynomial, i.e. the solutions of $(1 - x)(x^2 + x - 2) = 0$, that is $(1 - x)^2(2 + x) = 0$:

$$x_1 = 1, \qquad x_1 = -2$$

and such roots are the eigenvalues of A. We note that the eigenvalue 1 has algebraic multiplicity 2 and the eigenvalue -2 has algebraic multiplicity 1.

As in the previous exercise, we know that the V_{-2} eigenspace is 1, while to determine the dimension of V_1 we need to explicitly compute this eigenspace.

V_1 consists of the solutions of the homogeneous linear system associated with the $A - I$ matrix:

$$\begin{pmatrix} -8 - 1 & 18 & 2 \\ -3 & 7 - 1 & 1 \\ 0 & 0 & 1 - 1 \end{pmatrix}.$$

Reducing the matrix with the Gaussian algorithm we obtain

$$\begin{pmatrix} 1 & -2 & -1/9 \\ 0 & -12 & 1/3 \\ 0 & 0 & 0 \end{pmatrix},$$

so the system solutions depend on $3 - 2 = 1$ parameter. So V_1 has a dimension of 1.

In this case, A is not diagonalizable because we cannot have a basis of eigenvectors. This depends on the fact that the eigenvalue 1 has algebraic multiplicity 2, but V_1 has dimension 1 and not 2.

9.3.4 Determine for which values of k the matrix:

$$A = \begin{pmatrix} 1 & 1 & 1 + k \\ 2 & 2 & 2 \\ 0 & 0 & -k \end{pmatrix}$$

is diagonalizable. Also determine for which values of k we have that 5 is an eigenvalue of A.

Solution. We determine the characteristic polynomial of A:

$$p_A(x) = \det \begin{pmatrix} 1 - x & 1 & 1 + k \\ 2 & 2 - x & 2 \\ 0 & 0 & -k - x \end{pmatrix} = (x^2 - 3x)(-k - x).$$

The roots of the characteristic polynomial are:

$$x_1 = 0, \qquad x_2 = 3, \qquad x_3 = -k,$$

and such roots are the eigenvalues of A. If $k \neq 0$ and $k \neq -3$ we get that A has 3 distinct eigenvalues, so it is diagonalizable.

If $k = 0$ we get that

$$A = \begin{pmatrix} 1 & 1 & 1 \\ 2 & 2 & 2 \\ 0 & 0 & 0 \end{pmatrix},$$

and as we have just seen, we know that A has eigenvalues 0 and -3 with algebraic multiplicity 2 and 1, respectively. By Observation 9.2.20, we know that the eigenspace V_{-3} has dimension 1, while to determine the dimension of V_0 we need to explicitly compute this eigenspace.

V_0 consists of the solutions of the homogeneous linear system associated with the matrix $A - 0I = A$, i.e. $V_0 = \ker A$. Reducing the matrix with the Gaussian algorithm we obtain

$$\begin{pmatrix} 1 & 1 & 1 \\ 0 & 0 & 0 \\ 0 & 0 & 0 \end{pmatrix},$$

so the system solutions depend on $3 - 1 = 2$ parameters. So V_0 has dimension 2 and A is diagonalizable.

If $k = -3$ we get that

$$A = \begin{pmatrix} 1 & 1 & -2 \\ 2 & 2 & 2 \\ 0 & 0 & 3 \end{pmatrix},$$

and as we have just seen, we know that A has eigenvalues 0 and 3 with algebraic multiplicity 1 and 2, respectively. As before, to determine if A is diagonalizable we have to determine the dimension of the eigenspace relative to the eigenvalue of algebraic multiplicity 2, that is V_3.

We must solve the homogeneous linear system associated with the matrix:

$$A - 3I = \begin{pmatrix} -2 & 1 & -2 \\ 2 & -1 & 2 \\ 0 & 0 & 0 \end{pmatrix}.$$

Reducing the matrix with the Gaussian algorithm we obtain

$$\begin{pmatrix} 2 & -1 & 2 \\ 0 & 0 & 0 \\ 0 & 0 & 0 \end{pmatrix},$$

so the system solutions depend on $3 - 1 = 2$ parameters. So V_3 has dimension 2 and A is diagonalizable.

In summary, A is diagonalizable for every value of k.

Since the eigenvalues of A are $0, -3$ and $-k$, 5 is an eigenvalue of A if and only if $k = -5$.

9.4 SUGGESTED EXERCISES

9.4.1 Find eigenvalues and eigenvectors of the following matrices or linear transformations:

i) the matrix:

$$\begin{pmatrix} 2 & 1 & 0 \\ 0 & 1 & -1 \\ 0 & 2 & 4 \end{pmatrix}$$

ii) the linear transformation $L : \mathbb{R}^2 \to \mathbb{R}^2$ defined by:

$$L(x, y) = (2x + y, 2x + 3y)$$

iii) the linear transformation $L : \mathbb{R}^3 \to \mathbb{R}^3$ defined by:

$$L(x, y, z) = (x + y, x + z, y + z)$$

iv) the linear transformation $L : \mathbb{R}^2 \to \mathbb{R}^2$ defined by:

$$L(x, y) = (x - 3y, -2x + 6y)$$

v) the linear transformation $L : \mathbb{R}^2 \to \mathbb{R}^2$ defined by:

$$L(\mathbf{e}_1) = \mathbf{e}_1 - \mathbf{e}_2, \quad L(\mathbf{e}_2) = 2\mathbf{e}_1$$

9.4.2 a) Given the matrix:

$$A = \begin{pmatrix} 7 & 0 & 0 \\ 0 & 7 & -1 \\ 0 & 14 & -2 \end{pmatrix}$$

compute its eigenvalues and eigenvectors.
Is A diagonalizable? If so, determine a diagonal matrix A' similar to A.
b) Is it possible to find a matrix B such that $AB = I$ (where I is the identity matrix)?
Clearly motivate the answer.

9.4.3 Determine a linear transformation $T : \mathbb{R}^2 \longrightarrow \mathbb{R}^2$ that has $\mathbf{e}_1 - \mathbf{e}_2$ as an eigenvector of eigenvalue 2.

9.4.4 Consider the matrix:

$$A = \begin{pmatrix} 3 & 2 & -1 \\ 0 & 2 & 0 \\ -1 & -2 & 3 \end{pmatrix}.$$

Compute its eigenvalues and eigenvectors.
A is diagonalizable? If so, determine all diagonal matrices similar to A. Also, determine a diagonal matrix D that is not similar to A.

9.4.5 Consider the matrix:

$$A = \begin{pmatrix} -4 & -6 & 1 \\ 3 & 5 & -2 \\ 0 & 0 & -1 \end{pmatrix}.$$

a) Compute its eigenvalues and eigenvectors.
b) Is A diagonalizable? If so, write a diagonal matrix A' similar to A.
c) Is A invertible?

9.4.6 Let us consider the matrix:

$$A = \begin{pmatrix} 9 & 0 & 0 \\ 6 & 3 & 6 \\ 0 & 0 & 9 \end{pmatrix}.$$

a) Compute its eigenvalues and eigenvectors.
b) Determine whether A is diagonalizable and if so determine a diagonal matrix D similar to A and a matrix P such that $P^{-1}AP = D$. Is such P unique?

9.4.7 Consider the matrix:
$$A = \begin{pmatrix} 6 & -9 & -3 \\ 0 & -2 & 2 \\ 0 & -1 & 1 \end{pmatrix}.$$

a) Compute its eigenvalues and eigenvectors.
b) Determine, if possible, two distinct matrices P_1 and P_2 such that both $P_1^{-1}AP_1$ and $P_2^{-1}AP_2$ are diagonal.

9.4.8 Given the linear transformation $T : \mathbb{R}^3 \longrightarrow \mathbb{R}^3$ defined by: $T(\mathbf{e}_1) = \mathbf{e}_1 - 2\mathbf{e}_2$, $T(\mathbf{e}_2) = 2\mathbf{e}_1 + 6\mathbf{e}_2$, $T(\mathbf{e}_3) = 3\mathbf{e}_1 - \mathbf{e}_2 + 5\mathbf{e}_3$.
a) Compute the eigenvalues and an eigenspace of T.
b) Establish if T is diagonalizable.
c) Let $\mathcal{B} = \{3\mathbf{e}_2 + \mathbf{e}_3, 5\mathbf{e}_2 + \mathbf{e}_3, -\mathbf{e}_1 + \mathbf{e}_3\}$ be another basis of $\mathbb{R}^3$. Determine the matrix $A_{\mathcal{B}}$ associated with T with respect to the basis $\mathcal{B}$ (in domain and codomain).

9.4.9 Consider the linear transformation $T : \mathbb{R}^3 \longrightarrow \mathbb{R}^3$ defined by: $T(x, y, z) = (-3x + 6y, -x + 4y, x - 6y - 2z)$.
a) Compute the eigenvalues and an eigenspace of T.
b) Establish if T is diagonalizable and determine, if possible, a basis $\mathcal{B}$ of $\mathbb{R}^3$ such that the matrix $A_{\mathcal{B}}$ associated with T with respect to the basis $\mathcal{B}$ is diagonal.

9.4.10 Let $T : \mathbb{R}^3 \longrightarrow \mathbb{R}^3$ be the linear transformation defined by: $T(\mathbf{e}_1) = (k + 1)\mathbf{e}_1 - \mathbf{e}_3$, $T(\mathbf{e}_2) = k\mathbf{e}_2 + (k + 1)\mathbf{e}_3$, $T(\mathbf{e}_3) = k\mathbf{e}_3$ and let A be the matrix associated with T with respect to the canonical basis.
a) Determine for which values of k we have that T is diagonalizable.
b) Determine for which values of k we have that $2\mathbf{e}_1 - 2\mathbf{e}_3$ is an eigenvector of T.
b) For the values of k found in point a) determine, if possible, two distinct diagonal matrices D_1 and D_2 similar to A. Also determine a matrix P such that $P^{-1}AP = D_1$.

9.4.11 Given the matrix:
$$A = \begin{pmatrix} k & k & 2k - 1 \\ 0 & k & 0 \\ 0 & 0 & 1 \end{pmatrix}.$$

a) Determine for which values of k we have that A is invertible.
b) Determine for which values of k we have that A is diagonalizable.

9.4.12 Let $F : \mathbb{R}^3 \longrightarrow \mathbb{R}^3$ be the linear transformation defined by: $F(x, y, z) = (2x + 2y + z, 2x - y - 2z, kz)$, and let A be the matrix associated with F with respect to the canonical basis.
a) Determine for which values of k we have that F is diagonalizable.
b) Choose any value of k for which F is diagonalizable and determine all the diagonal matrices D which are similar to A.

9.4.13 Given the matrix:
$$A = \begin{pmatrix} k & 0 & 0 \\ k & 2 & 3 \\ k & 3 & 2 \end{pmatrix}.$$

a) Determine for which values of k we have that A is diagonalizable.

b) Determine for which values of k we have that $-2\mathbf{e}_1$ is an eigenvector of A.

c) When $k = 3$ determine, if possible, a matrix $B \in \mathrm{M}_3(\mathbb{R})$ having the same eigenvalues of A such that B is not similar to A.

CHAPTER 10

Scalar Products

In the definition of vector space, we have the two operations of sum of vectors and multiplication of a vector by a scalar (see Chapter 2). In this chapter, we want to introduce a new operation: the *scalar product* of two vectors. The result of this operation is a *scalar*, that is a real number. In addition to its vast importance in the applications to physics, we will see how the scalar product is essential in linear algebra for the solution of the problem of diagonalization of symmetric matrices, which we will discuss later.

10.1 BILINEAR FORMS

In this section, we want to introduce the concept of bilinear form or bilinear application and study the bijective correspondence between the set of bilinear forms on a vector space V of dimension n and the set of matrices of order n, once we fixed an ordered basis for V.

Definition 10.1.1 Let V be a vector space. A function $g : V \times V \longrightarrow \mathbb{R}$ is called a *bilinear form* or a *bilinear application* if:

1. $g(\mathbf{u} + \mathbf{u'}, \mathbf{v}) = g(\mathbf{u}, \mathbf{v}) + g(\mathbf{u'}, \mathbf{v})$,
   $g(\mathbf{u}, \mathbf{v} + \mathbf{v'}) = g(\mathbf{u}, \mathbf{v}) + g(\mathbf{u}, \mathbf{v'})$ for every $\mathbf{u}, \mathbf{u'}, \mathbf{v}, \mathbf{v'} \in V$.

2. $g(\lambda \mathbf{u}, \mathbf{v}) = \lambda g(\mathbf{u}, \mathbf{v})$,
   $g(\mathbf{u}, \mu \mathbf{v}) = \mu g(\mathbf{u}, \mathbf{v})$ for each $\mathbf{u}, \mathbf{v} \in V$ and for each $\lambda, \mu \in \mathbb{R}$.

In other words, for any fixed vector $\mathbf{u} \in V$ the functions $g(\mathbf{u}, \cdot) : V \longrightarrow \mathbb{R}$ and $g(\cdot, \mathbf{u}) : V \longrightarrow \mathbb{R}$ are linear applications, hence the term *bilinear*.

g is called *symmetric* if $g(\mathbf{u}, \mathbf{v}) = g(\mathbf{v}, \mathbf{u})$ for every $\mathbf{u}, \mathbf{v} \in V$. A symmetric bilinear form is called a *scalar product* on V and will be denoted with $< , >$.

We shall return to the definition of scalar product in Section 10.4, where we examine it in more detail.

Example 10.1.2 Let us define the function:

$$g : \mathbb{R}^2 \times \mathbb{R}^2 \longrightarrow \mathbb{R}, \qquad g((x_1, x_2), (y_1, y_2)) = x_1 y_1 + 2x_1 y_2.$$

Let us check property (1) of the previous definition:

$$g((x_1, x_2) + (x_1', x_2'), \quad (y_1, y_2)) = (x_1 + x_1')y_1 + 2(x_1 + x_1')y_2 =$$

$$= g((x_1, x_2), (y_1, y_2)) + g((x_1', x_2'), (y_1, y_2))$$

$$g(\lambda(x_1, x_2), (y_1, y_2)) = \lambda x_1 y_1 + 2\lambda x_1 y_2 = \lambda(g((x_1, x_2), (y_1, y_2))).$$

In a completely similar way, we can also verify the property (2). It is therefore a bilinear form. Note, however, that $g(\mathbf{e}_1, \mathbf{e}_2) = 2$ while $g(\mathbf{e}_2, \mathbf{e}_1) = 0$, so g is not a scalar product.

Let us look at another example, important in applications to physics.

Example 10.1.3 Let us define the function in $\mathbb{R}^n$:

$$g((x_1, \ldots, x_n), (y_1, \ldots, y_n)) = x_1 y_1 + \cdots + x_n y_n,$$

for every $\mathbf{u} = (x_1, \ldots, x_n)$ and $\mathbf{v} = (y_1, \ldots, y_n)$ in $\mathbb{R}^n$.
We leave to the reader to verify the properties (1) and (2) of Definition 10.1.1. So, we have a bilinear form. Since $g(\mathbf{u}, \mathbf{v}) = g(\mathbf{v}, \mathbf{u})$, g is a scalar product. This scalar product on $\mathbb{R}^n$ is called *Euclidean product* or *standard product*. We shall denote this product between two vectors $\mathbf{u}, \mathbf{v} \in \mathbb{R}^n$ as $< \mathbf{u}, \mathbf{v} >_e$ or also as $\mathbf{u} \cdot \mathbf{v}$.

As it happens for linear maps, a bilinear application is completely determined by the values it takes on pairs of elements of a fixed basis.

Proposition 10.1.4 *Let V be a vector space of dimension n and set in V a basis $\mathcal{B} = \{\mathbf{v}_1, \ldots, \mathbf{v}_n\}$. Let $c_{ij} \in \mathbb{R}$, $i, j = 1, \ldots, n$ be arbitrary scalars. Then, there exists a unique bilinear application g such that:*

$$g(\mathbf{v}_i, \mathbf{v}_j) = c_{ij}.$$

Proof. Let us first prove the existence of g. As each vector of V, is expressed in a unique way as a linear combination of the elements of the fixed basis $\mathcal{B}$, given $\mathbf{u}$ and $\mathbf{w}$ in V, we can write:

$$\mathbf{u} = \alpha_1 \mathbf{v}_1 + \cdots + \alpha_n \mathbf{v}_n, \qquad \mathbf{w} = \beta_1 \mathbf{v}_1 + \cdots + \beta_n \mathbf{v}_n. \tag{10.1}$$

We therefore define the function $g : V \times V \longrightarrow \mathbb{R}$ in the following way:

$$g(\mathbf{u}, \mathbf{w}) = \sum_{i,j=1}^{n} \alpha_i \beta_j c_{ij}, \tag{10.2}$$

where we used the symbol $\sum_{i,j=1}^{n}$ to indicate the sum for all possible $i, j = 1, \ldots n$. In full:

$$g(\mathbf{u}, \mathbf{w}) = \alpha_1 \beta_1 c_{11} + \alpha_1 \beta_2 c_{12} + \cdots +$$

$$+ \alpha_1 \beta_n c_{1n} + \alpha_2 \beta_1 c_{21} + \cdots + \alpha_n \beta_n c_{nn}.$$

We must now verify that it is a bilinear application, that is, that it satisfies the properties of Definition 10.1.1. We check the first of the conditions in (1) leaving the others by exercise.

Consider the three vectors in V:

$$\mathbf{u} = \alpha_1\mathbf{v}_1 + \cdots + \alpha_n\mathbf{v}_n, \ \mathbf{u}' = \alpha'_1\mathbf{v}_1 + \cdots + \alpha'_n\mathbf{v}_n, \ \mathbf{w} = \beta_1\mathbf{v}_1 + \cdots + \beta_n\mathbf{v}_n.$$

By definition of g we have that:

$$g(\mathbf{u} + \mathbf{u}', \mathbf{w}) = \sum_{i,j=1}^{n} (\alpha_i + \alpha'_i)\beta_j c_{ij} =$$

$$= \sum_{i,j=1}^{n} \alpha_i\beta_j c_{ij} + \sum_{i,j=1}^{n} \alpha'_i\beta_j c_{ij} =$$

$$= g(\mathbf{u}, \mathbf{w}) + g(\mathbf{u}', \mathbf{w}).$$

Similarly, we can also prove the other three properties.

We come now to uniqueness. Suppose that another form $\tilde{g}$ exists, and it is such that $g(\mathbf{v}_i, \mathbf{v}_j) = \tilde{g}(\mathbf{v}_i, \mathbf{v}_j)$. We want to prove that $g = \tilde{g}$, i.e. $g(\mathbf{u}, \mathbf{w}) = \tilde{g}(\mathbf{u}, \mathbf{w})$ for each $\mathbf{u}$, $\mathbf{w}$ in V. If we express $\mathbf{u}$ and $\mathbf{w}$ as linear combinations of the elements of the basis $\mathcal{B}$ as in (10.1), because of properties (1) and (2) of Definition 10.1.1, we can write:

$$\tilde{g}(\mathbf{u}, \mathbf{w}) = \tilde{g}(\alpha_1\mathbf{v}_1 + \cdots + \alpha_n\mathbf{v}_n, \beta_1\mathbf{v}_1 + \cdots + \beta_n\mathbf{v}_n) =$$

$$= \alpha_1\beta_1\tilde{g}(\mathbf{v}_1, \mathbf{v}_1) + \alpha_1\beta_2\tilde{g}(\mathbf{v}_1, \mathbf{v}_2) + \cdots + \alpha_1\beta_n\tilde{g}(\mathbf{v}_1, \mathbf{v}_n) +$$

$$+ \alpha_2\beta_1\tilde{g}(\mathbf{v}_2, \mathbf{v}_1) + \cdots + \alpha_n\beta_n\tilde{g}(\mathbf{v}_n, \mathbf{v}_n) =$$

$$= \sum_{i,j=1}^{n} \alpha_i\beta_j\tilde{g}(\mathbf{v}_i, \mathbf{v}_j).$$

Since $\tilde{g}(\mathbf{v}_i, \mathbf{v}_j) = g(\mathbf{v}_i, \mathbf{v}_j)$ by the very definition of $\tilde{g}$ (see (10.2)), we get $\tilde{g}(\mathbf{u}, \mathbf{w}) = g(\mathbf{u}, \mathbf{w})$

■

10.2 BILINEAR FORMS AND MATRICES

In this section, we want to prove that we have a one-to-one correspondence between the set of bilinear applications on a vector space V of dimension n and the set of matrices of order n, once we choose an ordered basis of V. This is the analog of what happens for a linear application, which is determined by its values on the vectors of an ordered basis. This correspondence is fundamental for any calculation concerning bilinear forms and scalar products.

Before proceeding, we observe a property of the matrices that will be particularly useful to us, whose verification is an easy calculation. The transpose of a product of matrices (rows by columns) is the product of the transposed matrices, with the factors order reversed.

In formulas, if $A \in M_{m,r}(\mathbb{R})$, $B \in M_{r,n}(\mathbb{R})$, then:

$$(AB) = B^T A^T. \tag{10.3}$$

Let us now continue our discussion on the one-to-one correspondence between bilinear forms and matrices, once fixed a basis of the given vector space.

Proposition 10.2.1 *Let V be a real vector space of dimension n and let $\mathcal{B} = \{\mathbf{v}_1, \ldots, \mathbf{v}_n\}$ be a fixed ordered basis. There is a one-to-one correspondence between the set of bilinear forms $g : V \times V \longrightarrow \mathbb{R}$ and the set $M_n(\mathbb{R})$. In this correspondence:*

- *The bilinear form $g : V \times V \longrightarrow \mathbb{R}$ is associated to the matrix:*

$$C = \begin{pmatrix} g(\mathbf{v}_1, \mathbf{v}_1) & \cdots & g(\mathbf{v}_1, \mathbf{v}_n) \\ \vdots & & \vdots \\ g(\mathbf{v}_n, \mathbf{v}_1) & \cdots & g(\mathbf{v}_n, \mathbf{v}_n) \end{pmatrix}.$$

- *The bilinear form*

$$g(\mathbf{u}, \mathbf{v}) = (\mathbf{u})_{\mathcal{B}}^T C(\mathbf{v})_{\mathcal{B}}, \qquad \text{for every } \mathbf{u}, \mathbf{v} \in V$$

is associated with the matrix $C \in M_n(\mathbb{R})$ where $(\mathbf{u})_{\mathcal{B}}$ denotes the coordinate column of the vector $\mathbf{u}$ relative to the basis $\mathcal{B}$.

Scalar products, i.e. symmetric bilinear forms, correspond to the symmetric matrices in $M_n(\mathbb{R})$.

Proof. The first point of this correspondence is a direct consequence of the previous proposition: to each bilinear application we can associate n^2 scalars $g(\mathbf{v}_i, \mathbf{v}_j)$.

Now let us see the second point, that is how to associate a bilinear application directly to a matrix.

We define

$$g(\mathbf{u}, \mathbf{v}) = (\mathbf{u})_{\mathcal{B}}^T C\, (\mathbf{v})_{\mathcal{B}} = (x_1 \ldots x_n) \begin{pmatrix} c_{11} & \cdots & c_{1n} \\ \vdots & & \vdots \\ c_{n1} & \cdots & c_{nn} \end{pmatrix} \begin{pmatrix} y_1 \\ \vdots \\ y_n \end{pmatrix},$$

where

$$(\mathbf{u})_{\mathcal{B}} = \begin{pmatrix} x_1 \\ \vdots \\ x_n \end{pmatrix}, \qquad (\mathbf{v})_{\mathcal{B}} = \begin{pmatrix} y_1 \\ \vdots \\ y_n \end{pmatrix}$$

are the coordinates of $\mathbf{u}$ and $\mathbf{v}$ with respect to the basis $\mathcal{B}$. It is immediate to verify that

$$g(\mathbf{v}_i, \mathbf{v}_j) = c_{ij},$$

therefore C is precisely the matrix associated with g as in point (1).

The bilinearity of g is almost immediate. Let us check the first part of condition (1) of Definition 10.1.1, leaving the other properties by exercise:

$$g(\mathbf{u} + \mathbf{u}', \mathbf{v}) = (\mathbf{u} + \mathbf{u}')_{\mathcal{B}}^{T} C(\mathbf{v})_{\mathcal{B}} = (\mathbf{u})_{\mathcal{B}}^{T} C(\mathbf{v})_{\mathcal{B}} + (\mathbf{u}')_{\mathcal{B}}^{T} C(\mathbf{v})_{\mathcal{B}} =$$

$$= g(\mathbf{u}, \mathbf{v}) + g(\mathbf{u}', \mathbf{v}).$$

We note now that g is symmetric if and only if the corresponding matrix $C = (c_{ij})$ is symmetric, i.e. $C = C^{T}$.

If g is symmetric then $c_{ij} = g(\mathbf{v}_i, \mathbf{v}_j) = g(\mathbf{v}_j, \mathbf{v}_i) = c_{ji}$ for every $i, j = 1, \ldots, n$, so C is symmetric.

Conversely, suppose that C is symmetric, that is, $C = C^{T}$. We observe that for every $\mathbf{u}, \mathbf{v} \in V$ we have that $g(\mathbf{u}, \mathbf{v})^{T} = g(\mathbf{u}, \mathbf{v})$, since $g(\mathbf{u}, \mathbf{v}) \in \mathbb{R}$.

For each $\mathbf{u}, \mathbf{v} \in V$ by (10.3), we have therefore that:

$$g(\mathbf{u}, \mathbf{v}) = (\mathbf{u})_{\mathcal{B}}^{T} C (\mathbf{v})_{\mathcal{B}} = ((\mathbf{u})_{\mathcal{B}}^{T} C (\mathbf{v})_{\mathcal{B}})^{T}$$

$$= (\mathbf{v})_{\mathcal{B}}^{T} C^{T} (\mathbf{u})_{\mathcal{B}} = (\mathbf{v})_{\mathcal{B}}^{T} C (\mathbf{u})_{\mathcal{B}} = g(\mathbf{v}, \mathbf{u}).$$

This shows that g is symmetric. ■

Let us see a concrete example of the correspondence described above.

Example 10.2.2 Observe that the matrix:

$$C = \begin{pmatrix} 3 & 1 \\ 1 & 2 \end{pmatrix},$$

is associated, with respect to the canonical basis of $\mathbb{R}^2$, to the scalar product $< (x_1, x_2), (y_1, y_2) >= 3x_1 y_1 + x_1 y_2 + x_2 y_1 2 + x_2 y_2$. Indeed:

$$\begin{pmatrix} x_1 & x_2 \end{pmatrix} \begin{pmatrix} 3 & 1 \\ 1 & 2 \end{pmatrix} \begin{pmatrix} y_1 \\ y_2 \end{pmatrix} = 3x_1 y_1 + x_1 y_2 + x_2 y_1 + 2x_2 y_2.$$

10.3 BASIS CHANGE

We now want to know how the matrix associated with a given bilinear form on $\mathbb{R}^n$, in the canonical basis, changes, if we decide to change basis. [1]

There is an analogy to what we saw for linear maps. We remind the reader that, as the basis changes, the matrix associated with a linear application can take various forms. For example, if we can find a basis of eigenvectors, the associated matrix is diagonal. However, *the linear application does not change!* The situation here is quite

---

[1]The fact that we take $V = \mathbb{R}^n$ is just for simplicity: all our reasoning applies in the same way to a generic space V of finite dimension, in which we fix a basis.

similar: the matrix associated to the given bilinear form can take very different forms and yet the bilinear form does not change.

Let $I_{B,C}$ be the basis change matrix; it is the matrix associated with the identity map, where we have fixed an ordered basis B in the domain and the canonical basis C in the codomain. We then have that, for each vector $\mathbf{u} \in \mathbb{R}^n$:

$$(\mathbf{u})_C = I_{B,C}(\mathbf{u})_B,$$

where $(\mathbf{u})_B$ denotes the column of coordinates of the vector $\mathbf{u}$ with respect to the ordered basis B.

Assume that C is the matrix associated with a given bilinear form, with respect to the canonical basis of $\mathbb{R}^n$. We want to determine the matrix C' associated with the same bilinear form with respect to the basis B. Let us replace $(\mathbf{u})_C$, $(\mathbf{v})_C$ using (10.3):

$$< \mathbf{u}, \mathbf{v} > = (\mathbf{u})_C^T C(\mathbf{v})_C = (I_{BC}(\mathbf{u})_B)^T C(I_{BC}(\mathbf{v})_B) = (\mathbf{u})_B^T I_{BC}^T C I_{BC}(\mathbf{v})_B.$$

We therefore obtained the following. Assume we fixed an arbitrary ordered basis B and let us denote with C' the corresponding matrix of a bilinear form g. Then C' is obtained from the matrix C associated with g with respect to the canonical basis through the formula:

$$C' = I_{B,C}^T C I_{B,C}. \tag{10.4}$$

More generally, as noted in Chapter 8, this formula also applies if C is replaced by an arbitrary basis B'; the proof of this statement remains the same. So formula (10.4) becomes:

$$C' = I_{B,B'}^T C I_{B,B'}, \qquad \text{with} \quad I_{B,B'} = I_{B',C}^{-1} I_{B,C}, \tag{10.5}$$

where for the last equality we used Theorem 8.3.1, considering the identity application of $\mathbb{R}^n$ as F.

Let us look at the previous example.

Example 10.3.1 Consider the scalar product $< , >$ associated to the matrix

$$C = \begin{pmatrix} 2 & 1 \\ 1 & 2 \end{pmatrix}$$

with respect to the the canonical basis. In particular, notice that scalar products are bilinear forms.

Suppose we choose $B = \{\mathbf{v}_1 = \mathbf{e}_1, \mathbf{v}_2 = \mathbf{e}_1 + \mathbf{e}_2\}$ as a basis, therefore:

$$I_{BC} = \begin{pmatrix} 1 & 1 \\ 0 & 1 \end{pmatrix}.$$

The matrix C' associated with the same scalar product with respect to the basis B is:

$$C' = \begin{pmatrix} 1 & 0 \\ 1 & 1 \end{pmatrix} \begin{pmatrix} 2 & 1 \\ 1 & 2 \end{pmatrix} \begin{pmatrix} 1 & 1 \\ 0 & 1 \end{pmatrix} = \begin{pmatrix} 2 & 3 \\ 3 & 3 \end{pmatrix}.$$

Consider the scalar product of $\mathbf{v}_1$ and $\mathbf{v}_2$, i.e. $< \mathbf{e}_1, \mathbf{e}_1 + \mathbf{e}_2 >$. Using first the canonical basis and then the basis $\mathcal{B}$, we verify that the result is the same:

$$< \mathbf{e}_1, \mathbf{e}_1 + \mathbf{e}_2 > = \begin{pmatrix} 1 & 0 \end{pmatrix} \begin{pmatrix} 2 & 1 \\ 1 & 2 \end{pmatrix} \begin{pmatrix} 1 \\ 1 \end{pmatrix} = 3$$

$$< \mathbf{e}_1, \mathbf{e}_1 + \mathbf{e}_2 > = \begin{pmatrix} 1 & 0 \end{pmatrix} \begin{pmatrix} 2 & 3 \\ 3 & 3 \end{pmatrix} \begin{pmatrix} 0 \\ 1 \end{pmatrix} = 3.$$

Observation 10.3.2 It is useful to compare the basis change formula for a linear application and that of the basis change for a bilinear form.

- Two matrices A and B represent the same linear application (with respect to different bases) if and only if they are similar, that is, there is an invertible matrix P such that $B = P^{-1}AP$.

- Two matrices A and B represent the same bilinear form (with respect to different bases) if and only if an invertible P matrix exists such that $B = P^T AP$.

It is clear by looking at these two formulas that matrices with the property $P^T = P^{-1}$, i.e. such that their transpose coincides with their inverse, are of particular importance. We will do a more detailed study of these matrices and their properties in later sections.

10.4 SCALAR PRODUCTS

From now on we will consider symmetric bilinear forms, i.e. scalar products only. Let us recall their definition, adding some remarks.

Definition 10.4.1 Let V be a vector space. The function $< , >: V \times V \longrightarrow \mathbb{R}$ is called a *scalar product* if:

1. $< \mathbf{u} + \mathbf{u}', \mathbf{v} > = < \mathbf{u}, \mathbf{v} > + < \mathbf{u}', \mathbf{v} >,$
   $< \mathbf{u}, \mathbf{v} + \mathbf{v}' > = < \mathbf{u}, \mathbf{v} > + < \mathbf{u}, \mathbf{v}' >$ for every $\mathbf{u}, \mathbf{u}', \mathbf{v}, \mathbf{v}' \in V$.

2. $< \lambda \mathbf{u}, \mathbf{v} > = \lambda < \mathbf{u}, \mathbf{v} >,$
   $< \mathbf{u}, \mu \mathbf{v} > = \mu < \mathbf{u}, \mathbf{v} >$, for each $\mathbf{u}, \mathbf{v} \in V$ and for each $\lambda, \mu \in \mathbb{R}$.

3. $< \mathbf{u}, \mathbf{v} > = < \mathbf{v}, \mathbf{u} >$ for every $\mathbf{u}, \mathbf{v} \in V$.

In other words, for any fixed vector $\mathbf{u} \in V$, the functions $< \mathbf{u}, \cdot >: V \longrightarrow \mathbb{R}$ and $< \cdot, \mathbf{u} >: V \longrightarrow \mathbb{R}$ are linear applications.

$<,>$ is *non-degenerate* when, if $< \mathbf{u}, \mathbf{v} > = 0$ for each $\mathbf{v} \in V$, then $\mathbf{u} = \mathbf{0}$.
$<,>$ is *positive definite* if $< \mathbf{u}, \mathbf{u} > \geq 0$ for each $\mathbf{u} \in V$ and $< \mathbf{u}, \mathbf{u} > = 0$ if and only if $\mathbf{u} = \mathbf{0}$. If a scalar product is positive definite, we say that $\sqrt{< \mathbf{u}, \mathbf{u} >}$ is the *norm* of the vector $\mathbf{u}$, and we denote it with $\|\mathbf{u}\|$.

As we will see, the positive definite scalar products are particularly important, and yet there are examples of nonpositive definite scalar products that have a fundamental importance in physics.

Example 10.4.2 We can immediately verify that the *standard scalar product* or *Euclidean scalar product* in $\mathbb{R}^n$, defined in the Example 10.1.3 given by:

$$< (x_1, \ldots, x_n), (x'_1, \ldots, x'_n) >_e = x_1 x'_1 + \cdots + x_n x'_n$$

is positive definite, indeed:

$$< (x_1, \ldots, x_n), (x_1, \ldots, x_n) >_e = x_1^2 + \cdots + x_n^2 \geq 0.$$

Also, $x_1^2 + \cdots + x_n^2 = 0$ if and only if $x_1 = x_2 = \cdots = x_n = 0$.

Note that the matrix associated with the Euclidean product with respect to the canonical basis of $\mathbb{R}^n$ is the identical matrix.

Definition 10.4.3 Let V be a vector space with a scalar product. We say that $\mathbf{u}, \mathbf{v} \in V$ are *perpendicular (orthogonal) to each other* if $< \mathbf{u}, \mathbf{v} >= 0$. We will also use the $\mathbf{u} \perp \mathbf{v}$ notation to indicate two vectors perpendicular to each other.

So we can reformulate the notions defined above in the following way:

A scalar product is non-degenerate if and only if there is no nonzero vector perpendicular to all the others. Furthermore, a positive definite scalar product is automatically not degenerate: in fact, being positive definite implies that there is no vector orthogonal to itself, while being degenerate requires that such a vector exists.

Observation 10.4.4 1. The notion of orthogonality depends on the scalar product chosen. For example, in $\mathbb{R}^2$ we consider the two scalar products:

$$< (x_1, x_2), (y_1, y_2) >_e \ = x_1 y_1 + x_2 y_2,$$

$$< (x_1, x_2), (y_1, y_2) >' \ = x_1 y_2 + x_2 y_1.$$

We see that the vectors $\mathbf{e}_1$ and $\mathbf{e}_2$ are perpendicular with respect to the Euclidean product $<, >_e$, but not with respect to the other scalar product, in fact:

$$< (1, 0), (0, 1) >'= 1.$$

2. The existence of a vector orthogonal to itself is a necessary, but not sufficient, condition to have a degenerate scalar product. In fact, if we consider in $\mathbb{R}^2$ the scalar product:

$$< (x_1, x_2), (y_1, y_2) >= x_1 y_1,$$

we immediately see that it is degenerate, since the vector $\mathbf{e}_2$ is orthogonal to any other vector.

On the other hand, if we consider in $\mathbb{R}^2$ the scalar product:

$$< (x_1, x_2), (y_1, y_2) >_m = x_1 y_1 - x_2 y_2$$

we have that the vector $(1, 1)$ is orthogonal to itself, in fact:

$$< (1, 1), (1, 1) >_m = 1 - 1 = 0.$$

However, this product is non-degenerate, in fact there is no nonzero vector that is orthogonal to all the others. If, by contradiction, such a vector (a, b) existed, it would be orthogonal to $\mathbf{e}_1$ and $\mathbf{e}_2$:

$$< (a, b), (1, 0) >_m = a = 0 \qquad < (a, b), (0, 1) >_m = -b = 0.$$

Hence, $(a, b) = (0, 0)$.

The latter scalar product is particularly important for physics and is called the *Minkowski scalar product*.

10.5 ORTHOGONAL SUBSPACES

In this section, we introduce the notion of *orthogonal subspace* $W^\perp$ to a subspace W of a vector space V.

Proposition 10.5.1 *Let* $W \subseteq V$ *be a vector subspace of* V *and* $<, >$ *a scalar product on* V. *Then the set:*

$$W^\perp = \{ \mathbf{u} \in V \mid < \mathbf{u}, \mathbf{w} >= 0, \text{ for all } \mathbf{w} \in W \}$$

is a vector subspace of V.

Proof. Let us check the three properties of Definition 10.4.1. We immediately see that $\mathbf{0}_V \in W^\perp$. Indeed, by property (2) of Definition 10.4.1:

$$< \mathbf{0}_V, \mathbf{w} >=< 0 \cdot \mathbf{0}_V, \mathbf{w} >= 0 \cdot < \mathbf{0}_V, \mathbf{w} >= 0.$$

Let us now verify that for every $\mathbf{u}_1, \mathbf{u}_2 \in W^\perp$, we have that $\mathbf{u}_1 + \mathbf{u}_2 \in W^\perp$. By property (1) of Definition 10.4.1:

$$< \mathbf{u}_1 + \mathbf{u}_2, \mathbf{w} >=< \mathbf{u}_1, \mathbf{w} > + < \mathbf{u}_2, \mathbf{w} >= 0.$$

Finally, we verify that for every scalar λ and for every $\mathbf{u} \in W^\perp$, we have that $\lambda \mathbf{u} \in W^\perp$. By properties (2) of Definition 10.4.1:

$$< \lambda \mathbf{u}, \mathbf{w} >= \lambda < \mathbf{u}, \mathbf{w} >= 0.$$

■

Definition 10.5.2 Given a vector subspace W of V and a scalar product $<, >$ on V, the *orthogonal subspace* to W is:

$$W^\perp = \{ \mathbf{u} \in V \mid < \mathbf{u}, \mathbf{w} >= 0, \text{ for all } \mathbf{w} \in W \}.$$

The observation that follows is essential for the exercises.

Observation 10.5.3 We observe that if $W = \langle \mathbf{w}_1, \ldots, \mathbf{w}_n \rangle \subseteq V$, then

$$W^\perp = \{ \mathbf{u} \in V \mid < \mathbf{u}, \mathbf{w}_i >= 0, \text{ for all } i = 1, \ldots, n \}. \tag{10.6}$$

In fact, if $\mathbf{u} \in W^\perp$ certainly $< \mathbf{u}, \mathbf{w}_i >= 0$, but the converse is also true. In fact, let $\mathbf{u} \in V$, such that $< \mathbf{u}, \mathbf{w}_i >= 0$. If $\mathbf{w} = \lambda_1 \mathbf{w}_1 + \cdots + \lambda_n \mathbf{w}_n \in W$, then

$$< \mathbf{u}, \mathbf{w} > \quad =< \mathbf{u}, \lambda_1 \mathbf{w}_1 + \cdots + \lambda_n \mathbf{w}_n >=$$

$$= \lambda_1 < \mathbf{u}, \mathbf{w}_1 > + \cdots + \lambda_n < \mathbf{u}, \mathbf{w}_n >= 0.$$

So, if we fix a basis of V, we can express conditions in (10.6) using linear equations, therefore the calculation of the subspace orthogonal to a given subspace amounts to the solution of an homogeneous linear system.

Example 10.5.4 Let us compute the orthogonal subspace to $W = \langle (1,1) \rangle$ with respect to the Euclidean scalar product and the Minkowski scalar product introduced in Observation 10.4.4. With respect to the Euclidean scalar product we have:

$$(W^\perp)_e = \{(x,y) \mid < (x,y), (1,1) >_e = 0\} = \{(x,y) \mid x + y = 0\} = \langle (1,-1) \rangle.$$

With respect to the Minkowski scalar product we have:

$$(W^\perp)_m = \{(x,y) \mid < (x,y), (1,1) >_m = 0\} = \{(x,y) \mid x - y = 0\} = \langle (1,1) \rangle.$$

Thanks to Observation 10.5.3, we are able to determine the dimension of the subspace orthogonal to $W \subseteq \mathbb{R}^n$ with respect to the Euclidean scalar product.

Proposition 10.5.5 *Let W be a vector subspace of $\mathbb{R}^n$ and let $W^\perp$ be the subspace orthogonal to W with respect to the Euclidean scalar product. Then:*

$$\dim(W) + \dim(W^\perp) = n.$$

Proof. Let $\mathcal{B} = \{\mathbf{v}_1, \ldots, \mathbf{v}_m\}$ be an ordered basis of W and let $A = (a_{ij}) \in M_{m,n}(\mathbb{R})$ be the matrix having the coordinates of the vectors $\mathbf{v}_1, \ldots, \mathbf{v}_m$ as rows. By Observation 10.5.3, we have that $(x_1, \ldots, x_n) \in W^\perp$ if and only if:

$$(a_{i1}, \ldots, a_{in}) \cdot (x_1, \ldots, x_n) = 0 \text{ for each } i = 1, \ldots, n.$$

In other words:

$$W^\perp = \{\mathbf{x} = (x_1, \ldots, x_n) \mid A\mathbf{x} = \mathbf{0}\} = \ker(L_A),$$

where L_A is the linear application associated with the matrix A with respect to the canonical basis.

By the Rank Nullity Theorem:

$$\dim(W^\perp) = n - \dim \mathrm{Im}(A).$$

We know that the dimension of the image of L_A is the rank of A, which is the dimension of the subspace generated by its columns or equivalently by its rows. Since the rows of A are given by the coordinates of the vectors of a basis of W, this dimension is just $\dim(W)$. Therefore:

$$\dim(W^\perp) = n - \dim(W).$$

■

10.6 GRAM-SCHMIDT ALGORITHM

Let V be a vector space with a positive definite scalar product. The Gram-Schmidt algorithm allows us to build a basis consisting of mutually orthogonal vectors.

Definition 10.6.1 Let V be a vector space with a scalar product $< , >$, which is positive definite and $\mathcal{B} = \{\mathbf{u}_1, \ldots, \mathbf{u}_n\}$ a basis of V. We say that $\mathcal{B}$ is an *orthogonal basis* if $\mathbf{u}_i \perp \mathbf{u}_j$ for every $i, j = 1, \ldots, n$ and that $\mathcal{B}$ is an *orthonormal basis* if:

$$< \mathbf{u}_i, \mathbf{u}_j >= \begin{cases} 1 & \text{for } i = j \\ 0 & \text{for } i \neq j \end{cases},$$

that is, if it is an orthogonal basis and each vector of the basis has norm equal to 1.

In this situation, it is useful to use compact notation:

$$< \mathbf{u}_i, \mathbf{u}_j >= \delta_{ij},$$

where the function δ_{ij} is called *Kronecker delta* and is defined by:

$$\Delta_{ij} = \begin{cases} 1 & \text{for } i = j \\ 0 & \text{for } i \neq j \end{cases}.$$

We observe that, given an orthogonal basis, we can immediately obtain an orthonormal basis by multiplying each vector by the inverse of its norm.

To obtain an orthogonal basis from a given basis, it is necessary to introduce the notion of *orthogonal projection*.

Definition 10.6.2 Let V be a vector space with a positive definite scalar product, and let $\mathbf{u}, \mathbf{v}$. The *orthogonal projection* of the vector $\mathbf{v}$ on the vector $\mathbf{u}$ is given by the vector:

$$\text{proj}_{\mathbf{u}} (\mathbf{v}) = \frac{< \mathbf{v}, \mathbf{u} >}{< \mathbf{u}, \mathbf{u} >} \mathbf{u}.$$

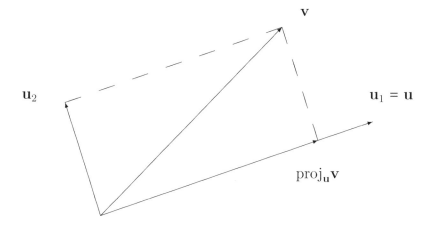

From the figure, where we chose the Euclidean product, we can see and verify with an easy calculation in the general setting, that if $\{\mathbf{u}, \mathbf{v}\}$ is a basis of $\mathbb{R}^2$, $\{\mathbf{u}_1 = \mathbf{u}, \mathbf{u}_2 = \mathbf{v} - \text{proj}_{\mathbf{u}_1}(\mathbf{v})\}$ is an orthogonal basis of $\mathbb{R}^2$. This process can be iterated and allows to build an orthogonal basis $\{\mathbf{u}_1, \ldots, \mathbf{u}_n\}$ starting from a generic basis $\{\mathbf{v}_1, \ldots, \mathbf{v}_n\}$ of V.

We can define then, in a similar way to what we have seen for $\mathbb{R}^2$:

$$\mathbf{u}_1 = \mathbf{v}_1, \qquad\qquad\qquad \mathbf{f}_1 = \frac{1}{\|\mathbf{u}_1\|}\mathbf{u}_1$$

$$\mathbf{u}_2 = \mathbf{v}_2 - \text{proj}_{\mathbf{u}_1}(\mathbf{v}_2), \qquad\qquad \mathbf{f}_2 = \frac{1}{\|\mathbf{u}_2\|}\mathbf{u}_2$$

$$\mathbf{u}_3 = \mathbf{v}_3 - \text{proj}_{\mathbf{u}_1}(\mathbf{v}_3) - \text{proj}_{\mathbf{u}_2}(\mathbf{v}_3), \qquad \mathbf{f}_3 = \frac{1}{\|\mathbf{u}_3\|}\mathbf{u}_3 \qquad (10.7)$$

$$\vdots \qquad\qquad\qquad\qquad\qquad \vdots$$

$$\mathbf{u}_k = \mathbf{v}_k - \text{proj}_{\mathbf{u}_1}(\mathbf{v}_k)\cdots - \text{proj}_{\mathbf{u}_{k-1}}(\mathbf{v}_k), \quad \mathbf{f}_k = \frac{1}{\|\mathbf{u}_k\|}\mathbf{u}_k,$$

where $\|\mathbf{u}_i\| = \sqrt{<\mathbf{u}_i, \mathbf{u}_i>}$.

With the procedure described above, called *Gram-Schmidt algorithm*, we immediately obtained a set of mutually orthogonal vectors.

Theorem 10.6.3 Gram-Schmidt algorithm. *Let V be a vector space with a positive definite scalar product $<,>$. Let $\{\mathbf{v}_1, \ldots, \mathbf{v}_n\}$ be a basis of V. Then the set of vectors $\mathbf{f}_1, \ldots, \mathbf{f}_n$ obtained through (10.7) is an orthonormal basis of V.*

Proof. The fact that the vectors $\mathbf{f}_1, \ldots, \mathbf{f}_n$ are mutually orthogonal and their norm is equal to 1 is easy to verify. Since the dimension of V is n, in order to show $\mathbf{f}_1, \ldots, \mathbf{f}_n$ is a basis, it is enough to verify linear independence. If

$$\lambda_1 \mathbf{f}_1 + \cdots + \lambda_n \mathbf{f}_n = \mathbf{0},$$

then:

$$< \mathbf{f}_i, \lambda_1 \mathbf{f}_1 + \cdots + \lambda_n \mathbf{f}_n >= \lambda_i = 0 \text{ for all } i = 1, \ldots, n.$$

So $\mathbf{f}_1, \ldots, \mathbf{f}_n$ are linearly independent and form an orthonormal basis of V. ■

We conclude with an observation that establishes the importance of choosing coordinates with respect to an orthonormal basis for a given positive definite scalar product. This gives a nice explicit description of the product.

Observation 10.6.4 Let V be a finite dimensional vector space with a positive definite scalar product $<,>$ and let $\mathcal{B} = \{\mathbf{v}_1, \ldots, \mathbf{v}_n\}$ be an orthonormal basis of V. Such a basis always exists by the Gram-Schmidt algorithm.

By definition, the matrix $C = (< \mathbf{v}_i, \mathbf{v}_j >)_{i,j=1,\cdots,n}$ is associated with the scalar product $<,>$, if we fix the basis $\mathcal{B}$. Thus C coincides with the identity matrix I.

Moreover, if we express two vectors $\mathbf{u}$ and $\mathbf{w}$ as linear combinations of the vectors of the basis $\mathcal{B}$, $\mathbf{u} = \alpha_1 \mathbf{v}_1 + \cdots + \alpha_n \mathbf{v}_n$, $\mathbf{w} = \beta_1 \mathbf{v}_1 + \cdots + \beta_n \mathbf{v}_n$, we have that:

$$< \mathbf{u}, \mathbf{w} > \quad =< \alpha_1 \mathbf{v}_1 + \cdots + \alpha_n \mathbf{v}_n, \beta_1 \mathbf{v}_1 + \cdots + \beta_n \mathbf{v}_n >=$$

$$= \alpha_1 \beta_1 < \mathbf{v}_1, \mathbf{v}_1 > + \alpha_1 \beta_2 < \mathbf{v}_1, \mathbf{v}_2 > + \cdots + \alpha_n \beta_n < \mathbf{v}_n, \mathbf{v}_n >=$$

$$= \alpha_1 \beta_1 + \cdots + \alpha_n \beta_n,$$

where we used the bilinearity properties of the scalar product and the fact that $< \mathbf{v}_i, \mathbf{v}_j >= \delta_{ij}$. Hence, if we choose an orthonormal basis $\mathcal{B}$, the matrix associated to the given positive definite scalar product is the identity, just as it happens for the standard scalar product in $\mathbb{R}^n$ and the canonical basis, and the scalar product of two vectors $\mathbf{v}$ and $\mathbf{w}$ coincides with the Euclidian product of their coordinates with respect to the basis $\mathcal{B}$.

10.7 EXERCISES WITH SOLUTIONS

10.7.1 Consider the bilinear form $< (x_1, x_2), (y_1, y_2) >= x_1 y_1 - 2x_1 y_2 - 2x_1 y_2$. Determine the matrix associated with it with respect to the canonical basis, and show that it is a scalar product. Write the matrix associated with the same scalar product with respect to the basis $\mathcal{B} = \{\mathbf{e}_1 + \mathbf{e}_2, 2\mathbf{e}_1 - \mathbf{e}_2\}$.

Solution. We can immediately write the associated matrix to the given bilinear form with respect to the canonical basis:

$$C = \begin{pmatrix} 1 & -2 \\ -2 & 0 \end{pmatrix}.$$

It is a scalar product since the matrix is symmetric. The matrix C' associated with the same scalar product with respect to the basis $\mathcal{B}$ is:

$$C' = \begin{pmatrix} 1 & 2 \\ 1 & -1 \end{pmatrix}^T \begin{pmatrix} 1 & -2 \\ -2 & 0 \end{pmatrix} \begin{pmatrix} 1 & 2 \\ 1 & -1 \end{pmatrix} = \begin{pmatrix} -3 & 0 \\ 0 & 12 \end{pmatrix}.$$

10.7.2 Consider the vector subspace W of $\mathbb{R}^3$ generated by the vectors $\mathbf{w}_1 = \mathbf{e}_1 + \mathbf{e}_2 - 3\mathbf{e}_3$ and $\mathbf{w}_2 = -\mathbf{e}_1 + 2\mathbf{e}_2 - 3\mathbf{e}_3$. Determine a basis for $W^\perp$, computed with respect to the Euclidean scalar product in $\mathbb{R}^3$. Also determine an orthonormal basis for W and an orthonormal basis for $W^\perp$.

Solution. $W^\perp$ consists of the vectors $(x, y, z) \in \mathbb{R}^3$, such that:

$$(1, 1, -3) \cdot (x, y, z) = 0, \qquad (-1, 2, -3) \cdot (x, y, z) = 0.$$

We must therefore solve the linear system:

$$\begin{cases} x + y - 3z = 0 \\ -x + 2y - 3z = 0. \end{cases}$$

We immediately get that $W^\perp = \{(z, 2z, z) \mid z \in \mathbb{R}\}$, therefore a basis for $W^\perp$ is given by the vector $(1, 2, 1)$.

To determine an orthonormal basis of W it is necessary to use the Gram-Schmidt algorithm described in (10.7). We therefore have:

$$\mathbf{u}_1 = \mathbf{w}_1 = (1, 1, -3), \quad \mathbf{f}_1 = \frac{1}{\|\mathbf{u}_1\|}\mathbf{u}_1 = (1/\sqrt{11}, 1/\sqrt{11}, -3/\sqrt{11})$$

$$\mathbf{u}_2 = \mathbf{w}_2 - \frac{\langle \mathbf{w}_2, \mathbf{u}_1 \rangle}{\langle \mathbf{u}_1, \mathbf{u}_1 \rangle}\mathbf{u}_1, \quad \mathbf{f}_2 = \frac{1}{\|\mathbf{u}_2\|}\mathbf{u}_2 = (-7/\sqrt{66}, 2\sqrt{2/33}, -1/\sqrt{66}).$$

An orthonormal basis for W is therefore given by:

$$\{(1/\sqrt{11}, 1/\sqrt{11}, -3/\sqrt{11}), (-7/\sqrt{66}, 2\sqrt{2/33}, -1/\sqrt{66})\}.$$

An orthonormal basis for $W^\perp$ is obtained by taking a generator, for example $(1, 2, 1)$, and dividing it by its norm:

$$\{(1/\sqrt{6}, 2/\sqrt{6}, 1/\sqrt{6})\}.$$

10.7.3 Consider the vector subspace W of $\mathbb{R}^4$ consisting of the solutions of the linear system:

$$\begin{cases} x + y + z - t = 0 \\ x + 2y - z + t = 0. \end{cases}$$

Determine a basis for $W^\perp$, calculated with respect to Euclidean scalar product in $\mathbb{R}^3$.

Solution. We observe that we can write the equations that define W in the following way:

$$(1, 1, 1, -1) \cdot (x, y, z, t) = 0, \qquad (1, 2, -1, 1) \cdot (x, y, z, t) = 0.$$

Therefore the vectors $(1, 1, 1, -1)$ and $(1, 2, -1, 1)$ belong to $W^\perp$ as they are perpendicular to all vectors of W. As $\dim(W^\perp) = 4 - \dim(W) = 2$ by Proposition 10.5.1, it follows that they form a basis of $W^\perp$.

10.8 SUGGESTED EXERCISES

10.8.1 Consider the bilinear form in $\mathbb{R}^2$ given by $g((x_1, x_2), (y_1, y_2)) = 2x_1 y_1 - x_1 y_2 + 3x_2 y_2$.

a) Write the matrix associated with it with respect to the canonical basis.

b) Write the matrix associated with it with respect to the basis $\mathcal{B} = \{\mathbf{v}_1 = \mathbf{e}_1 + \mathbf{e}_2, \mathbf{v}_2 = -2\mathbf{e}_2\}$.

10.8.2 Let W be the vector subspace of $\mathbb{R}^4$ defined from the equation $x + y + 2z - t = 0$ and consider the Euclidian scalar product of $\mathbb{R}^4$.

a) Determine a basis for $W^\perp$.

b) Determine an orthogonal basis of W.

10.8.3 Let W be the vector subspace of $\mathbb{R}^4$ generated by the vectors $e_1 + e_4$, $e_2 - 2e_3 + e_4$.

a) Determine a basis for $W^\perp$.

b) Determine an orthogonal basis of W.

10.8.4 Let W be the vector subspace of $\mathbb{R}^4$ defined by the following equations:

$$\begin{cases} x + y + z = 0 \\ -x + y + z + w = 0 \\ y + z + \frac{1}{2}w = 0. \end{cases}$$

Calculate an orthonormal basis for W relative to the Euclidean scalar product of $\mathbb{R}^4$.

10.8.5 Let W be the vector subspace of $\mathbb{R}^3$ generated by the vectors $(1, 2, -1)$, $(-1/2, -1, -1)$, $(1, 2, 1)$.

a) Find an orthonormal basis of W with respect to the Euclidean scalar product.

b) Find an orthonormal basis for $W^\perp$.

10.8.6 Consider the vector subspace W of $\mathbb{R}^5$ described by the solutions of the homogeneous linear system:

$$\begin{cases} x_1 + 2x_2 + x_5 = 0 \\ x_1 - x_2 - x_3 = 0 \\ 2x_1 - x_3 - x_4 = 0 \end{cases}$$

a) Determine an orthogonal basis of W with respect to the Euclidean scalar product.

b) Determine $W^\perp$. What is the relation between $W^\perp$ and the matrix of the system?

10.8.7 Let W be the following vector subspace of $\mathbb{R}^4$:

$$W = \langle (1, -1, 0, 1), (2, -1, 0, 1), (1, 1, 1, 0) \rangle.$$

a) Determine $W^\perp$.

b) Determine an orthogonal basis for W and an orthogonal basis for $W^\perp$ with respect to the standard scalar product in $\mathbb{R}^4$.

c) Determine whether the basis found in point (b) remains an orthogonal basis also with respect to the scalar product:

$$< (x_1, x_2, x_3, x_4), (y_1, y_2, y_3, y_4) > = x_1y_1 + 2x_2y_2 + 3x_3y_3 + 4x_4y_4.$$

d) Determine the matrix associated with the scalar product of point (c) with respect to the ordered basis

$$\mathcal{B} = \{(1, 1, 1, -1), (0, 1, 0, 1), (0, 0, 1, 0), (1, 0, 1, 2)\}.$$

10.8.8   a) Show that, in a vector space V of finite dimension, a scalar product associated with a diagonal matrix D, relative to a given basis, is non-degenerate if and only if all elements on the diagonal of D are nonzero.

b) Prove that, in a vector space V of finite dimension, a scalar product associated with a diagonal matrix D, relative to a given basis, is positive definite if and only if all the elements on the diagonal of D are positive.

10.8.9 Let V be a vector space with a scalar product $<,>$.

a) If $<,>$ is a positive definite scalar product, then is it not degenerate?

b) If $<,>$ is a real nondegenerate scalar product, is it positive definite?

CHAPTER 11

Spectral Theorem

The Spectral Theorem represents one of the most important results of elementary linear algebra. In Chapter 9, we examined the problem of calculating eigenvalues and eigenvectors and the question of diagonalizability for a square matrix A of order n with real entries. We have seen that it is not always possible to find a diagonal matrix similar to the given matrix, because sometimes we do not have a basis of eigenvectors of A for the space $\mathbb{R}^n$. However, if the matrix A is symmetric, i.e. it coincides with its transpose, the Spectral Theorem guarantees that it is diagonalizable. Furthermore, not only is A similar to a diagonal matrix, but, if we denote with $< , >$ the scalar product associated with it, there is a basis consisting of eigenvectors of A such that the matrix associated with $< , >$ with respect to this basis is diagonal.

In order to prove all these results, it is necessary to introduce the concepts of orthogonal linear transformation and symmetric linear transformation.

11.1 ORTHOGONAL LINEAR TRANSFORMATIONS

In this section we want to define both orthogonal linear transformations and a set of matrices which are extremely important for a deeper understanding of the concept of scalar product: the *orthogonal matrices*. Let us start with the first of these concepts.

Definition 11.1.1 Let V be a real vector space with a positive definite scalar product. We say that a linear transformation $U : V \longrightarrow V$ is *orthogonal* if

$$< U(\mathbf{u}), U(\mathbf{v}) > = < \mathbf{u}, \mathbf{v} >, \qquad \text{for all} \quad \mathbf{u}, \mathbf{v} \in V$$

In other words, the linear transformation U preserves the scalar product given in V.

Proposition 11.1.2 *Let V be a real vector space of dimension n with a positive definite scalar product $<,>$. Let $U : V \longrightarrow V$ be a linear map and let $\mathbf{u}, \mathbf{v} \in V$. The following statements are equivalent.*

1. *U is orthogonal, i.e. $< U(\mathbf{u}), U(\mathbf{v}) > = < \mathbf{u}, \mathbf{v} >$ for each $\mathbf{u}, \mathbf{v} \in V$.*

2. *U preserves the norm, i.e.: $< U(\mathbf{u}), U(\mathbf{u}) > = < \mathbf{u}, \mathbf{u} >$ for each $\mathbf{u} \in V$.*

3. *If $\mathcal{B} = \{\mathbf{v}_1, \ldots, \mathbf{v}_n\}$ is an orthonormal basis of V (with respect to the scalar product $<,>$) then $\{U(\mathbf{v}_1), \ldots, U(\mathbf{v}_n)\}$ is also an orthonormal basis of V.*

Proof. Let us see (2) $\implies$ (1). Let $\mathbf{u}, \mathbf{v} \in V$ and consider:

$$< U(\mathbf{u} - \mathbf{v}), U(\mathbf{u} - \mathbf{v}) > = < (\mathbf{u} - \mathbf{v}), (\mathbf{u} - \mathbf{v}) >$$

Thus:

$$< U(\mathbf{u}), U(\mathbf{u}) > -2 < U(\mathbf{u}), U(\mathbf{v}) > + < U(\mathbf{v}), U(\mathbf{v}) > =$$

$$= < \mathbf{u}, \mathbf{u} > -2 < \mathbf{u}, \mathbf{v} > + < \mathbf{v}, \mathbf{v} >$$

Using (2) we immediately see that $< U(\mathbf{u}), U(\mathbf{v}) > = < \mathbf{u}, \mathbf{v} >$.

(1) $\implies$ (3). If $\mathcal{B}$ is an orthonormal basis then $< \mathbf{v}_i, \mathbf{v}_j > = \delta_{ij}$, where δ_{ij} is the Kronecker delta, i.e. $\delta_{ij} = 1$ if $i = j$ and $\delta_{ij} = 0$ if $i \neq j$.

By (1) we have that $< \mathbf{v}_i, \mathbf{v}_j > = < U(\mathbf{v}_i), U(\mathbf{v}_j) > = \delta_{ij}$ so $\{U(\mathbf{v}_1), \dots, U(\mathbf{v}_n)\}$ is also an orthonormal basis of V.

(3) $\implies$ (2). Let $\mathbf{u} = \alpha_1 \mathbf{v}_1 + \cdots + \alpha_n \mathbf{v}_n$ a vector of V expressed as a linear combination of the elements of $\mathcal{B}$. Then

$$< \mathbf{u}, \mathbf{u} > = < \alpha_1 \mathbf{v}_1 + \cdots + \alpha_n \mathbf{v}_n, \alpha_1 \mathbf{v}_1 + \cdots + \alpha_n \mathbf{v}_n > =$$

$$= \alpha_1^2 < \mathbf{v}_1, \mathbf{v}_1 > + \alpha_1 \alpha_2 < \mathbf{v}_1, \mathbf{v}_2 > + \cdots + \alpha_n^2 < \mathbf{v}_n, \mathbf{v}_n >$$

using the bilinearity of the scalar product. Since $\mathcal{B}$ is an orthonormal basis

$$< \mathbf{u}, \mathbf{u} > = \alpha_1^2 + \cdots + \alpha_n^2 \tag{11.1}$$

Now let us compute $< U(\mathbf{u}), U(\mathbf{u}) >$:

$$< U(\mathbf{u}), U(\mathbf{u}) > = < \alpha_1 U(\mathbf{v}_1) + \cdots + \alpha_n U(\mathbf{v}_n),$$

$$\alpha_1 U(\mathbf{v}_1) + \cdots + \alpha_n U(\mathbf{v}_n) > =$$

$$= \alpha_1^2 < U(\mathbf{v}_1), U(\mathbf{v}_1) > + \alpha_1 \alpha_2 < U(\mathbf{v}_1), U(\mathbf{v}_2) > + \cdots +$$

$$+ \alpha_n^2 < U(\mathbf{v}_n), U(\mathbf{v}_n) >$$

using the bilinearity of the scalar product and the linearity of U. We recall that by hypothesis also $\{U(\mathbf{v}_1), \dots U(\mathbf{v}_n)\}$ is an orthonormal basis, i.e. $< U(\mathbf{v}_i), U(\mathbf{v}_j) > = \delta_{ij}$, we get:

$$< U(\mathbf{u}), U(\mathbf{u}) > = \alpha_1^2 + \cdots + \alpha_n^2$$

So, by (11.1), we have $< U(\mathbf{u}), U(\mathbf{u}) > = < \mathbf{u}, \mathbf{u} >$.

This concludes the proof. ■

11.2 ORTHOGONAL MATRICES

We now want to define the set of *orthogonal matrices* and establish the relation between orthogonal matrices and orthogonal linear transformations.

Let us consider the vector space $\mathbb{R}^n$ with the standard scalar product $< , >_e$:

$$< \mathbf{u}, \mathbf{v} >_e = (\alpha_1 \dots \alpha_n) \begin{pmatrix} \beta_1 \\ \vdots \\ \beta_n \end{pmatrix} = \alpha_1 \beta_1 + \dots + \alpha_n \beta_n \qquad (11.2)$$

where $\mathbf{u} = (\alpha_1 \dots \alpha_n)^T$, $\mathbf{v} = (\beta_1 \dots \beta_n)^T$ are (column) vectors in $\mathbb{R}^n$.

Definition 11.2.1 We say that a matrix $A \in M_n(\mathbb{R})$ is *orthogonal* if it preserves the standard scalar product, that is, if:

$$< A\mathbf{u}, A\mathbf{v} >_e = < \mathbf{u}, \mathbf{v} >_e \qquad \text{for all} \quad \mathbf{u}, \mathbf{v} \in \mathbb{R}^n$$

where we denote with $A\mathbf{u}$ the product rows by columns of the matrix A by the column vector $\mathbf{u}$.

Recall now that, given a matrix $A \in M_n(\mathbb{R})$, we denote by $L_A : \mathbb{R}^n \longrightarrow \mathbb{R}^n$ the linear transformation associated with it, if we fix the canonical bases in domain and codomain; if $\mathbf{u} \in \mathbb{R}^n$, $L_A(\mathbf{u}) = A\mathbf{u}$.

The relation between orthogonal matrices and orthogonal linear transformations is expressed by the following proposition, whose proof is immediate if we compare the Definitions 11.1.1 and 11.2.1.

Proposition 11.2.2 *A matrix $A \in M_n(\mathbb{R})$ is orthogonal if and only if the linear transformation $L_A : \mathbb{R}^n \longrightarrow \mathbb{R}^n$ is orthogonal with respect to the standard scalar product in $\mathbb{R}^n$ (see formula (11.2)).*

Orthogonal matrices satisfy many properties, which we summarize in a proposition.

Proposition 11.2.3 *Let $A \in M_n(\mathbb{R})$ be a square matrix. The following statements are all equivalent.*

a) A is orthogonal, i.e. $< A\mathbf{u}, A\mathbf{v} >_e = < \mathbf{u}, \mathbf{v} >_e$ for each $\mathbf{u}, \mathbf{v} \in \mathbb{R}^n$.

b) A preserves the vector norm, i.e .: $< A\mathbf{u}, A\mathbf{u} >_e = < \mathbf{u}, \mathbf{u} >_e$ for each $\mathbf{u}, \mathbf{v} \in \mathbb{R}^n$.

c) $A^T A = I = A^T A$ where I is the identity matrix. In particular, $A^{-1} = A^T$.

d) The columns and rows of A form two orthonormal bases.

Proof. The equivalence between (a) and (b) is immediate by Proposition 11.1.2. Before proceeding with the proof, we recall that the standard scalar product of two column vectors $\mathbf{u} = (\alpha_1 \dots \alpha_n)^T$, $\mathbf{v} = (\beta_1 \dots \beta_n)^T$, can be expressed as the product rows by columns:

$$\mathbf{u}^T \mathbf{v} = (\alpha_1 \dots \alpha_n) \begin{pmatrix} \beta_1 \\ \vdots \\ \beta_n \end{pmatrix}$$

So we have that:

$$< A\mathbf{u}, A\mathbf{v} >_e = (A\mathbf{u})^T (A\mathbf{v}) = \mathbf{u}^T (A^T A)\mathbf{v} \qquad (11.3)$$

remembering that if X and Y are matrices, we have $(XY)^T = Y^T X^T$.

$(a) \implies (c)$. By formula (11.3) and by our assumptions, it follows that

$$\mathbf{u}^T (A^T A) \mathbf{v} = \mathbf{u}^T \mathbf{v}$$

for each $\mathbf{u}, \mathbf{v} \in \mathbb{R}^n$. Let $A^T A = C = (c_{ij})_{i,j=1,\dots,n}$ and choose $\mathbf{u} = \mathbf{e}_i$ and $\mathbf{v} = \mathbf{e}_j$ in the above formula. We see that

$$\mathbf{e}_i^T (A^T A) \mathbf{e}_j = c_{ij},$$

while

$$\mathbf{e}_i^T \mathbf{e}_j = \delta_{ij}$$

where δ_{ij} is the Kronecker delta. So we have

$$A^T A = I$$

since we proved that the matrices $C = A^T A$ and I have the same entries.
In particular, it follows that the determinant of A is nonzero, so A is invertible. Multiplying both sides of the equality by the matrix A^{-1}, we obtain that $A^T = A^{-1}$, so $AA^T = I$.

$(c) \implies (a)$. Suppose that $AA^T = I$. Then by formula (11.3), we have that

$$< A\mathbf{u}, A\mathbf{v} >_e = (A\mathbf{u})^T (A\mathbf{v}) = \mathbf{u}^T (A^T A) \mathbf{v} = \mathbf{u}^T \mathbf{v} = < \mathbf{u}, \mathbf{v} >_e,$$

as we wanted to prove.

$(c) \iff (d)$. If $\mathbf{v}_1, \dots, \mathbf{v}_n$ are the column vectors of A the equation $A^T A = I$ is equivalent to:

$$\mathbf{v}_i \cdot \mathbf{v}_j = 0, \quad i \neq j, \quad \mathbf{v}_i \cdot \mathbf{v}_i = 1.$$

that is, it is equivalent to the fact that these columns are orthonormal vectors. For the row vectors we argue similarly. The equation $AA^T = I$ is equivalent to the fact that the rows of A are orthonormal vectors. We conclude by remembering that A is an invertible matrix if and only if its rows (columns) form a basis of $\mathbb{R}^n$. ■

Remark 11.2.4 We observe that in the previous proposition, condition (c), $AA^T = I = A^T A$, is equivalent to only one of the two equalities. In fact, in general, given $X, Y \in M_n(\mathbb{R})$, if $XY = I$ then, by Binet theorem, $\det(XY) = \det(X)\det(Y) = 1$. Hence both matrices are invertible and they are one the inverse of the other (see Observation 7.4.3 of Chapter 7).
Similarly in part (d), the two conditions " the columns of A form a basis" and " the rows of A form a basis" are equivalent. This is a consequence of the general fact that the row rank of a matrix is equal to its column rank.

Because of the previous remark and the above discussion we have the following proposition.

Proposition 11.2.5 *Let V be a real vector space of finite dimension with a positive definite scalar product. Let $U : V \longrightarrow V$ be a linear map and let A be the matrix associated with it, with respect to an orthonormal $\mathcal{B}$ basis. Then U is orthogonal if and only if the matrix A is orthogonal.*

We conclude this section with some properties of orthogonal matrices.

Observation 11.2.6   1. The determinant of an orthogonal matrix can only be ± 1. Indeed

$$\det(A^T A) = \det(A^T)\det(A) = \det(A)^2 = \det(I) = 1$$

because the determinant of a matrix is equal to the determinant of its transpose, by Corollary 7.9.8.

2. Also note that, if A is an orthogonal matrix, we have that $A^{-1} = A^T$ is orthogonal. This immediately follows from property (c) of Proposition 11.2.3. To verify that $(A^T)^T A^T = I$ just recall that $(A^T)^T = A$.

3. If A and B are orthogonal matrices then AB is an orthogonal matrix. In fact, we can directly verify that:

$$(AB)^T(AB) = B^T A^T AB = B^T IB = B^T B = I$$

Let us see some examples of orthogonal matrices. We leave to the reader the easy verification that $A^{-1} = A^T$.

Example 11.2.7 1. Rotations in $\mathbb{R}^2$:

$$A = \begin{pmatrix} cos(t) & sin(t) \\ -sin(t) & cos(t) \end{pmatrix}$$

As the reader can easily verify, this matrix is associated with a rotation of the plane by an angle t, centered on the origin of the Cartesian axes.

2. Reflection with respect to the line $x = y$ in $\mathbb{R}^2$:

$$A = \begin{pmatrix} 0 & 1 \\ 1 & 0 \end{pmatrix}$$

As the reader can easily verify, this matrix it is associated with a reflection, i.e. the points that belong to the line $x = y$ are fixed, and every other point of the plane is sent to the his symmetric with respect to that line.

3. Rotations in $\mathbb{R}^3$. Let us consider a linear transformation that rotates any vector applied in the origin around a given axis passing through the origin, of a certain fixed angle t. Such a linear transformation preserves the Euclidean norm and therefore is an orthogonal transformation (with respect to the Euclidean scalar product). A famous theorem by Euler states that every orthogonal transformation of $\mathbb{R}^3$ whose matrix has determinant equal to 1 is of this type.

11.3   SYMMETRIC LINEAR TRANSFORMATIONS

In this section we want to define symmetric linear maps and see which matrices are associated with them, once we fix suitable bases in domain and codomain.

Definition 11.3.1 Let V be a real vector space with a positive definite scalar product $<,>$. We say that the linear transformation $T : V \longrightarrow V$ is *symmetric* if

$$< T(\mathbf{u}), \mathbf{v} >=< \mathbf{u}, T(\mathbf{v}) > \qquad \text{for each} \quad \mathbf{u}, \mathbf{v} \in V \qquad (11.4)$$

Let us now look at the particular case of a linear transformation $T : \mathbb{R}^n \longrightarrow \mathbb{R}^n$ and the Euclidean scalar product. If we fix the canonical basis $\mathcal{C}$, we can write:

$$< \mathbf{u}, \mathbf{v} >_e = \mathbf{u}^T I \mathbf{v} = \mathbf{u}^T \mathbf{v} \qquad \text{for all} \quad \mathbf{u}, \mathbf{v} \in \mathbb{R}^n$$

where $<,>_e$ denotes the standard scalar product in $\mathbb{R}^n$ and $\mathbf{u}$ denotes a *column* vector in $\mathbb{R}^n$. This notation identifies a vector in $\mathbb{R}^n$ with the column of its coordinates with respect to the canonical basis.

In other words, the Euclidean product associated with the canonical basis corresponds to the identity matrix. Let us now consider the matrix A associated with the linear transformation T, with respect to the canonical basis, i.e. $T = L_A$.

We can therefore write:

$$< T(\mathbf{u}), \mathbf{v} >_e = (A\mathbf{u})^T \mathbf{v} = \mathbf{u}^T A^T \mathbf{v} = \mathbf{u}^T A^T \mathbf{v} \qquad \text{for each} \quad \mathbf{u}, \mathbf{v} \in \mathbb{R}^n \qquad (11.5)$$

If A is a symmetric matrix, i.e. $A = A^T$, we have that T is a symmetric linear map. In fact from (11.5):

$$< T(\mathbf{u}), \mathbf{v} >_e =< A\mathbf{u}, \mathbf{v} >_e = \mathbf{u}^T A^T \mathbf{v} = \mathbf{u}^T A \mathbf{v} =< \mathbf{u}, A\mathbf{v} >_e =< \mathbf{u}, T(\mathbf{v}) >_e$$

Conversely, if T is symmetric, we see that the A matrix is symmetric. Indeed:

$$< T(\mathbf{e}_i), \mathbf{e}_j >_e = (A\mathbf{e}_i)^T \mathbf{e}_j = \begin{pmatrix} a_{1i} & \cdots & a_{ni} \end{pmatrix} \begin{pmatrix} 1 & 0 & \cdots & 0 \\ 0 & 1 & \cdots & 0 \\ \vdots & & & \vdots \\ 0 & & \cdots & 1 \end{pmatrix} \begin{pmatrix} 0 \\ \vdots \\ 1 \\ \vdots \\ 0 \end{pmatrix} = a_{ji}$$

$$< \mathbf{e}_i, T(\mathbf{e}_j) >_e = \mathbf{e}_i^T A\mathbf{e}_j = \begin{pmatrix} 0 \ldots 1 \ldots 0 \end{pmatrix} \begin{pmatrix} 1 & 0 & \cdots & 0 \\ 0 & 1 & \cdots & 0 \\ \vdots & & & \vdots \\ 0 & & \cdots & 1 \end{pmatrix} \begin{pmatrix} a_{1j} \\ a_{2j} \\ \vdots \\ a_{nj} \end{pmatrix} = a_{ij}$$

and therefore by the condition (11.4), we have $a_{ij} = a_{ji}$, that is, the matrix A is symmetric.

The same happens for the general case: every symmetric linear map is associated to a symmetric matrix, if we fix an orthonormal basis in domain and codomain.

Proposition 11.3.2 *Let V be a real vector space, with a positive definite scalar product $<,>$ and let $\mathcal{B} = \{\mathbf{v}_1, \ldots, \mathbf{v}_n\}$ be an orthonormal basis. Let $T : V \longrightarrow V$ be a linear map and let $A = (a_{ij})$ be the matrix associated with T with respect to the basis $\mathcal{B}$. Then T is symmetric if and only if the matrix A is symmetric.*

Proof. We first prove that the matrix associated to the symmetric linear map T with respect to the basis $\mathcal{B}$ is symmetric. We have:

$$T(\mathbf{v}_j) = a_{1j}\mathbf{v}_1 + \cdots + a_{nj}\mathbf{v}_n$$

from which:

$$< T(\mathbf{v}_j), \mathbf{v}_i > \ = a_{1j} < \mathbf{v}_1, \mathbf{v}_i > + \cdots + a_{ij} < \mathbf{v}_i, \mathbf{v}_i > + \ldots$$

$$\cdots + a_{nj} < \mathbf{v}_n, \mathbf{v}_j >= a_{ij}$$

and similarly

$$< \mathbf{v}_j, T(\mathbf{v}_i) > \ = a_{j1} < \mathbf{v}_j, \mathbf{v}_1 > + \cdots + a_{ji} < \mathbf{v}_j, \mathbf{v}_i > + \ldots$$

$$\cdots + a_{nj} < \mathbf{v}_n, \mathbf{v}_j >= a_{ji}$$

So we have that:

$$a_{ij} =< T(\mathbf{v}_j), \mathbf{v}_i >=< \mathbf{v}_j, T(\mathbf{v}_i) >= a_{ji} \tag{11.6}$$

On the other hand, equation (11.6) proves that if the matrix A associated with T with respect to an orthonormal basis is symmetric, then $< T(\mathbf{v}_j), \mathbf{v}_i >=< \mathbf{v}_j, T(\mathbf{v}_i) >$. As each $\mathbf{u}, \mathbf{v} \in V$ is a linear combination of $\mathbf{v}_1, \ldots, \mathbf{v}_n$ we have that the previous equality gives immediately $< T(\mathbf{u}), \mathbf{v} >=< \mathbf{u}, T(\mathbf{v}) >$ for each $\mathbf{u}, \mathbf{v} \in \mathbb{R}^n$. This concludes the proof. ■

Remark 11.3.3 Note that in the preliminary observations to the previous proposition we have proved that if A is a $n \times n$ matrix with real entries then:

$$< A\mathbf{u}, \mathbf{v} >_e =< \mathbf{u}, A^T\mathbf{v} >_e$$

where $<,>_e$ is the Euclidean scalar product and $\mathbf{u}, \mathbf{v} \in \mathbb{R}^n$ are column vectors.

11.4   THE SPECTRAL THEOREM

In this section, we want to state and prove one of the most important results of linear algebra: the spectral theorem.

Let us start with two lemmas followed by some immediate observations. The first lemma tells us that every symmetric matrix admits at least one real eigenvalue, the second lemma, which is actually a consequence of the first, tells us that eigenvectors of distinct eigenvalues of a symmetric matrix are always perpendicular to each other. These are the key steps to the proof of the spectral theorem.

Lemma 11.4.1 *Let $A \in M_n(\mathbb{R})$ be a symmetric matrix. Then A admits a real eigenvalue.*

Proof. See Appendix 11.7. ■

We observe that once proven that a matrix A with real entries admits an eigenvalue $\lambda \in \mathbb{R}$, we immediately have that the eigenspace $V_\lambda = \ker(A - \lambda I) \subseteq \mathbb{R}^n$ contains a nonzero vector and therefore there exists also a *real* eigenvector relative to the eigenvalue λ.

Let us now establish another result that will be fundamental in the proof of the spectral theorem.

Lemma 11.4.2 *Let* $A \in M_n(\mathbb{R})$ *be a symmetric matrix,* λ *a real eigenvalue and* $\mathbf{u} \in \mathbb{R}^n$ *an eigenvector of eigenvalue* λ*. Let* $\mathbf{w}$ *be a vector perpendicular to* $\mathbf{u}$ *with respect to the Euclidean scalar product. Then,* $\mathbf{u}$ *is perpendicular to* $A\mathbf{w}$*.*

Proof. Since $\mathbf{u}$ is perpendicular to $\mathbf{w}$ and $A = A^T$, we have

$$0 = \lambda <\mathbf{u}, \mathbf{w}>_e = <\lambda\mathbf{u}, \mathbf{w}>_e = <A\mathbf{u}, \mathbf{w}>_e = <\mathbf{u}, A\mathbf{w}>_e,$$

and therefore also $A\mathbf{w}$ is perpendicular to $\mathbf{u}$. ■

We almost immediately have a particularly important corollary.

Corollary 11.4.3 *Let* λ *and* μ *be distinct eigenvalues of a symmetric matrix* A *and* $\mathbf{u}$*,* $\mathbf{w}$ *two corresponding eigenvectors. Then* $\mathbf{u}$ *is perpendicular to* $\mathbf{w}$*.*

Proof. We need to show that $<\mathbf{u}, \mathbf{w}>_e = 0$. We have

$$\lambda <\mathbf{u}, \mathbf{w}>_e = <A\mathbf{u}, \mathbf{w}>_e = <\mathbf{u}, A\mathbf{w}>_e = \mu <\mathbf{u}, \mathbf{w}>_e,$$

So $(\lambda - \mu) <\mathbf{u}, \mathbf{w}>_e = 0$ and since $\lambda \neq \mu$, we get $<\mathbf{u}, \mathbf{w}>_e = 0$. ■

Let us summarize with a corollary what we proved for symmetric matrices in terms of symmetric linear transformations.

Corollary 11.4.4 *Let* V *be a real vector space of finite dimension with a positive definite scalar product, and let* $T : V \longrightarrow V$ *be a symmetric linear map. Then:*

1. *T admits at least one real eigenvalue.*

2. *If $\mathbf{w} \in V$ is perpendicular to an eigenvector $\mathbf{u}$ of T then $T(\mathbf{w})$ is also perpendicular to $\mathbf{u}$.*

3. *Eigenvectors of T relative to distinct eigenvalues are perpendicular to each other.*

Proof. Let $\mathcal{B}$ be an orthonormal basis for the positive definite scalar product. Such a basis exists thanks to the Gram-Schmidt algorithm. By Proposition 11.3.2, the matrix A associated to T with respect to the basis $\mathcal{B}$ is symmetric. Therefore, the statements of the corollary immediately follow from Lemmas 11.4.1, 11.4.2 and from Corollary 11.4.3. ■

We can finally state the spectral theorem for real symmetric matrices and symmetric linear maps at the same time.

Theorem 11.4.5 *Let V be a real vector space of dimension n with a positive definite scalar product. Let $T : V \longrightarrow V$ be a symmetric linear transformation and let $A \in M_n(\mathbb{R})$ be the symmetric matrix associated with T with respect to an orthonormal basis $\mathcal{B}$.*
Then:

- *T is diagonalizable, and there exists an orthonormal basis $\mathcal{N}$ of eigenvectors of T.*

- *A is diagonalizable by an orthogonal matrix, that is, there exists an orthogonal matrix P, such that $D = P^{-1}AP$ is diagonal.*

Before starting with the proof, we observe that the two statements of the theorem are completely equivalent. The orthonormal basis $\mathcal{N}$ is a basis of mutually perpendicular eigenvectors of T that have norm 1. The existence of this basis of eigenvectors is equivalent to the existence of an orthogonal matrix P, such that $P^{-1}AP$ is diagonal. This matrix has as columns the coordinates of the eigenvectors with respect to the basis $\mathcal{B}$.

Proof. Let λ_1 be a real eigenvalue of T, and let $\mathbf{u}_1 \in V$ be an eigenvector of norm 1 of eigenvalue λ_1. We know that such λ_1 and $\mathbf{u}_1$ exist by Lemma 11.4.1. Let $W_1 = \langle \mathbf{u}_1 \rangle^{\perp}$. Then, we have $\dim(W_1) = n - 1$. Let us now consider the linear map $T_1 = T|_{W_1}$, that is, let us look at the restriction of T_1 to the subspace W_1, then $T_1 : W_1 \longrightarrow V$. For Lemma 11.4.2, since $\mathbf{u}$ is also perpendicular to $T_1(\mathbf{w})$ for each $\mathbf{w} \in W_1$, we have $\mathrm{Im}(T_1) \subseteq W_1 = \langle \mathbf{u}_1 \rangle^{\perp}$, therefore we can write $T_1 : W_1 \longrightarrow W_1$. Let us now repeat all the arguments we have given for $T : V \longrightarrow V$ for the symmetric transformation $T_1 : W_1 \longrightarrow W_1$. Such a transformation has a real eigenvalue λ_2 and a real eigenvector, $\mathbf{u}_2 \in W_1 \subseteq V$ of norm 1. Clearly since T_1 is nothing else than T restricted to W_1, the eigenvalue λ_2 and the corresponding eigenvector $\mathbf{u}_2$ will also be an eigenvalue and an eigenvector of T, respectively. So, reasoning as above, we define $W_2 = \langle \mathbf{u}_2 \rangle^{\perp} \subseteq W_1$, $\dim(W_2) = n - 2$. We note that every vector of W_2 is also perpendicular to $\mathbf{u}_1$ as $W_2 \subseteq W_1$. It is clear that, proceeding in this way, after n steps, we obtain n eigenvectors $\mathbf{u}_1, \ldots, \mathbf{u}_n$ of T of norm 1, which are mutually perpendicular. Therefore, they form an orthonormal basis of eigenvectors and so T is diagonalizable. ■

Example 11.4.6 Let us consider the real matrix

$$A = \begin{pmatrix} 3 & -2 & 4 \\ -2 & 6 & 2 \\ 4 & 2 & 3 \end{pmatrix}.$$

The eigenvalues are $\lambda = 7$ with algebraic multiplicity 2 and $\lambda = -2$ with algebraic multiplicity 1. The eigenspaces are:

$$V_7 = \langle \mathbf{v}_1 = (1, 0, 1), \mathbf{v}_2 = (-1/2, 1, 0) \rangle,$$

$$V_{-2} = \langle \mathbf{v}_3 = (-1, -1/2, 1) \rangle,$$

The matrix Q, which has as columns the eigenvectors of the basis $\{v_1, v_2, v_3\}$, diagonalizes the matrix A, however it is not orthogonal:

$$D = \begin{pmatrix} 7 & 0 & 0 \\ 0 & 7 & 0 \\ 0 & 0 & -2 \end{pmatrix} = Q^{-1}AQ, \qquad Q = \begin{pmatrix} 1 & -1/2 & -1 \\ 0 & 1 & -1/2 \\ 1 & 0 & 1 \end{pmatrix}.$$

If we want to diagonalize A via an orthogonal basis change, we have to orthogonalize the basis of eigenvectors of each eigenspace with the Gram-Schmidt algorithm. The orthogonal matrix we are looking for is:

$$P = \begin{pmatrix} 1/\sqrt{2} & -1/\sqrt{5} & -2/3 \\ 0 & 2/\sqrt{5} & -1/3 \\ 1/\sqrt{2} & 0 & 2/3 \end{pmatrix}.$$

and we have $D = P^{-1}AP$.

11.5   EXERCISES WITH SOLUTIONS

11.5.1 Consider the linear map $T : \mathbb{R}^2 \longrightarrow \mathbb{R}^2$ defined by: $T(e_1) = e_1 + 2e_2$, $T(e_2) = 2e_1 + e_2$. Check that this is a symmetric linear transformation and determine an orthonormal basis $\mathcal{B}$ with respect to which T is associated with a diagonal matrix.

Solution. The matrix associated with T with respect to the canonical basis (in domain and codomain) is:

$$A = \begin{pmatrix} 1 & 2 \\ 2 & 1 \end{pmatrix}.$$

Since A is a symmetric matrix, T is a symmetric linear transformation.
The eigenvalues of A and T are: $-1, 3$. We compute the eigenspaces:

$$V_{-1} = \langle (1, -1) \rangle, \qquad V_3 = \langle (1, 1) \rangle.$$

To get an orthonormal basis, we have to choose eigenvectors with norm equal to 1. Therefore:

$$\mathcal{B} = \{(1/\sqrt{2}, -1/\sqrt{2}), (1/\sqrt{2}, 1/\sqrt{2})\}$$

is a basis that satisfies our requirements.

11.5.2 Let $u = e_1 - 2e_3 \in \mathbb{R}^3$ and consider the linear transformation $\text{proj}_u : \mathbb{R}^3 \longrightarrow \mathbb{R}^3$ that associates to each vector v of $\mathbb{R}^3$ its orthogonal projection on the vector u with respect to the Euclidean scalar product, i.e.

$$\text{proj}_u(v) = \frac{\langle v, u \rangle_e}{\langle u, u \rangle_e} u$$

(see Definition 10.6.2). Verify that this is a symmetric linear transformation and determine an orthonormal basis $\mathcal{B}$ with respect to which proj_u is associated to a diagonal matrix D. Then write D explicitly.

Solution. We determine the A matrix associated with $\text{proj}_\mathbf{u}$ with respect to the canonical basis (in domain and codomain). We have:

$\text{proj}_\mathbf{u}(\mathbf{e}_1) = \frac{<\mathbf{e}_1,\mathbf{u}>_e}{<\mathbf{u},\mathbf{u}>_e}(\mathbf{e}_1 - 2\mathbf{e}_3) = \frac{1}{5}(\mathbf{e}_1 - 2\mathbf{e}_3) = \frac{1}{5}\mathbf{e}_1 - \frac{2}{5}\mathbf{e}_3$

$\text{proj}_\mathbf{u}(\mathbf{e}_2) = \frac{<\mathbf{e}_2,\mathbf{u}>_e}{<\mathbf{u},\mathbf{u}>_e}(\mathbf{e}_1 - 2\mathbf{e}_3) = 0(\mathbf{e}_1 - 2\mathbf{e}_3) = 0$

$\text{proj}_\mathbf{u}(\mathbf{e}_3) = \frac{<\mathbf{e}_3,\mathbf{u}>_e}{<\mathbf{u},\mathbf{u}>_e}(\mathbf{e}_1 - 2\mathbf{e}_3) = -\frac{2}{5}(\mathbf{e}_1 - 2\mathbf{e}_3) = -\frac{2}{5}\mathbf{e}_1 + \frac{4}{5}\mathbf{e}_3.$

Therefore:

$$A = \begin{pmatrix} \frac{1}{5} & 0 & -\frac{2}{5} \\ 0 & 0 & 0 \\ -\frac{2}{5} & 0 & \frac{4}{5} \end{pmatrix}.$$

Since A is a symmetric matrix, $\text{proj}_\mathbf{u}$ is a symmetric linear transformation.

We want to determine an orthonormal basis $\mathcal{B}$ with respect to which $\text{proj}_\mathbf{u}$ is associated with a diagonal matrix. We can proceed as in the previous exercise or observe that $\mathbf{u}$ is an eigenvector of $\text{proj}_\mathbf{u}$ relative to the eigenvalue 1, in fact: $\text{proj}_\mathbf{u}(\mathbf{u}) = \frac{<\mathbf{u},\mathbf{u}>_e}{<\mathbf{u},\mathbf{u}>_e}\mathbf{u} = \mathbf{u}$. Moreover, if $\mathbf{v}$ is any nonzero vector orthogonal to $\mathbf{u}$, i.e. $< \mathbf{v}, \mathbf{u} >_e = 0$, we have that $\text{proj}_\mathbf{u}(\mathbf{v}) = 0\mathbf{u}$, so $\mathbf{v}$ is an eigenvector of $\text{proj}_\mathbf{u}$ relative to the eigenvalue 0.

Let us now consider an orthonormal basis $\mathcal{B}$ obtained by applying the Gram-Schmidt algorithm to basis $\{\mathbf{u}, \mathbf{e}_1, \mathbf{e}_2\}$.

Using the notation of Theorem 10.6.3, we have:

$\mathbf{u}_1 = \mathbf{u},$ $\qquad\qquad\qquad\qquad\qquad f_1 = \frac{1}{\|\mathbf{u}\|}\mathbf{u} = \frac{1}{\sqrt{5}}\mathbf{e}_1 - \frac{2}{\sqrt{5}}\mathbf{e}_3$

$\mathbf{u}_2 = \mathbf{e}_1 - \text{proj}_\mathbf{u}(\mathbf{e}_1) = \frac{4}{5}\mathbf{e}_1 + \frac{2}{5}\mathbf{e}_3, \qquad f_2 = \frac{1}{\|\mathbf{u}_2\|}\mathbf{u}_2 = \frac{2}{\sqrt{5}}\mathbf{e}_1 + \frac{1}{\sqrt{5}}\mathbf{e}_3$

$\mathbf{u}_3 = \mathbf{e}_2 - \text{proj}_\mathbf{u}(\mathbf{e}_2) - \text{proj}_{\mathbf{u}_2}(\mathbf{e}_2) = \mathbf{e}_2, \quad f_3 = \frac{1}{\|\mathbf{u}_3\|}\mathbf{u}_3 = \mathbf{e}_2.$

The vector f_1 is a multiple of $\mathbf{u}$ so it is an eigenvector of $\text{proj}_\mathbf{u}$ of eigenvalue 1, the vectors f_2, f_3 are perpendicular to $\mathbf{u}$ by construction, so they are eigenvectors of $\text{proj}_\mathbf{u}$ of eigenvalue 0. A basis with the required properties is: $\mathcal{B} = \{f_1, f_2, f_3\}$ and the matrix D is:

$$D = \begin{pmatrix} 1 & 0 & 0 \\ 0 & 0 & 0 \\ 0 & 0 & 0 \end{pmatrix}.$$

11.6 SUGGESTED EXERCISES

11.6.1 Say which of the following matrices are orthogonal and give reasons for the answer:

$$a) \begin{pmatrix} 2 & 0 \\ 0 & 1/2 \end{pmatrix}, \quad b) \begin{pmatrix} 1 & 1 \\ 1 & -1 \end{pmatrix}, \quad c) \begin{pmatrix} 0 & 1 & 0 \\ 0 & 0 & -1 \\ 1 & 0 & 0 \end{pmatrix}, \quad d) \begin{pmatrix} -1/\sqrt{5} & 4/\sqrt{45} & -2/3 \\ 2/\sqrt{5} & 2/\sqrt{5} & -1/3 \\ 0 & 5/\sqrt{5} & 2/3 \end{pmatrix}.$$

11.6.2 Consider the matrix:

$$A = \begin{pmatrix} 1/2 & 0 & \sqrt{3}/2 \\ 0 & 1 & 0 \\ a & b & c \end{pmatrix}.$$

If possible, determine values for a, b, c that make it an orthogonal matrix.

11.6.3   a) Determine if there is an angle value t for which the rotation in $\mathbb{R}^2$ described in Example 11.2.7 is a symmetric linear transformation.

  b) Determine the linear transformation that associates to each point of the plane, its symmetric with respect to the x-axis. Establish whether it is an orthogonal linear transformation or not.

  c) Determine whether the orthogonal projection described in Exercise 11.5.2 is an orthogonal linear transformation.

11.6.4 Diagonalize by an orthogonal transformation the following matrices:

$$\begin{pmatrix} 3 & 1 & 1 \\ 1 & 3 & 1 \\ 1 & 1 & 3 \end{pmatrix}, \quad \begin{pmatrix} 2 & -1 \\ -1 & 1 \end{pmatrix}, \quad \begin{pmatrix} 3 & 2 & 0 \\ 2 & 6 & 0 \\ 0 & 0 & 3 \end{pmatrix}, \quad \begin{pmatrix} 2 & 0 & 0 \\ 0 & 1 & -2 \\ 0 & -2 & 4 \end{pmatrix}.$$

11.6.5 Determine the linear transformation $\mathrm{proj}_{\mathbf{u}} : \mathbb{R}^2 \to \mathbb{R}^2$ associating to each vector its projection on the vector $\mathbf{u} = \mathbf{e}_1 - \mathbf{e}_2$. Write explicitly the matrix T associated to it with respect to the canonical basis. Say if it is a symmetric and/or orthogonal linear transformation. Determine (if possible) a diagonal matrix similar to T.

11.6.6 Let $\mathrm{proj}_{\mathbf{u}} : \mathbb{R}^2 \to \mathbb{R}^2$ be the linear transformation associating to any vector its projection on the vector $\mathbf{u} = 3\mathbf{e}_1 - 4\mathbf{e}_2$. Let A be the matrix associated with it with respect to the canonical basis. Determine if there is an orthogonal matrix P, such that $P^{-1}AP = \begin{pmatrix} 0 & 0 \\ 0 & 1 \end{pmatrix}$.

11.7 APPENDIX: THE COMPLEX CASE

In this appendix, we want to give the proof of Lemma 11.4.1, and so we will introduce *hermitian products*, which represent a generalization of scalar products to the complex case. This appendix requires familiarity with complex numbers. We invite the reader to refer to Appendix A.

We define $\mathbb{C}^n = \{(x_1, \dots, x_n) \mid x_i \in \mathbb{C}\}$ as the set of n-tuples of complex numbers. In analogy with what happens for $\mathbb{R}^n$, we can define the sum and the multiplication by a scalar $\lambda \in \mathbb{C}$ so that properties (1) through (8) of Definition 2.3.1 are verified. We thus obtain that $\mathbb{C}^n$ is a *complex vector space* (see Appendix A). All the theory we developed in Chapters (1) through (9) for real vector spaces also holds for complex vector spaces, that is for sets satisfying properties (1) through (8) of Definition 2.3.1, where $\mathbb{C}$ is substituted for $\mathbb{R}$ as the set of scalars.

Let us now see an important generalization of the notion of scalar product.

Definition 11.7.1 Let V be a complex vector space. The function $<,>: V \times V \longrightarrow \mathbb{C}$ is called a *hermitian product* if:

  1. $< \mathbf{u} + \mathbf{u}', \mathbf{v} > = < \mathbf{u}, \mathbf{v} > + < \mathbf{u}', \mathbf{v} >$,
     $< \mathbf{u}, \mathbf{v} + \mathbf{v}' > = < \mathbf{u}, \mathbf{v} > + < \mathbf{u}, \mathbf{v}' >$ for each $\mathbf{u}, \mathbf{u}', \mathbf{v}, \mathbf{v}' \in V$.

2. $< \lambda\mathbf{u}, \mathbf{v} >= \lambda < \mathbf{u}, \mathbf{v} >,$

 $< \mathbf{u}, \mu\mathbf{v} >= \overline{\mu} < \mathbf{u}, \mathbf{v} >$ for each $\mathbf{u}, \mathbf{v} \in V$ and for each $\lambda, \mu \in \mathbb{C}$.

3. $< \mathbf{u}, \mathbf{v} >= \overline{< \mathbf{v}, \mathbf{u} >}$ for each $\mathbf{u}, \mathbf{v} \in V$.

$<,>$ is *non-degenerate* when, if $< \mathbf{u}, \mathbf{v} >= 0$ for each $\mathbf{v} \in V$, then $\mathbf{u} = \mathbf{0}$.
$<,>$ is *positive definite* if $< \mathbf{u}, \mathbf{u} > \geq 0$ for each $\mathbf{u} \in V$ and $< \mathbf{u}, \mathbf{u} >= 0$ if only if $\mathbf{u} = \mathbf{0}$.

The difference between a hermitian product and a scalar product is that for a hermitian product we require the linearity of the function $< \mathbf{u}, \cdot >: V \longrightarrow \mathbb{C}$, but *the antilinearity* of the function $< \cdot, \mathbf{u} >: V \longrightarrow \mathbb{C}$, i.e. for $\mathbf{u} \in V$ fixed we have $< \mathbf{u}, \mu\mathbf{v} >= \overline{\mu} < \mathbf{u}, \mathbf{v} >$.

We are particularly interested in the following example.

Example 11.7.2 In the complex vector space $\mathbb{C}^n$ we define:

$$< (x_1, \ldots, x_n), (x_1', \ldots, x_n') >_h = x_1\overline{x_1'} + \cdots + x_n\overline{x_n'}$$

We leave to the reader to verify this is an hermitian product. This product is called the *standard hermitian product* in $\mathbb{C}^n$. It is immediate to verify that it is positive definite, indeed:

$$< (x_1, \ldots, x_n), (x_1, \ldots, x_n) >_h = x_1\overline{x_1} + \cdots + x_n\overline{x_n} = |x_1|^2 + \cdots + |x_n|^2 \geq 0,$$

and $|x_1|^2 + \cdots + |x_n|^2 = 0$ if and only if $x_1 = x_2 = \cdots = x_n = 0$. Therefore, the standard hermitian product is also non-degenerate and positive definite.
Let us consider the canonical basis $\mathcal{C} = \{\mathbf{e}_1, \ldots, \mathbf{e}_n\}$ of $\mathbb{C}^n$, where $\mathbf{e}_i$ is the vector having 1 in the i-th position and 0 elsewhere. We observe that, as we did for scalar products, to each hermitian product $<,>$ in $\mathbb{C}^n$ we can associate a C matrix, with $c_{ij} =< \mathbf{e}_i, \mathbf{e}_j >$, such that

$$< (x_1, \ldots, x_n), (x_1', \ldots, x_n') >= (x_1, \ldots, x_n)C\begin{pmatrix} \overline{x_1'} \\ \vdots \\ \overline{x_n'} \end{pmatrix}.$$

In the case of the standard hermitian product, the matrix associated with it is the identity matrix I since $< \mathbf{e}_i, \mathbf{e}_j >_h = \delta_{ij}$. It is not difficult to prove, in complete analogy with the case of the scalar products, that a matrix C is associated with a hermitian product if and only if $C = \overline{C}^T$, that is, it coincides with its complex conjugate transpose.

The following observation is crucial in the proof of Lemma 11.4.1.

Observation 11.7.3 Let us consider the complex vector space $\mathbb{C}^n$ with the standard hermitian product, described in the previous example.

If $(x_1, \ldots, x_n), (x'_1, \ldots, x'_n) \in \mathbb{R}^n \subseteq \mathbb{C}^n$, since $\overline{x'_i} = x'_i$ we have that:

$$< (x_1, \ldots, x_n), (x'_1, \ldots, x'_n) >_h \ = x_1 x'_1 + \cdots + x_n x'_n$$

$$=< (x_1, \ldots, x_n), (x'_1, \ldots, x'_n) >_e .$$

In other words, the hermitian product of vectors in $\mathbb{R}^n$ coincides with the usual Euclidean product in $\mathbb{R}^n$.

If A is a matrix with real entries, we have that:

$$< Au, v >_h =< u, A^T v >_h, \qquad \text{for each } u, v \in \mathbb{C}^n. \tag{11.7}$$

In fact, (11.7) is true for $u = e_i$, $v = e_j$ (where the e_i are the vectors of the canonical basis), therefore, by the linearity and antilinearity of the hermitian product, it is not difficult to verify that it is true also for generic vectors $u, v \in \mathbb{C}^n$.

We are therefore ready to state and prove Lemma 11.4.1.

Lemma 11.7.4 *Let $A \in M_n(\mathbb{R})$ be a symmetric matrix. Then A admits a real eigenvalue.*

Proof. A is a matrix with real entries, however, as real numbers are contained in the complex field we also have that $A \in M_n(\mathbb{C})$. By the Fundamental Theorem of algebra (see Appendix A), the characteristic polynomial of A, $\det(A - \lambda I)$, is equal to zero for at least one complex number, $\lambda_0 \in \mathbb{C}$. We want to prove that λ_0 is real, that is $\lambda_0 = \overline{\lambda_0}$. Let $u \in \mathbb{C}^n$ be an eigenvector of eigenvalue λ_0, and let $<, >_h$ be the standard hermitian product in $\mathbb{C}^n$, i.e .:

$$< u, v >_h = u^T \overline{v}, \qquad \text{for all} \quad u, v \in \mathbb{C}^n,$$

where $u = (\alpha_1, \ldots, \alpha_n)^T, v = (\beta_1, \ldots, \beta_n)^T \in \mathbb{C}^n$ are column vectors and $\overline{v} = (\overline{\beta_1}, \ldots, \overline{\beta_n})^T$.

We have:

$$< Au, u >_h = (Au)^T \overline{u} = u^T A^T \overline{u} = u^T A \overline{u} = u^T (\overline{Au}) =< u, Au >_h$$

since $A = A^T$ and, because A is a matrix with real entries, $\overline{A} = A$. Therefore,

$$\lambda_0 < u, u >_h =< Au, u >_h =< u, Au >_h = \overline{\lambda_0} < u, u >_h .$$

Hence:

$$(\lambda_0 - \overline{\lambda_0}) < u, u >_h = 0.$$

From the fact that $u \neq 0$, since it is an eigenvector, it follows that $< u, u >_h \neq 0$, then $\lambda_0 = \overline{\lambda_0}$. ■

In the rest of this appendix, we want to revisit the results we have stated in this chapter for a real vector space with a positive definite scalar product, for the case of a complex vector space with a positive definite hermitian product. As we will see all the main theorems, including the Spectral Theorem, have statements and proofs similar to those seen, which we will therefore leave by exercise.

Reading this part is not necessary for understanding the real case; we include it for completeness.

In analogy with the real case, we can give the following definitions. In the complex case, the notions *unitary* and *hermitian* linear maps or matrices replace the corresponding notions of real symmetric and orthogonal ones, respectively.

Definition 11.7.5 Let V be a complex vector space with a positive definite hermitian product $< , >_h$. We say that a linear transformation $T : V \longrightarrow V$ is *unitary* if

$$< T\mathbf{u}, T\mathbf{v} >_h = < \mathbf{u}, \mathbf{v} >_h .$$

We say that a linear transformation $T : V \longrightarrow V$ is *hermitian* if

$$< T\mathbf{u}, \mathbf{v} >_h = < \mathbf{u}, T\mathbf{v} >_h .$$

We say that a complex coefficient matrix A is *unitary* if $A^{-1} = A^*$, where $A^* = \overline{A^T}$; instead we say it is *hermitian* if $A = A^*$, that is, A coincides with its transpose complex conjugate. We note that these two operations, that is, transposition of A and conjugation of each entry of A, can be interchanged; that is, the result is independent of which one we choose to do first.

It is easy to verify that the hermitian condition on A corresponds to the fact that with respect to the hermitian scalar product in $\mathbb{C}^n$ we have:

$$< A\mathbf{u}, \mathbf{v} >_h = < \mathbf{u}, A\mathbf{v} >_h .$$

Also note that if A is a real symmetric matrix we have immediately that it is also an hermitian matrix. In fact, it satisfies the condition $A = A^*$, as the conjugate complex of a number real is the real number itself.

We can state the analogue of Proposition 11.7.6, whose proof is the same as in the real case.

Given a complex vector space V with an hermitian scalar product $< , >_h$, we say that $\mathbf{u}, \mathbf{v} \in V$ are *perpendicular (orthogonal)* if $< \mathbf{u}, \mathbf{v} >_h = 0$.
If V is finite dimensional and the hermitian scalar product is positive definite, with the same calculations as in Section 10.6 it can be proved that there exists a basis $\mathcal{B}$ of V consisting of vectors of norm 1 such that any two of them are orthogonal, where the norm of a vector $\mathbf{u}$ is defined as $\|\mathbf{u}\| = \sqrt{< \mathbf{u}, \mathbf{u} >_h}$.

Proposition 11.7.6 *Let V be a complex vector space of finite dimension, with a positive definite hermitian product $< , >_h$. Let $T : V \longrightarrow V$ be a linear transformation, and let $\mathcal{B}$ be an orthonormal basis. Then:*

1. *T is hermitian if and only if its associated matrix with respect to the basis $\mathcal{B}$ is an hermitian matrix.*

2. *T is unitary if and only if its associated matrix with respect to the basis $\mathcal{B}$ is a unitary matrix.*

Similarly to the real case, we can also state and prove the following results.

Lemma 11.7.7 *Let A be an hermitian matrix. Then:*

1. *A admits at least one real eigenvalue.*

2. *If **u** is the eigenvector of A, and **w** is perpendicular to **u** then A**w** is perpendicular to **u**.*

3. *Eigenvectors of A relative to distinct eigenvalues are perpendicular to each other.*

Finally, we can state the Spectral Theorem for hermitian linear maps and equivalently for hermitian matrices. The proof is the same as the one for the real case.

Theorem 11.7.8 *Let V be a complex vector space of dimension n with a positive definite hermitian product. Let $T : V \longrightarrow V$ be an hermitian linear transformation and let $A \in M_n(\mathbb{C})$ be the hermitian matrix associated with T with respect to an orthonormal basis $\mathcal{B}$ of V.*
Then:

- *T is diagonalizable and furthermore there exists an orthonormal basis $\mathcal{N}$ of V consisting of eigenvectors of T.*

- *A is diagonalizable by means of a unitary matrix; that is, there exists a unitary matrix P such that $D = P^{-1}AP$ is diagonal.*

CHAPTER 12

Applications of Spectral Theorem and Quadratic Forms

In this chapter, we want to study some consequences of the Spectral Theorem for scalar products and quadratic forms associated with them.

12.1 DIAGONALIZATION OF SCALAR PRODUCTS

Let us now revisit the problem of the basis change for scalar products. We wonder if it is possible to determine a basis for the vector space, such that the matrix associated with a given scalar product is as simple as possible, i.e. diagonal. We have already solved this problem in the case of a positive definite scalar product using Gram-Schmidt Theorem 10.6.3. We are now interested in the more general case, together with some extra conditions on the basis.

Given a vector space V and a scalar product $<,>$, once we fix an ordered basis $\mathcal{A} = \{\mathbf{u}_1, \ldots, \mathbf{u}_n\}$, we can associate to $<,>$ a unique matrix C whose coefficients are $c_{ij} = <\mathbf{u}_i, \mathbf{u}_j>$, as we saw in Chapter 10.

If we choose a different basis, $\mathcal{B} = \{\mathbf{v}_1, \ldots, \mathbf{v}_n\}$, we will have that the matrix associated with the scalar product changes according to the following formula (see (10.5)):

$$C' = I_{\mathcal{B},\mathcal{A}}^T C I_{\mathcal{B},\mathcal{A}}, \tag{12.1}$$

where $I_{\mathcal{B},\mathcal{A}}$ is the basis change matrix between the bases $\mathcal{B}$ and $\mathcal{A}$.

Now suppose we have a vector space V with a positive definite scalar product $<,>_V$. As we saw in Observation 10.6.4, if we fix an orthonormal basis, $<,>_V$ is associated with the identity matrix. Therefore, if we write the vectors of V using the coordinates with respect to the chosen orthonormal basis, we can identify V with $\mathbb{R}^n$ and $<,>_V$ with the standard scalar product.

Now, consider an arbitrary scalar product $<,>$ in V (not necessarily positive definite or non-degenerate). We will see shortly that, thanks to the Spectral Theorem, it is possible choose a basis $\mathcal{N}$, orthonormal with respect to $<,>_V$, such that the matrix associated with the scalar product $<,>$ with respect to the basis $\mathcal{N}$ is diagonal. Hence, $\mathcal{N}$ will be an orthogonal basis (not necessarily orthonormal) also with respect to $<,>$. This will allow us to immediately to determine some fundamental properties of the scalar product $<,>$. For example, we can determine if the product is non-degenerate or positive definite, simply by looking at the signs of the elements on the diagonal (which are, in fact, the eigenvalues) of the matrix associated to $<,>$ with respect to $\mathcal{N}$.

We start with an equivalent statement of the Spectral Theorem.

Theorem 12.1.1 *Let V be a vector space of dimension n with a positive definite scalar product $<,>_V$. Let $<,>$ be another scalar product in V. Then, there exists an orthonormal basis $\mathcal{N}$ for $<,>_V$, which is also orthogonal for $<,>$.*

Proof. By the Gram-Schmidt Theorem 10.6.3, there exists a basis $\mathcal{A}$, which is orthonormal for $<,>_V$. Let C be the matrix associated with $<,>$ with respect to the basis $\mathcal{A}$ and let $T : V \longrightarrow V$ be the linear application associated with C with respect to the basis $\mathcal{A}$ in both the domain and codomain. By the Spectral Theorem, there exists a basis $\mathcal{N}$, orthonormal for the positive definite scalar product $<,>_V$, consisting of eigenvectors of T. If $P = I_{\mathcal{N},\mathcal{A}}$ is the matrix of basis change between $\mathcal{N}$ and $\mathcal{A}$, we have, again by the Spectral Theorem, that P is orthogonal, i.e. $P^{-1} = P^T$. Therefore, since $\mathcal{N}$ is a basis of eigenvectors, we can write:

$$D = P^{-1}CP = P^T CP, \tag{12.2}$$

where D is a diagonal matrix, with the eigenvalues of C on the diagonal.

By formula (12.2), we have that the diagonal matrix D is the matrix associated with the scalar product $<,>$ with respect to the basis $\mathcal{N}$, which is then an orthogonal basis for $<,>$. This concludes the proof. ■

Remark 12.1.2 Formula (12.2) tells us a surprising fact: given a vector space V with a positive definite scalar product and a fixed orthonormal basis $\mathcal{A}$, there exists an orthonormal basis $\mathcal{N}$, such that we can write the basis change formula for a symmetric linear application T as the basis change formula for a scalar product $<,>$ in the same way!

Hence, we can use the theory of diagonalization of linear applications, that we studied in the Chapter 9, to solve the diagonalizability problem for scalar products. We need to keep in mind two things:

1. Unlike what happens for linear applications, scalar products are always diagonalizable. This happens because, with a fixed-ordered basis, a scalar product is associated to a symmetric matrix and the Spectral Theorem guarantees us the diagonalizability of such matrices, via an orthogonal matrix P.

2. Formula (12.2) solves the problem of diagonalizability of a scalar product *using orthogonal transformations (changes of basis)*. If we relax this request it is possible to prove in a more general way the result of the diagonalizability of the scalar products, through Sylvester's Theorem, which, however, will not be

dealt with. We want to emphasize that orthogonal basis changes are particularly useful in applications, especially in physics.

From the previous theorem, we immediately have a corollary, which is very important for applications.

Corollary 12.1.3 *Let $<,>$ be a scalar product in $\mathbb{R}^n$. Then, there exists a basis $\mathcal{N}$, orthonormal for the Euclidean scalar product, such that the matrix associated with $<,>$ is diagonal.*

Let us now see how the above results can be applied to determine if an arbitrary scalar product is positive definite and non-degenerate.

Proposition 12.1.4 *Let V be a vector space of dimension n with a scalar product $<,>$ associated with a diagonal matrix D with respect to a given basis $\mathcal{N}$. Then:*

1. *$<,>$ is non-degenerate if and only if all the elements on the diagonal of D are non zero;*

2. *$<,>$ is positive definite if and only if all elements on the diagonal of D are positive.*

Proof. Given two vectors $\mathbf{u}, \mathbf{v} \in V$, let $(\mathbf{u})_{\mathcal{N}} = (\alpha_1, \ldots, \alpha_n)^T$, $(\mathbf{v})_{\mathcal{N}} = (\beta_1, \ldots, \beta_n)^T$ be their coordinates with respect to the ordered basis $\mathcal{N} = \{\mathbf{w}_1, \ldots, \mathbf{w}_n\}$. Then we have:

$$< \mathbf{u}, \mathbf{v} >= (\mathbf{u})_{\mathcal{N}}^T D (\mathbf{v})_{\mathcal{N}} = \lambda_1 \beta_1 \alpha_1 + \cdots + \lambda_n \beta_n \alpha_n, \qquad (12.3)$$

where $\lambda_1, \ldots, \lambda_n$ are the elements on the diagonal of the diagonal matrix D. Therefore:

$$< \mathbf{w}_i, \mathbf{w}_i >= \lambda_i. \qquad (12.4)$$

Let us prove the claims.

(1). Suppose $<,>$ is non-degenerate and suppose by contradiction that $\lambda_i = 0$ for some i. Then from formula (12.4) it follows that the vector $\mathbf{w}_i$ is orthogonal to any other vector of the space V, and this gives us a contradiction.

Let us now look at the other implication. We assume, by contradiction, that $<,>$ is degenerate; then there is a non zero vector $\mathbf{u}$ orthogonal to all vectors of the space V, in particular to all vectors of $\mathcal{N}$. Let us set $(\mathbf{u})_{\mathcal{N}} = (\alpha_1, \ldots, \alpha_n)$; therefore:

$$< \mathbf{u}, \mathbf{w}_i >= (\mathbf{u})_{\mathcal{N}}^T D (\mathbf{w}_i)_{\mathcal{N}} = \lambda_i \alpha_i = 0$$

for each i. Since by hypothesis $\lambda_i \neq 0$ we have $\alpha_i = 0$ for each i, which gives a contradiction.

Let us see the other implication. By contradiction, we assume that there is a non zero vector $\mathbf{u}$ such that $< \mathbf{u}, \mathbf{u} >\leq 0$. Then, from (12.3), we have:

$$< \mathbf{u}, \mathbf{u} >= \lambda_1 \alpha_1^2 + \cdots + \lambda_n \alpha_n^2 \leq 0.$$

This implies that at least one of the λ_i is negative or null, getting the contradiction.

■

Let us see an example of an application of this result.

Example 12.1.5 In $\mathbb{R}^2$, we consider the scalar product defined as:

$$< (x_1, y_1), (x_2, y_2) > = -4x_1y_1 + 2x_1y_2 + 2x_2y_1 - 4x_2y_2$$

The matrix associated with it with respect to the canonical basis is:

$$C = \begin{pmatrix} -4 & 2 \\ 2 & -4 \end{pmatrix}$$

The Spectral Theorem guarantees us that such a matrix can be diagonalized through an orthogonal change of basis. With an easy calculation, we see that the eigenvalues of C are $\lambda_1 = -2, \lambda_2 = -6$, and its eigenspaces are: $V_{-2} = \langle (1, 1) >, V_{-6} = \langle (1, -1) >$. By the Spectral Theorem we immediately have that

$$\begin{pmatrix} -2 & 0 \\ 0 & -6 \end{pmatrix} = P^{-1}CP = P^TCP =$$

$$= \begin{pmatrix} 1/\sqrt{2} & 1/\sqrt{2} \\ 1/\sqrt{2} & -1/\sqrt{2} \end{pmatrix} \begin{pmatrix} -4 & 2 \\ 2 & -4 \end{pmatrix} \begin{pmatrix} 1/\sqrt{2} & 1/\sqrt{2} \\ 1/\sqrt{2} & -1/\sqrt{2} \end{pmatrix}.$$

By the previous proposition, with respect to the ordered orthonormal basis $\mathcal{N} = ((1/\sqrt{2}, 1/\sqrt{2}), (1/\sqrt{2}, -1/\sqrt{2}))$, the scalar product $< , >$ is associated to the diagonal matrix:

$$D = \begin{pmatrix} -2 & 0 \\ 0 & -6 \end{pmatrix}.$$

We can then conclude that $< , >$ is non-degenerate but not positive definite.

Thanks to the previous example, we can make an easy observation, which is very important for exercises.

Observation 12.1.6 Let $< , >$ be a scalar product in a vector space V of dimension n, and let C be the associated matrix, with respect to any basis of V. Then:

1. $< , >$ is non-degenerate if and only if C has no null eigenvalues.

2. $< , >$ is positive definite if and only if C has no negative or null eigenvalues.

In fact, by Corollary 12.1.3 there exists a basis $\mathcal{N}$, such that the matrix associated to $< , >$ is diagonal. Looking at the proof of Theorem 12.1.1, we see that this matrix has the eigenvalues of C on its diagonal. Hence, our claims descend from Proposition 12.1.4.

Let us see an example.

Example 12.1.7 In $\mathbb{R}^3$ we define the scalar product:

$$< (x_1, y_1), (x_2, y_2) > = -2x_1y_1 + 2x_1y_2 + 2x_2y_1 - x_2y_3 - x_3y_2 + x_3y_3.$$

We wonder if this product is non-degenerate and positive definite.
The matrix associated with $<,>$ with respect to the canonical basis is:

$$C = \begin{pmatrix} -2 & 2 & 0 \\ 2 & 0 & -1 \\ 0 & -1 & 1 \end{pmatrix}.$$

The eigenvalues are:

$$\lambda_1 = 1/2(-3 - \sqrt{13}), \qquad \lambda_2 = 2, \qquad \lambda_3 = 1/2(\sqrt{13} - 3).$$

By the previous observation we can conclude *without further calculations*, that the given product is non-degenerate, but not positive definite, since all eigenvalues are non zero, but the eigenvalue λ_1 is negative.

12.2   QUADRATIC FORMS

In this section we want to discuss quadratic forms. We can use the information we have about scalar products to diagonalize a quadratic form and, in the case of two variables, to draw curves in the plane.

Definition 12.2.1 Let V be a real vector space and let $<,>$ be a scalar product in V. We define the *real quadratic form* q associated with $<,>$ as the function $q : V \longrightarrow \mathbb{R}$ such that $q(\mathbf{v}) = <\mathbf{v}, \mathbf{v}>$.

For example in $\mathbb{R}^n$, given the Euclidean product, the quadratic form associated with it is the function that associates to a vector its norm squared: $q(\mathbf{v}) = \|\mathbf{v}\|^2 = \alpha_1^2 + \cdots + \alpha_n^2$, for each $\mathbf{v} = (\alpha_1, \ldots, \alpha_n)$.

Observation 12.2.2 If q is a quadratic form, then q uniquely determines the scalar product that defines it. Indeed, such product on two arbitrary vectors is given by:

$$< \mathbf{u}, \mathbf{v} >= (1/2)[q(\mathbf{u} + \mathbf{v}) - q(\mathbf{u}) - q(\mathbf{v})]. \tag{12.5}$$

So that:
$$(1/2)[q(\mathbf{u} + \mathbf{v}) - q(\mathbf{u}) - q(\mathbf{v})] =$$

$$= (1/2)[< \mathbf{u} + \mathbf{v}, \mathbf{u} + \mathbf{v} > - < \mathbf{u}, \mathbf{u} > - < \mathbf{v}, \mathbf{v} >]$$

$$= (1/2)[2 < \mathbf{u}, \mathbf{v} >].$$

Definition 12.2.3 Given a quadratic form q on a vector space V we say that q is *non-degenerate*, or *positive definite*, if the scalar product associated with it has the same properties.

Given a quadratic form q, and a basis $\mathcal{B}$ for V, we can therefore write:

$$q(\mathbf{v}) = (\mathbf{v})_{\mathcal{B}}^T C(\mathbf{v})_{\mathcal{B}},$$

where C is the matrix associated with the scalar product corresponding to q, with respect to the basis $\mathcal{B}$, and $(\mathbf{v})_{\mathcal{B}}$ represents the column of the coordinates of $\mathbf{v}$ with respect to the basis $\mathcal{B}$.

Observation 12.2.4 In $\mathbb{R}^n$ consider the following function $q : \mathbb{R}^n \longrightarrow \mathbb{R}$:

$$q(x_1, \ldots, x_n) = a_{11}x_1^2 + a_{12}x_1x_2 + a_{13}x_1x_3 + \ldots a_{1n}x_1x_n +$$

$$+ a_{22}x_2^2 + a_{23}x_2x_3 + \cdots + a_{2n}x_2x_n + \cdots + a_{nn}x_n^2.$$

We construct the symmetric matrix C as follows:

$$C = \begin{pmatrix} a_{11} & \frac{a_{12}}{2} & \frac{a_{13}}{2} & \cdots & \frac{a_{1n}}{2} \\ \frac{a_{12}}{2} & a_{22} & \frac{a_{23}}{2} & \cdots & \frac{a_{2n}}{2} \\ \cdots & & & & \\ \frac{a_{1n}}{2} & \frac{a_{2n}}{2} & \frac{a_{3n}}{2} & \cdots & a_{nn} \end{pmatrix}. \tag{12.6}$$

It is easy to verify that

$$q(x_1, \ldots, x_n) = (x_1, \ldots, x_n)C \begin{pmatrix} x_1 \\ \vdots \\ x_n \end{pmatrix}.$$

So the function q represents a quadratic form in $\mathbb{R}^n$ and we can immediately check using (12.6) that every quadratic form in $\mathbb{R}^n$ is of this form.

Let us see an example.

Example 12.2.5 Consider $q : \mathbb{R}^3 \longrightarrow \mathbb{R}$ given by: $q(x, y, z) = x^2 + 2xy + 3zy - 2z^2$. We have that q is a quadratic form, and the matrix associated with it is given by:

$$q(x, y, z) = \begin{pmatrix} x & y & z \end{pmatrix} \begin{pmatrix} 1 & 1 & 0 \\ 1 & 0 & 3/2 \\ 0 & 3/2 & -2 \end{pmatrix} \begin{pmatrix} x \\ y \\ z \end{pmatrix}.$$

We leave to the reader to verify the above equality.

The Spectral Theorem and its version for scalar products given by Theorem 12.1.1 have an immediate consequence for quadratic forms.

Corollary 12.2.6 (Principal Axes Theorem). *Let $q : \mathbb{R}^n \longrightarrow \mathbb{R}$ be a quadratic form associated with the matrix C with respect to the canonical basis of $\mathbb{R}^n$. Then there exists an orthonormal basis $\mathcal{N}$ of $\mathbb{R}^n$, consisting of eigenvectors of C, such that the matrix associated with q takes a diagonal form. We can therefore write:*

$$q(x_1, \ldots, x_n) = \lambda_1 x_1^2 + \cdots + \lambda_n x_n^2,$$

where $(x_1, \ldots, x_n)$ are the coordinates with respect to basis $\mathcal{N}$ and $\lambda_1, \ldots, \lambda_n$ are the eigenvalues of the matrix C.

We conclude this section with a fundamental definition for the classification of real quadratic forms. Despite of its importance, we shall not develop this topic any further here.[1]

---

[1] Sylvester's Theorem, which we do not study here, states that two quadratic forms, or equivalently two scalar products in a finite dimensional vector space, coincide up to a change of basis if and only if they have the same signature.

Definition 12.2.7 Let q be a quadratic form on $\mathbb{R}^n$. We define the *signature* of q as the pair (r, s), where r and s are the number of positive and negative eigenvalues respectively, of the matrix C associated with q with respect to the canonical basis, each counted with multiplicity.

We note that, in this definition, instead of the canonical basis, we may choose any orthonormal basis to determine C. In fact, we know that all matrices associated with q have the same eigenvalues.

12.3 QUADRATIC FORMS AND CURVES IN THE PLANE

As an application of the results on quadratic forms, we want to describe the curve consisting of the points satisfying the equation

$$q(x, y) = c, \tag{12.7}$$

in the plane, where q is a non-degenerate quadratic form and $c \in \mathbb{R}$ is a positive constant.

Let us start with considering the particular case

$$q(x, y) = \lambda_1 x^2 + \lambda_2 y^2 \tag{12.8}$$

with $\lambda_1, \lambda_2 \neq 0$. We immediately see that if $\lambda_1, \lambda_2 < 0$, no point of the plane has coordinates that satisfy (12.7). We divide the other cases according to the sign of λ_1 and λ_2 obtaining the following classification:

- $\lambda_1, \lambda_2 > 0$ ellipse;

- $\lambda_1 < 0, \lambda_2 > 0$ (or $\lambda_1 > 0, \lambda_2 < 0$) hyperbola.

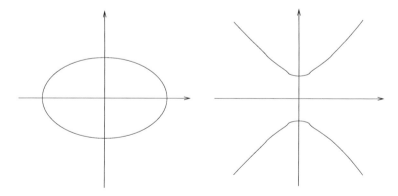

Thanks to the Principal Axis Theorem 12.2.6, we can treat the general case as one of these two geometric figures. We will therefore say that a quadratic form is in *canonical form*, if it takes the expression (12.8). Let us see an example.

***Example* 12.3.1** We want to draw the curve in the plane whose points have coordinates satisfying the equation: $5x^2 - 4xy + 5y^2 = 48$.

The matrix associated with the quadratic form $q(x, y) = 5x^2 - 4xy + 5y^2$ is:

$$A = \begin{pmatrix} 5 & -2 \\ -2 & 5 \end{pmatrix}$$

The eigenvalues of A are 3 and 7, the corresponding eigenvectors of unit length are $\mathbf{u}_1 = (1/\sqrt{2}, 1/\sqrt{2})$, $\mathbf{u}_2 = (-1/\sqrt{2}, 1/\sqrt{2})$. Using the coordinates with respect to the basis $\mathcal{B} = \{\mathbf{u}_1, \mathbf{u}_2\}$, we have $q(x', y') = 3(x')^2 + 7(y')^2$. So $q(x, y) = 48$ is an ellipse that we can immediately draw:

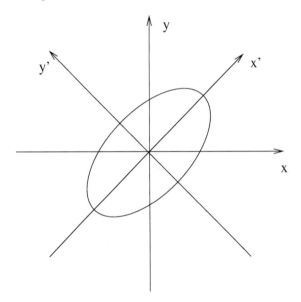

Let us now look at another example related to hyperbola.

Example 12.3.2 We want to draw the curve in the plane whose points have coordinates satisfying the equation: $x^2 - 8xy - 5y^2 = 16$. The matrix associated with the quadratic form $q(x, y) = x^2 - 8xy - 5y^2$ is:

$$A = \begin{pmatrix} 1 & -4 \\ -4 & -5 \end{pmatrix}.$$

The eigenvalues of A are -7 and 3, the corresponding eigenvectors of unit length are $\mathbf{u}_1 = (1/\sqrt{5}, 2/\sqrt{5})$, $\mathbf{u}_2 = (-2/\sqrt{5}, 1/\sqrt{5})$. Using coordinates with respect to the basis $\mathcal{B} = \{\mathbf{u}_1, \mathbf{u}_2\}$, we have $q(x', y') = -7(x')^2 + 3(y')^2$. So $q(x, y) = 16$ is a hyperbola, which we can immediately draw, similarly to what we have done previously.

12.4   EXERCISES WITH SOLUTIONS

12.4.1 Consider the quadratic form in $\mathbb{R}^3$ given by:

$$q(x, y, z) = x^2 + 2xz - y^2 - 2yz.$$

Say if it is non-degenerate, positive definite and give its signature. Write the scalar product associated with it with respect to the canonical basis.

Also, determine a basis with respect to which this scalar product is associated with a diagonal matrix.

Solution. First we write the matrix associated with the given quadratic form, in the canonical basis:

$$C = \begin{pmatrix} 1 & 0 & 1 \\ 0 & -1 & -1 \\ 1 & -1 & 0 \end{pmatrix}.$$

Let us compute the eigenvalues:

$$\lambda_1 = -\sqrt{3}, \qquad \lambda_2 = \sqrt{3}, \qquad \lambda_3 = 0.$$

The quadratic form is degenerate as one of the eigenvalues of the associated matrix is equal to zero. It is not positive definite as it is degenerate. The signature is $(1,1)$.

The scalar product associated with q with respect to the canonical basis is given by:

$$< (x,y,z), (x',y',z') >= (x,y,z) \begin{pmatrix} 1 & 0 & 1 \\ 0 & -1 & -1 \\ 1 & -1 & 0 \end{pmatrix} \begin{pmatrix} x' \\ y' \\ z' \end{pmatrix}.$$

A basis $\mathcal{N}$ with respect to which the matrix C is diagonal consists of normalized eigenvectors of C. Indeed if we compute the eigenvectors:

$$\mathbf{v}_1 = (-(2 - \sqrt{3})/(-1 + \sqrt{3}), 1/(-1 + \sqrt{3}), 1)$$

$$\mathbf{v}_2 = (-(-2 - \sqrt{3})/(1 + \sqrt{3}), -1/(1 + \sqrt{3}), 1)$$

$$\mathbf{v}_3 = (-1, -1, 1).$$

we have $\mathcal{N} = (\mathbf{v}_1/\|\mathbf{v}_1\|, \mathbf{v}_2/\|\mathbf{v}_2\|, \mathbf{v}_3/\|\mathbf{v}_3\|)$.

12.4.2 Consider the curve consisting of the points in the plane that satisfy the equation:

$$3x^2 - 2xy - y^2 = 1.$$

Determine what curve it is and give a sketch in the cartesian plane.

Solution. The matrix associated with the quadratic form $q(x,y) = 3x^2 - 2xy - y^2$ with respect to the canonical basis is:

$$C = \begin{pmatrix} 3 & -1 \\ -1 & -1 \end{pmatrix}.$$

Let us compute the eigenvalues:

$$\lambda_1 = 1 + \sqrt{5}, \qquad \lambda_2 = 1 - \sqrt{5}.$$

We can see immediately, without the need for further calculations, that it is a hyperbola, whose canonical form is given by:

$$q(x',y') = (1 + \sqrt{5})(x')^2 + (1 - \sqrt{5})(y')^2.$$

However, if we want to draw it, it is necessary to compute the eigenvectors, as their directions correspond to the x' and y' axes of the new coordinate system. We note that in this case it is not necessary to normalize the eigenvectors, that is to divide them by the norm, as we are only interested in the direction of the axes and not in the basis change.

Let us compute the eigenvectors:

$$\mathbf{v}_1 = (-2 - \sqrt{5}, 1), \qquad \mathbf{v}_2 = (-2 + \sqrt{5}, 1).$$

We can then draw the hyperbola. It is useful to compute the intersections with the axes x' and y': for $y' = 0$, $x' = \pm 1/(1 + \sqrt{5})$.

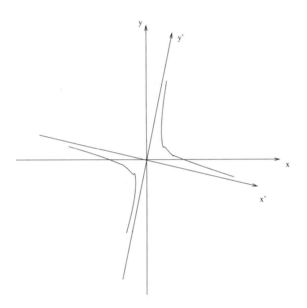

12.5   SUGGESTED EXERCISES

12.5.1 Given the quadratic form $q(x_1, x_2) = 5x_1^2 - 4x_1x_2 + 5x_2^2$.

a) Write the matrix associated with it with respect to the canonical basis.

b) Write the quadratic form in canonical form $q(x_1, x_2) = ax_1^2 + bx_2^2$ for appropriate a and b.

c) Draw the curve described by the equation $q(x_1, x_2) = 48$ in the cartesian plane.

12.5.2 Given the quadratic form $q(x_1, x_2) = 3x_1^2 + 2x_2^2 + x_3^2 + 4x_1x_2 + 4x_2x_3$.

a) Write the matrix A associated with it with respect to the canonical basis.

b) Say if q is positive definite.

c) Find (if possible) a matrix P such that $D = P^{-1}AP$ is diagonal.

d) Write the quadratic form q_1 associated with D and establish the relation between q_1 and q.

12.5.3 Given the matrix:

$$C = \begin{pmatrix} 4 & 2 & 3 \\ 2 & 0 & 2 \\ 3 & 2 & 4 \end{pmatrix}$$

a) Write the scalar product and the quadratic form associated with it with respect to the canonical basis of $\mathbb{R}^3$.

b) Determine whether the scalar product in (a) is positive definite and/or non-degenerate.

c) Compute the signature of the quadratic form in (a).

d) Determine a basis in which the scalar product given in (a) is associated with a diagonal matrix.

[Help: 8 and -1 are eigenvalues of C.]

12.5.4 Consider the quadratic form: $q(x, y) = x^2 + 5xy$.

1) Write the scalar product associated with it and determine if it is non-degenerate or positive definite. Compute the signature of q.

2) Given the curve in the plane consisting of the points with coordinates satisfying the equation $x^2 + 5xy = 1$, say what curve it is.

12.5.5 Draw the curve $x_1^2 - 8x_1x_2 - 5x_2^2 = 16$.

12.5.6 Given the curve described by the equation

$$5x^2 + 5y^2 - 6xy - 16 = 0,$$

find the canonical form and give a drawing of it.

12.5.7 Given the equation:

$$x^2 - 4xy + y^2 = 4$$

I) Say which curve it describes.
II) Determine its canonical form.
III) Draw the curve in the plane.

12.5.8 Given the curve described by the equation $x^2 + y^2 - 16xy = 1$;

1) Find its canonical form and give a drawing.

2) Write the scalar product associated with it and say if it is non-degenerate and if it is positive definite.

CHAPTER 13

Lines and Planes

In this chapter, we introduce three-dimensional geometry in an elementary way, and we focus on the study of lines and planes in $\mathbb{R}^3$.

Hence, we need to introduce the concept of *dot product* and *cross product* in $\mathbb{R}^3$. Since this chapter is self-contained, we discuss such notions without any reference to definitions in Chapters 10, 11, 12. We warn the reader is that the dot product defined below is an example of the more general notion of standard scalar product as defined in Chapter 11.

13.1 POINTS AND VECTORS IN $\mathbb{R}^3$

Consider $\mathbb{R}^3$, the set of ordered triples of real numbers:

$$\mathbb{R}^3 = \{(x, y, z) \mid x, y, z \in \mathbb{R}\}$$

We can represent the elements of $\mathbb{R}^3$ as points, where:

- We have a point O, called the *origin*, corresponding to the element $(0, 0, 0)$.

- We choose three lines through O perpendicular to each other, and we call them *coordinate axes*, denoted respectively as the x, y and z *axis*. We usually think of the x and y axes as *horizontal* and to the z axis as *vertical* (see Fig. 13.1).

- We set an orientation of the axes according to the *right-hand rule* (see also Fig. 13.1). We first choose a direction for the x-axis and one for the y-axis and we call these *positive directions*. For the z axis the positive direction is identified as follows: if we wrap the fingers of the right hand (pointer, middle, ring finger, little finger) around the z-axis from the x-axis in the positive direction, to the y-axis in the positive direction, the thumb points in the direction that we call positive for the z axis.

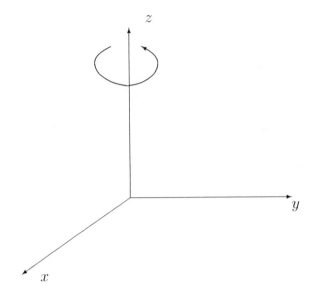

Fig. 13.1

We can then associate with a point P the three ordered real numbers (a, b, c) that we call *coordinates* of P (see Fig. 13.2). From now on, we will identify $\mathbb{R}^3$ with the points of space through this representation.

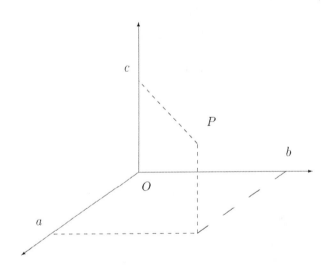

Fig. 13.2

Given the points $P = (a, b, c)$ and $Q = (a', b', c') \in \mathbb{R}^3$, we can define the distance between them as:

$$d(P, Q) = \sqrt{(a - a')^2 + (b - b')^2 + (c - c')^2}. \tag{13.1}$$

We leave to the reader as an easy exercise to prove that actually $d(P, Q)$ represents the length of the segment which connects P with Q.

We now define the notion of *vector*, which is extremely important for our discussion. In this appendix we will use round brackets for the points of $\mathbb{R}^3$ and square brackets for vectors, in order to mark the difference between these two notions.

Given a point $P = (a, b, c) \in \mathbb{R}^3$, we define the *position vector* of P as:

$$\mathbf{v} = \overrightarrow{OP} = [a, b, c].$$

We represent the position vector as an arrow from the origin with its tip in P.

Given the points $P = (a, b, c), Q = (a', b', c') \in \mathbb{R}^3$, we define the vector $\overrightarrow{PQ}$ like:

$$\mathbf{w} = \overrightarrow{PQ} = [((a - a'), (b - b'), (c - c')].$$

We can represent the vector $\mathbf{w}$ as an arrow starting from the origin and parallel to the segment PQ.

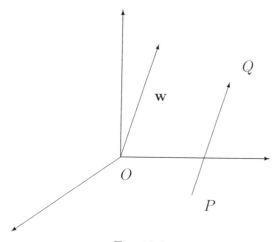

Fig. 13.3

Given two vectors $\mathbf{u} = [u_1, u_2, u_3]$, $\mathbf{v} = [v_1, v_2, v_3]$ we can define their sum as follows:

$$\mathbf{u} + \mathbf{v} = [u_1 + v_1, u_2 + v_2, u_3 + v_3].$$

Similarly we can define multiplication by a scalar, i.e. by a real number λ:

$$\lambda \mathbf{u} = [\lambda u_1, \lambda u_2, \lambda u_3].$$

The two operations of sum between vectors and multiplication of a vector by a scalar equip the set of vectors with the structure of vector space, that is, the 8 properties of Definition 2.3.1 of Chapter 2 apply.

We leave the easy verification to the reader.

13.2   SCALAR PRODUCT AND VECTOR PRODUCT

We now want to define the dot product, also called scalar product between two vectors. Let $\mathbf{u} = [u_1, u_2, u_3]$, $\mathbf{v} = [v_1, v_2, v_3]$ be two vectors. We define their *dot product* as:

$$\mathbf{u} \cdot \mathbf{v} = u_1 v_1 + u_2 v_2 + u_3 v_3.$$

The dot product has the following properties, which we leave as an easy exercise for the reader. For each vector $\mathbf{u}, \mathbf{v}, \mathbf{w}$ and for each scalar λ we have:

- Commutativity:

$$\mathbf{u} \cdot \mathbf{v} = \mathbf{v} \cdot \mathbf{u}$$

- Compatibility with multiplication by a scalar:

$$(\lambda \mathbf{u}) \cdot \mathbf{v} = \lambda (\mathbf{u} \cdot \mathbf{v}) = \mathbf{u} \cdot (\lambda \mathbf{v})$$

- Distributivity:

$$\mathbf{u} \cdot (\mathbf{v} + \mathbf{w}) = \mathbf{u} \cdot \mathbf{v} + \mathbf{u} \cdot \mathbf{w}$$

Note that $\sqrt{\mathbf{u} \cdot \mathbf{u}} = \sqrt{u_1^2 + u_2^2 + u_3^2}$ represents the distance of the point $P = (u_1, u_2, u_3)$ from the origin and therefore represents the length of the segment OP. We define as *length* of the vector $\mathbf{u}$ the number $\sqrt{\mathbf{u} \cdot \mathbf{u}}$, which we also denote as $\|\mathbf{u}\|$.

Let us now remind the reader of an elementary result of Euclidean geometry, the *cosine theorem*, which allows us to find the length of an edge of a triangle with vertices A, B, C, knowing the lengths of the other two edges and the angle θ between them (see Fig. 13.4). This theorem states that:

$$\overline{BC}^2 = \overline{AC}^2 + \overline{AB}^2 - 2 \cdot \overline{AC} \cdot \overline{AB} \cos\theta,$$

where θ is the angle corresponding to the vertex C.

This theorem allows us to prove a result that is fundamental for our discussion.

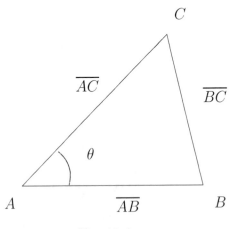

Fig. 13.4

Theorem 13.2.1 *Let $\mathbf{u}$ and $\mathbf{v}$ be two nonzero vectors and let θ be the angle having as sides the half-lines determined by the two vectors. Then*

$$\mathbf{u} \cdot \mathbf{v} = \|\mathbf{u}\| \, \|\mathbf{v}\| \cos\theta.$$

Proof. With an easy calculation, we have that:

$$\|\mathbf{u} - \mathbf{v}\|^2 = (\mathbf{u} - \mathbf{v}) \cdot (\mathbf{u} - \mathbf{v}) = \|\mathbf{u}\|^2 + \|\mathbf{v}\|^2 - 2\mathbf{u} \cdot \mathbf{v}. \qquad (13.2)$$

By the cosine theorem, if we consider the triangle with a vertex in the origin and sides of length $\|\mathbf{u}\|$, $\|\mathbf{v}\|$ and $\|\mathbf{u} - \mathbf{v}\|$, we immediately have that:

$$\|\mathbf{u} - \mathbf{v}\|^2 = \|\mathbf{u}\|^2 + \|\mathbf{v}\|^2 - \|\mathbf{u}\|\,\|\mathbf{v}\|^2 \cos(\theta). \qquad (13.3)$$

Equalities (13.2) and (13.3) immediately give us our result. ■

From the previous theorem, we can easily obtain the following corollary, which establishes when two vectors are perpendicular, that is, when the angle between them is $\pi/2$. This result will be very useful for exercises.

Corollary 13.2.2 *Two vectors* $\mathbf{u}$ *and* $\mathbf{v}$ *are perpendicular to each other if and only if* $\mathbf{u} \cdot \mathbf{v} = 0$.

Let us now turn to another extremely important product for our discussion: the *cross product*, also called *vector product*. Let $\mathbf{u} = [u_1, u_2, u_3]$, $\mathbf{v} = [v_1, v_2, v_3]$ be two vectors. We define their vector product as:

$$\mathbf{u} \times \mathbf{v} = (u_2 v_3 - u_3 v_2, u_3 v_1 - u_1 v_3, u_1 v_2 - u_2 v_1).$$

The vector product has the following properties that we leave for exercise. For each vector $\mathbf{u}, \mathbf{v}, \mathbf{w}$ and for each scalar λ:

- The vector product of a vector by itself is the zero vector.

$$\mathbf{u} \times \mathbf{u} = \mathbf{0}$$

- Anticommutativity

$$\mathbf{u} \times \mathbf{v} = -(\mathbf{v} \times \mathbf{u})$$

- Distributivity with respect to addition

$$\mathbf{u} \times (\mathbf{v} + \mathbf{w}) = (\mathbf{u} \times \mathbf{v}) + (\mathbf{u} \times \mathbf{w})$$

- Compatibility with multiplication by a scalar:

$$(\lambda \mathbf{u}) \times \mathbf{v} = \mathbf{u} \times (\lambda \mathbf{v}) = \lambda (\mathbf{u} \times \mathbf{v})$$

We conclude the section with a result of great importance for the exercises. It is an immediate consequence of Corollary 13.2.2.

Proposition 13.2.3 *The cross product between two vectors* $\mathbf{u}$ *and* $\mathbf{v}$ *is perpendicular to both* $\mathbf{u}$ *and* $\mathbf{v}$.

13.3   LINES IN $\mathbb{R}^3$

In the space $\mathbb{R}^3$, we want to describe the points laying on a straight line r, using equations. We require our line to pass through a point $P_0 = (x_0, y_0, z_0)$, we ask that its direction is determined by the vector $\mathbf{v} = [v_1, v_2, v_3]$. In short, with the help of a drawing, we can immediately write the equation for $P = (x, y, z)$, the generic point of the line r:

$$\overrightarrow{OP} = \overrightarrow{OP_0} + t[v_1, v_2, v_3]. \tag{13.4}$$

Therefore:

$$[x, y, z] = [x_0, y_0, z_0] + t[v_1, v_2, v_3]. \tag{13.5}$$

We call equation (13.5) a *vector parametrization or vector equation* of the line r.

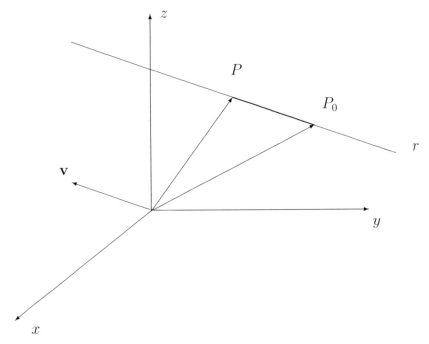

In general, the line r is the set points with coordinates $(x, y, z) \in \mathbb{R}^3$ expressed as:

$$\begin{cases} x = x_0 + tv_1 \\ y = y_0 + tv_2 \\ z = z_0 + tv_3 \end{cases} \tag{13.6}$$

as the parameter $t \in \mathbb{R}$ changes. The equations in (13.6) are called *parametric equations* of the line r. The vector $\mathbf{v}$ is called the *direction vector* of r. Note that the vector $\mathbf{v}$ identifies the *direction* of r; however, we can use any nonzero vector, multiple of $\mathbf{v}$, to define the same line.

Let us see a concrete example.

Example 13.3.1 We want to write parametric equations of the line r through the point $P_0 = (1, 0 - 1)$ with direction given by the vector $\mathbf{v} = [2, 1, -1]$. We also want to know if r is parallel to the line r' given by parametric equations $[x, y, z] = [2, 0, -3] + t'[-4, -2, 2]$. Substituting in formula (13.6), we have immediately the parametric equations of the line:

$$\begin{cases} x = 1 + 2t \\ y = t \\ z = -1 - t. \end{cases} \tag{13.7}$$

The two lines r and r' have the same direction, since the direction vector of r, $\mathbf{v} = [2, 1, -1]$, is a multiple of the direction vector of r', $\mathbf{v}' = [-4, -2, 2]$. Therefore, the two lines are parallel or they coincide. To establish their mutual position it is sufficient to check if the point P_0 belongs to the line r', that is, if it exists a value of the parameter t' such that $[1, 0, -1] = [2, 0, -3] + t'[-4, -2, 2]$. We leave to the reader to verify that this value does not exist. Therefore, the two given lines are parallel and they do not coincide.

Let us look at another example.

Example 13.3.2 We want to write parametric equations of the line $\hat{r}$ through the points $P_1 = (3, 1, -2)$ and $P_2 = (5, 2, -3)$. A direction of $\hat{r}$ is obtained by taking the difference between the coordinates of P_2 and those of P_1: $\mathbf{v} = [2, 1, -1]$. So $\hat{r}$ is given by:

$$\begin{cases} x = 3 + 2\hat{t} \\ y = 1 + \hat{t} \\ z = -2 - \hat{t}, \end{cases} \tag{13.8}$$

where we have chosen $P_0 = P_1$ in formula (13.6) (we very well could have chosen P_2 instead).

We now ask whether the line $\hat{r}$ is parallel or coincident with the line r of the previous example, since they have the same direction. A quick calculation shows that Q belongs to r, and therefore the two lines are coincident.

These examples show that a parametric form of a given line is not unique: we can in fact change the point $P_0 = (x_0, y_0, z_0)$ used in the representation (13.6), choosing it arbitrarily among all the infinite points of the line, or we can multiply the parameter t for an arbitrary nonzero constant: in both the cases the line does not change, even though its parametric equations can take a different form.

Let us now see an equivalent way of describing the points of a line in $\mathbb{R}^3$ without using a parameter. For any set of parametric equations of a line r, it is always possible to get the parameter t from one of the equations and, replacing it in the other two, we obtain a linear system of two equations in the unknowns x, y, z. Such equations are called *Cartesian equations* of the line r. It is obvious that if the coordinates of a point verify parametric equations of the line r then they also verify the Cartesian equations, the vice-versa will be clear at the end of next section. In fact, we will see that the two linear equations of the system we obtain represent two planes both containing the line r. The solutions of the linear system are the points that lie in the intersection of two planes, that is, the points of a line, and this line must be indeed r. Let us see an example.

Example 13.3.3 We want to write the line r of the previous example in Cartesian form (see 13.3.1). In this case, it is very simple; since $t = y$, just substitute y instead of t directly in the other equations:

$$\begin{cases} x = 1 + 2y \\ z = -1 - y \end{cases} \tag{13.9}$$

So $x - 2y - 1 = 0$ and $y + z + 1 = 0$ are the Cartesian equations of the line r. We can then think of the points of the line as the set of solutions of the linear system (13.9).

In the next section, we will see how the previous example can be reinterpreted to see a line as the intersection of two planes in $\mathbb{R}^3$.

13.4 PLANES IN $\mathbb{R}^3$

We now want to determine the equation that describes the points of a plane perpendicular to a given line with direction $\mathbf{n} = [a, b, c]$ and passing through a fixed point $P_0 = (x_0, y_0, z_0)$. By Corollary 13.2.2, we have that two vectors are perpendicular if and only if their dot product is zero. Therefore, the set of points of the plane containing $P_0 = (x_0, y_0, z_0)$, perpendicular to the vector $\mathbf{n}$, is obtained by imposing that the generic point $P = (x, y, z)$ on the plane satisfies the equation:

$$\mathbf{n} \cdot \overrightarrow{P_0 P} = 0.$$

We write this equation as:

$$a(x - x_0) + b(y - y_0) + c(z - z_0) = 0. \tag{13.10}$$

Equation (13.10) is called a *Cartesian equation* of the given plane and the vector $\mathbf{n}$ is said *normal vector* to the plane.

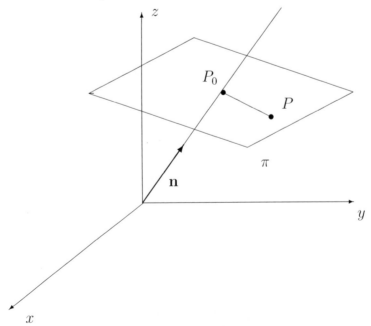

Example 13.4.1 We want to determine the plane through the point $P_0 = (1, -1, 2)$ and with normal vector $\mathbf{n} = [2, -3, -1]$. Substituting in equation (13.10) we immediately have:

$$2(x - 1) - 3(y + 1) - (z - 2) = 0.$$

Thus the plane consists of all the points (x, y, z) that satisfy the equation $2x - 3y - z = 3$.

It is easy to verify that, on the other hand, any linear equation of the type:

$$ax + by + cz = d$$

represents a plane perpendicular to the vector $\mathbf{n} = [a, b, c]$.

Similarly to what we saw in the previous section, we can also write parametric equations that represent the points belonging to a plane containing the point $P_0 = (x_0, y_0, z_0)$ and with directions given by $\mathbf{u} = [u_1, u_2, u_3]$ and $\mathbf{v} = [v_1, v_2, v_3]$, which are not one multiple of the other.

We write:

$$[x, y, z] = [x_0, y_0, z_0] + t[u_1, u_2, u_3] + s[v_1, v_2, v_3]$$

or more extensively:

$$\begin{cases} x = x_0 + tu_1 + sv_1 \\ y = y_0 + tu_2 + sv_2 \\ z = z_0 + tu_3 + sv_3. \end{cases} \tag{13.11}$$

The equations in (13.11) are called *parametric equations* of the given plane. Note that while the parametric equations of a line (see (13.6)) depend on a single parameter t, the parametric equations of a plane depend on two parameters t and s. Intuitively, this corresponds to the fact that, while along a line we can move in one direction only, in the plane we have two degrees of freedom, and therefore we can move according to all linear combinations of the two vectors $\mathbf{u}$ and $\mathbf{v}$.

In a similar way to what we have seen for the lines, we can transform parametric equations of a plane deriving t and s from two equations and replacing them in the third, thus obtaining a single linear equation, which is precisely the equation of a plane. It is uniquely determined up to a multiple.

Let us see some examples.

Example 13.4.2 We want to determine, in parametric form and in Cartesian form, the plane passing through the point $P = (3, 1, 0)$ with normal vector $\mathbf{n} = [1, 2, -5]$. From fomula (13.10) we immediately obtain the equation of the plane in Cartesian form:

$$1(x - 3) + 2(y - 1) - 5z = 0 \qquad \Longrightarrow \qquad x + 2y - 5z = 11.$$

To obtain the equations of the plane in parametric form we proceed in a similar way to what we have seen for the lines: we assign to two variables (arbitrarily) the values of the parameters t and s and replace them in the Cartesian equation:

$$\begin{cases} x = 11 - 2t + 5s \\ y = t \\ z = s. \end{cases}$$

Let us look at another more complicated example.

Example 13.4.3 We want to determine, in parametric form and in Cartesian form, the plane passing through the three points: $P = (1, 0, -1)$, $Q = (2, 2, 1)$, $R = (4, 1, 2)$. First we determine the vectors giving the directions:

$$\overrightarrow{PQ} = Q - P = [1, 2, 2], \qquad \overrightarrow{PR} = R - P = [3, 1, 3].$$

Note that we could very well have chosen for instance $\overrightarrow{PR}$, $\overrightarrow{QR}$: the plane thus obtained would be the same, we invite the student to verify this fact by doing the calculations. At this point a parametric form is immediate:

$$\begin{cases} x = 1 + t + 3s \\ y = 0 + 2t + s \\ z = -1 + 2t + 3s. \end{cases} \tag{13.12}$$

To determine the Cartesian form, we could obtain t and s from two equations and replace them into the third. However, from formula (13.10) we know that, to determine the plane, it is enough know the coordinates of a point in the plane and a vector normal to it. To determine a vector normal to the plane, we can take the *vector product* of the two direction vectors. Such product is in fact always perpendicular to both vectors. Let us see the calculation:

$$\mathbf{n} = \overrightarrow{PQ} \times \overrightarrow{PR} = [4, 3, -5].$$

Therefore the plane is immediately given by the Cartesian equation:

$$4(x - 1) + 3y - 5(z + 1) = 0.$$

i.e. $4x + 3y - 5z = 9$.

We end the section by giving the definition of skew lines.

Definition 13.4.4 We say that two distinct lines r and r' in $\mathbb{R}^3$ are *skew* lines, if they are neither parallel nor they intersect.

13.5 EXERCISES WITH SOLUTIONS

13.5.1 Consider the plane π passing through the point $P = (1,0,1)$ with normal vector $\mathbf{n} = [1,-2,4]$. Determine the line r perpendicular to π and passing through the point $P = (1,-2,3)$.

Solution. The plane is given by the Cartesian equation:

$$x - 2y + 4z = 5.$$

The line we need to find has as a direction vector the vector normal to the plane and therefore we can immediately write the equations of the line r in parametric form:

$$\begin{cases} x = 1 + t \\ y = -2 - 2t \\ z = 3 + 4t, \end{cases} \tag{13.13}$$

which correspond to the equations in Cartesian form:

$$\begin{cases} 4x - z = 1 \\ 2x + y = 0. \end{cases} \tag{13.14}$$

13.5.2 Determine the distance of the point $P = (2,-1,3)$ from the plane passing through $Q = (1,-1,-8)$ with normal vector $\mathbf{n} = (2,-2,-1)$.

Solution. We immediately write the Cartesian equation of the plane:

$$2x - 2y - z = 12$$

and then a set of parametric equations of the line passing through P and with direction vector $\mathbf{n}$ is:

$$\begin{cases} x = 2 + 2t \\ y = -1 - 2t \\ z = 3 - t. \end{cases}$$

We then calculate the point R of intersection between this line and the given plane by substituting the generic point obtained from the parametric equations of the line in the equation of the plane:

$$2(2 + 2t) - 2(-1 - 2t) - (3 - t) = 12,$$

We obtain $t = 1$, thus $R = (4,-3,2)$. Using the distance formula (13.1), we get that the distance between P and R is $\sqrt{17}$.

13.5.3 Consider the line r through the two points $P = (1,0,-2)$ and $Q = (-1,1,-1)$ and the line r' through the point $R = (3,1,-5)$ with direction vector $\mathbf{v}' = (0,-1,1)$. Say if r is perpendicular to r' and compute the distance between Q and r'.

Solution. We compute a direction vector: $\mathbf{v} = [-2, 1, 1]$ and therefore the line r has parametric equations:

$$\begin{cases} x = 1 - 2t \\ y = 0 + t \\ z = -2 + t. \end{cases}$$

We immediately see that $\mathbf{v} \cdot \mathbf{v}' = 0$, however two lines are perpendicular if they intersect and if they have perpendicular direction vectors. So we have to check if they intersect. We write parametric equations of r':

$$\begin{cases} x = 3 \\ y = 1 - t' \\ z = -5 + t'. \end{cases}$$

At this point, to check whether or not they intersect we solve the system:

$$\begin{cases} 3 = 1 - 2t \\ 1 - t' = 0 + t \\ -5 + t' = -2 + t. \end{cases}$$

We see that this system admits solution for $t = -1$ and $t' = 2$, that is, the point $S = (3, -1, -3)$ belongs to both lines. Therefore, r and r' are perpendicular.

We now want to calculate the distance between Q and r'. As the lines are perpendicular and intersect at the point S this distance will be given by the distance between the points Q and S which we can immediately calculate using formula (13.1):

$$\sqrt{16 + 4 + 4} = 2\sqrt{6}.$$

13.5.4 Determine the intersection of the two planes:

$$x + y - z = 0, \qquad y + 2z = 6. \tag{13.15}$$

Solution. Before proceeding, let us note a very important fact: the expression of a line in Cartesian form, that is, given by a system of two equations in the unknowns x, y, z, corresponds to the intersection of two planes each identified by its own Cartesian equation. The direction of the line is uniquely identified as it is perpendicular to both normal directions of the two planes. So, to find a vector perpendicular to both the normal vectors of the planes, we calculate the vector product of the two normal vectors: $\mathbf{n}_1 = [1, 1, -1]$ and $\mathbf{n}_2 = [0, 1, 2]$:

$$\mathbf{n}_1 \times \mathbf{n}_2 = [3, -2, 1].$$

Let us now choose an arbitrary point on the line, that is, a point that satisfies both equations (13.15). We can for example set $z = 0$ and obtain values for x and y from the equations: $y = 6$ and $x = -6$. Hence, parametric equations of the line intersection of the two planes are:

$$\begin{cases} x = -6 + 3t \\ y = 6 - 2t \\ z = t. \end{cases}$$

13.6   SUGGESTED EXERCISES

13.6.1 Consider the following lines and determine what is their reciprocal position (they are parallel lines, they intersect or they are skew).

$$r_1: \qquad x = 1 + t, \quad y = t, \quad z = 2 - 5t$$

$$r_2: \qquad x + 1 = y - 2 = 1 - z,$$

$$r_3: \qquad x = 1 + t, \quad y = 4 + t, \quad z = 1 - t$$

$$r_4: \qquad x = 2 + 2t, \quad y = 1 + 2t, \quad z = -3 - 10t$$

Also, in case of parallel lines, compute the distance between them.

13.6.2 a) Let the π be the plane given by:

$$\begin{cases} x = \quad 1 + s + 2t \\ y = \quad 3s \\ z = \quad 2 + t. \end{cases}$$

Determine the line r perpendicular to π and passing through the point $P = (2, 1, 0)$.

b) Determine the plane π' containing $Q = (1, 0, -1)$ and r, both in parametric and in Cartesian form.

13.6.3 Find parametric equations for the intersection of the planes $x + y - z = 2$, $3x - 4y + 5z = 6$. Also calculate the angle formed by the two planes, that is, by the two lines normal to the planes.

13.6.4 In $\mathbb{R}^3$, given the plane π, $x + y - z = 0$ and the line r $x = 2t$, $y = t$, $z = 2t + 1$ calculate (if it exists) the line through $(0, 0, 0)$ and $\pi \cap r$.

13.6.5 Give in $\mathbb{R}^3$ the lines r, s of Cartesian equations

$$r: \begin{cases} x + y - 1 = 0 \\ z - 1 = 0 \end{cases} \qquad s: \begin{cases} x + y - 2z - 2 = 0 \\ z + 1 = 0; \end{cases}$$

a) Show that r, s belong to the same plane.

b) Determine the equation of the plane that contains r, s.

c) Determine the distance between r and s.

13.6.6 In $\mathbb{R}^3$, given the planes $\pi_1 : x - y + 1 = 0$, $\pi_2 : x + y + 3z = 0$:

a) Find parametric equations for the line r intersection of π_1 and π_2.

b) Determine the Cartesian equation of the plane π containing r and the point $P = (1, 0, 1)$.

c) Find parametric equations for the line s passing through P and orthogonal to π.

13.6.7 In $\mathbb{R}^3$, given the plane π with Cartesian equation $\pi : x + y - 2z + 4 = 0$ and the line r with Cartesian equations

$$\begin{cases} x - 2z + 12 = 0 \\ y - 4 = 0 \end{cases}$$

a) Find parametric equations for r.

b) Determine the relative position of π, r.

c) Determine the distance between π and r.

13.6.8 In $\mathbb{R}^3$, given the plane π with Cartesian equation

$$\pi : x - y + z + 1 = 0$$

and the point $Q = (1, 1, 0)$:

a) Find a Cartesian equation of the plane passing through Q that is parallel to π.

b) Determine the distance between these two planes.

c) Find Cartesian equations for the line s passing through Q and parallel to the vector $v = [1, 0, 1]$.

13.6.9 Consider two generic vectors $\mathbf{u}$, $\mathbf{v}$ laying on the plane xy. Prove that the norm of $\|\mathbf{u} \times \mathbf{v}\|$ is the area of the parallelogram of sides $\mathbf{u}$ and $\mathbf{v}$.

13.6.10 Consider three generic vectors $\mathbf{u}$, $\mathbf{v}$ $\mathbf{w}$ in $\mathbb{R}^3$. Prove that $\|\mathbf{u} \times \mathbf{v} \cdot \mathbf{w}\|$ is the volume of the parallelepiped with sides $\mathbf{u}$, $\mathbf{v}$ and $\mathbf{w}$.

CHAPTER 14

Introduction to Modular Arithmetic

In this chapter, we want to study the arithmetic of the integers. We will start with the principle of induction, a result of fundamental importance that has several applications in various areas of mathematics. We will then continue with the division algorithm, Euclid's algorithm, arriving to congruences, the most important topic of elementary discrete mathematics.

14.1 THE PRINCIPLE OF INDUCTION

The principle of induction is the main technique for proving statements regarding natural numbers that is, the set $\mathbb{N} = \{0, 1, 2, \ldots\}$. Let start with an example.

Assume we want to prove that the sum of the integers between 0 and n is $n(n+1)/2$:

$$\sum_{k=0}^{n} k = \frac{n(n+1)}{2}. \tag{14.1}$$

We will denote this statement with $P(n)$. First of all, we check its validity for $n = 0$: we need to show that the sum of natural numbers between 0 and 0 is 0. In fact, $0 = 0(0+1)/2 = 0$. Let us now verify that $P(1)$ is true: the sum of the integers between 0 and 1 is $0+1 = 1 = 1(1+1)/2$. Similarly, $P(2)$: $0+1+2 = 2(2+1)/2 = 3$. It is clear that, with a little patience, we could go on like this by verifying formula (14.1) when n is very large, but we want to prove that it is true *for all natural numbers n*. The induction principle helps us by allowing us to prove the validity of a statement for all natural numbers.

First of all, we state the axiom of good ordering. We will see later that the principle of induction and the axiom of good ordering are equivalent to each other. However in the theory we will illustrate we shall take one as an axiom and then prove the other.

Axiom of good ordering. *Each non-empty subset of the set of natural numbers contains an element that is smaller than all the others.*

Theorem 14.1.1 (**Principle of induction**). *Assume that a statement $P(n)$ is given for each natural number. If it occurs that:*

1) *$P(0)$ is true;*

2) *if $P(k)$ is true then $P(k+1)$ is true;*

then $P(n)$ is true for every $n \in \mathbb{N}$.

Proof. Let S be the subset of $\mathbb{N}$ for which the statement $P(n)$ does not hold. We want to show that $S = \varnothing$. Assume by contradiction that S is not empty. So, by the axiom of the good ordering, S contains a natural number m smaller than all other elements in S. In particular $P(m)$ is not true and since $P(0)$ is true by hypothesis, $m \neq 0$. On the other hand, being m the smallest element of S, $m-1$ does not belong to S, ie $P(m-1)$ is true. By hypothesis 2), however, $P(m-1+1) = P(m)$ is true and therefore we have obtained a contradiction. ■

Example 14.1.2 Let us see how to use the principle of induction to prove the formula (14.1), which is the following statement $P(n)$:

the sum of the first n natural numbers is $\frac{n(n+1)}{2}$.

We have already seen that this statement is true for $n = 0$, that is, hypothesis 1) of the principle of induction is true. Now assume that $P(k)$ is true, that is:

$$0 + 1 + 2 + 3 + \cdots + k = \frac{k(k+1)}{2}.$$

We have to show that $P(k) \implies P(k+1)$ (hypothesis 2) of the induction principle, i.e. that:

$$0 + 1 + 2 + 3 + \cdots + k + (k+1) = \frac{(k+1)(k+2)}{2}.$$

Since $P(k)$ is true, we have:

$$0 + 1 + \cdots + k + (k+1) \quad = \frac{k(k+1)}{2} + k + 1 = \frac{k^2+k+2k+2}{2} =$$

$$= \frac{(k+1)(k+2)}{2},$$

so $P(k+1)$ is true. At this point the principle of induction guarantees that the statement $P(n)$ holds for every n.

The use of the statement $P(k)$ to prove $P(k+1)$ takes the name of *inductive hypothesis*.

We now state a variant of the principle of induction, known as the principle of complete induction, that is very useful in applications. Despite this statement appears weaker than the induction principle, at the end of this section we will see that the principle of induction and the principle of complete induction are equivalent.

Theorem 14.1.3 (**Principle of Complete Induction**). *Assume that a statement $P(n)$ is given for each natural number. If it occurs that:*

1. $P(0)$ *is true;*

2. *if $P(j)$ is true for every $j < k$ then $P(k)$ is true;*

then $P(n)$ is true for every $n \in \mathbb{N}$.

Proof. It is an immediate consequence of the principle of induction. ■

Theorem 14.1.4 *The following are equivalent:*

1. *The axiom of good ordering.*

2. *The principle of induction.*

3. *The principle of complete induction.*

Proof. We have already seen that $(1) \implies (2) \implies (3)$. It is therefore sufficient to show that $(3) \implies (1)$. The axiom of good ordering can be formulated as follows: "If a set $S \subset \mathbb{N}$ does not have a smallest element then $S = \varnothing$" or, equivalently "If a set $S \subset \mathbb{N}$ does not have a smallest element then $n \notin S$ for every $n \in \mathbb{N}$". We denote with $P(n)$ this last statement and we show that it is true using the principle of complete induction. So S is a subset of $\mathbb{N}$ that has no smallest element. $P(0)$ is true, i.e. $0 \notin S$, otherwise 0 would be the smallest element of S (being 0 the smallest element of $\mathbb{N}$). Assume that $P(j)$ is true for every $j \leq k - 1$, that is, assume that $0, 1, 2, \ldots, k - 1 \notin S$. We want to show that $k \notin S$. But if $k \in S$ then that k would be the minimum of S (because all natural numbers smaller than k are not in S) and this would be absurd because we are assuming that S has no minimum element. So, by the complete induction principle, $P(n)$ holds for every $n \in \mathbb{N}$, that is, the axiom of good order is true. ■

This apparently convoluted proof is, however, instructive because it proves the equivalence of three statements, which look quite different from each other.

14.2   THE DIVISION ALGORITHM AND EUCLID'S ALGORITHM

The division algorithm formalizes a procedure that we know very well since elementary school. However it is necessary to understand that a procedure, to be well defined, must be proved, so that its validity becomes absolute and not confined to the possible examples that we can build.

Theorem 14.2.1 (Division algorithm). *Consider $n, b \in \mathbb{N}$, $b \neq 0$. Then there are two unique integers q and r, respectively called quotient and remainder, such that:*

$$n = qb + r, \qquad \text{with} \quad 0 \leq r < b.$$

Proof. Let us first prove the existence of q and r with the principle of complete induction. $P(0)$ it is true because $n = 0 = 0b + 0$, with $q = r = 0$. Assume that $P(j)$ is true for every j such that $0 \leq j < k$ and we want to show $P(k)$ (that is, we want to verify hypothesis 2) of the principle of complete induction). If $k < b$ then:

$$k = 0b + k, \qquad \text{with remainder } r = k, \quad 0 \leq k < b,$$

so $P(k)$ is true for $k < b$. Now assume that $k \geq b$. Since $0 \leq k - b < k$, by applying the inductive hypothesis to $k - b$ we have:

$$k - b = q_1 b + r_1, \qquad \text{with} \quad 0 \leq r_1 < b,$$

from which we deduce:

$$k = (q_1 + 1)b + r_1, \qquad \text{with} \quad 0 \leq r_1 < b,$$

which shows what we want.

We now show that q and r are unique. Assume $n = q_1 b + r_1 = q_2 b + r_2$, with $0 \leq r_1 \leq r_2 < b$. Then it follows that $r_2 - r_1 = (q_1 - q_2)b$. Now in the first member of the equality there is a non-negative integer less than b, in the second member there is a multiple of b, so they both must be zero, i.e. $r_1 = r_2$ and $q_1 = q_2$. ■

Observation 14.2.2 With a similar proof, which uses the axiom of good ordering, we obtain that the division algoritm holds for any two integers a and b, not necessarily belonging to natural numbers. More precisely we have that:

Theorem 14.2.3 *Given two integers n and b with $b \neq 0$, there exists a unique pair of integers p and q, respectively called quotient and remainder, such that*

$$n = qb + r, \qquad \text{with} \quad 0 \leq r < |b|,$$

where $|b|$ indicates the absolute value of b.

We now want to introduce the concept of *divisibility* and of *greatest common divisor*.

Definition 14.2.4 Let a and b be two integers. We say that b *divides* a if there is an integer c such that $a = bc$ and we write $b|a$. We say that d is the *greatest common divisor* between two numbers a and b if it divides them both and it is the largest integer with this property. We will denote the greatest common divisor between a and b with $\gcd(a, b)$.

Observation 14.2.5 Note that if a, b and c are integers and a divides both b and c, then a also divides $b + c$ and $b - c$. We leave to the reader the easy verification of this property that we will use several times later.

We now want to find an efficient algorithm to determine the greatest common divisor between two integers.

Theorem 14.2.6 (Euclid's algorithm) *Let a and b be two positive integers such that $b \leq a$ and b does not divide a. Then we have:*

$$a = q_0 b + r_0, \qquad \text{where} \qquad 0 \leq r_0 < b$$

$$b = q_1 r_0 + r_1, \qquad \text{where} \qquad 0 \leq r_1 < r_0$$

$$r_0 = q_2 r_1 + r_2, \qquad \text{where} \qquad 0 \leq r_2 < r_1$$

$$\vdots$$

$$r_{t-2} = q_t r_{t-1} + r_t, \qquad \text{where} \qquad 0 < r_t < r_{t-1}$$

$$r_{t-1} = q_{t+1} r_t$$

and the last nonzero remainder r_t is the greatest common divisor between a and b.

Proof. By Theorem 14.2.1 we can write: $a = q_0 b + r_0$. Now we want to show that $\gcd(a, b) = \gcd(b, r_0)$. In fact if $c|a$ and $c|b$ then $c|r_0$, since $r_0 = a - q_0 b$. Similarly, if $c|b$ and $c|r_0$ then $c|a = q_0 b + r_0$. So the set of integers that divide both a and b coincides with the set of integers that divide both b and r_0. Therefore the greatest common divisor of the two pairs (a, b) and (b, r_0) is the same. Once established that, the result follows immediately from the chain of equalities:

$$\gcd(a, b) = \gcd(b, r_0) = \gcd(r_0, r_1) = \cdots = \gcd(r_{t-1}, r_t) = r_t.$$

■

Let us see concretely how to use this algorithm to determine the greatest common divisor of two given numbers.

Example 14.2.7 We want to compute $\gcd(603, 270)$. We use Euclid's algorithm (Theorem 14.2.6):

$$603 = 2 \cdot 270 + 63$$

$$270 = 4 \cdot 63 + 18$$

$$63 = 3 \cdot 18 + 9$$

$$18 = 2 \cdot 9.$$

So $\gcd(603, 270) = 9$.

The following theorem is a consequence of Euclid's algorithm and will be the fundamental tool for the resolution of congruences, which we will study in Section 14.4.

Theorem 14.2.8 (Bézout Identity). *Let a, b be positive integers and let $d = \gcd(a, b)$. Then, there are two integers u, v (not unique) such that:*

$$d = ua + vb.$$

Proof. The proof of this result uses Euclid's algorithm 14.2.6. We show that at each step there exist $u_i, v_i \in \mathbb{Z}$ such that $r_i = u_i a + v_i b$.

For r_0 the result is true, with $u_0 = 1$ and $v_0 = -q_0$, indeed

$$r_0 = a - q_0 b.$$

Then we have that:

$$r_1 = b - r_0 q_1 = b - (u_0 a + v_0 b) q_1 = -u_0 a q_1 + (1 - v_0 q_1) b$$

and the result is also true for r_1, just take $u_1 = -u_0 q_1$ and $v_1 = 1 - v_0 q_1$. In general, after the step $i - 1$ we know $u_{i-2}, u_{i-1}, v_{i-2}, v_{i-1} \in \mathbb{Z}$ such that:

$$r_{i-2} = u_{i-2} a + v_{i-2} b, \quad r_{i-1} = u_{i-1} a + v_{i-1} b.$$

So we have that:

$$r_i = r_{i-2} - r_{i-1} q_i = u_{i-2} a + v_{i-2} b - (u_{i-1} a + v_{i-1} b) q_i$$

$$= (u_{i-2} - u_{i-1} q_i) a + (v_{i-2} - v_{i-1} q_i) b,$$

so the result is true for r_i, with $u_i = u_{i-2} - u_{i-1} q_i$ and $v_i = v_{i-2} - v_{i-1} q_i$. Since $\gcd(a, b)$ is the last nonzero remainder r_t, after the step t we know u_t and v_t such that $r_t = \gcd(a, b) = u_t a + v_t b$ and $u = u_t$, $v = v_t$ are the integers we were looking for. ■

Observation 14.2.9 In the previous theorem, the existence of two numbers u, v such that $d = ua + vb$ does not guarantee that $d = \gcd(a, b)$. For example $10 = 15 \cdot 14 - 25 \cdot 8$ but $\gcd(14, 8) = 2$.

Example 14.2.10 In Example 14.2.7, we computed $\gcd(603, 270) = 9$. We now want to compute two numbers u and v such that $u\,603 + v\,270 = 9$. We proceed backwards by carefully replacing the remainders in the sequence of equations obtained in Example 14.2.7.

$$9 = 63 - 3 \cdot 18 =$$

$$= 63 - 3 \cdot [270 - 4 \cdot 63] = 63 - 3 \cdot 270 + 12 \cdot 63 =$$

$$= (-3) \cdot 270 + 13 \cdot 63 = (-3) \cdot 270 + 13 \cdot [603 - 2 \cdot 270] =$$

$$= (-3) \cdot 270 + 13 \cdot 603 + (-26) \cdot 270 =$$

$$= 13 \cdot 603 + (-29) \cdot 270.$$

So $9 = 13 \cdot 603 + (-29) \cdot 270$, i.e. we have $u\,603 + v\,270 = 9$ with $u = 13$ and $v = -29$.

We conclude this section with a result that we do not prove and that we will not use later, but which plays a fundamental role for the theory of the integer numbers and whose generalizations are extremely important in number theory (see [2]).

Definition 14.2.11 Let us say that a positive integer p is *prime* if its only divisors are $\pm p$ and ± 1.

Theorem 14.2.12 (Fundamental Theorem of Arithmetic). *Each integer greater than 1 is the product of primes in a unique way up to reordering:*

$$n = p_1 p_2 \ldots p_r,$$

where $p_1, \ldots, p_r$ are prime numbers (not necessarily distinct).

14.3 CONGRUENCE CLASSES

Congruence classes represent a way of counting, adding up and multiplying numbers different from the ones we know from the arithmetic of integers and that we use every day. For example, if we ask someone to meet precisely in 10 hours and now it is 23 o'clock, this person knows very well that the meeting will take place at 9 o'clock, not at 33, as the sum of integers would suggest. Similarly if our clock marks 8 o'clock and we have an appointment in 6 hours, we know that the appointment will be at 2 o'clock.

We want to formalize this way of carrying out operations, which in the cases examined simply consists of doing the sum of the integer numbers, divide the result by a certain integer (which in the first example is 24, in the second 12) and then take the remainder of the division.

Definition 14.3.1 Let a, b and n be three integers, with $n > 0$. We say that a is *congruent* to b modulo n and we write $a \equiv_n b$ if $n | a - b$.

The congruence relationship between integers has the following properties:

(i) Reflexive property: $a \equiv_n a$, in fact $n | a - a = 0$.

(ii) Symmetric property: if $a \equiv_n b$ then $b \equiv_n a$. Indeed if $n | a - b$ then $n | b - a$.

(iii) Transitive property: if $a \equiv_n b$ and $b \equiv_n c$ then $a \equiv_n c$. In fact, if $n | a - b$ and $n | b - c$ then $n | a - b + b - c = a - c$.

A relationship satisfying the reflexive property, the symmetric property and the transitive property is called *equivalence relation*. We will not study equivalence relations in general, however we will use the three aforementioned properties of congruences.

We now state a result that will be useful later and whose proof we leave for exercise.

Proposition 14.3.2 *Let $a, b, c, d, n \in \mathbb{Z}$, with $n > 0$. If $a \equiv_n b$ and $c \equiv_n d$ then:*

1. $a + c \equiv_n b + d$,

2. $ac \equiv_n bd$.

Let us now define the congruence classes, i.e. the sets that contain all integer numbers which are congruent to each other modulo a certain integer n.

Definition 14.3.3 Let $a, n \in \mathbb{Z}$. The *congruence class of a modulo n*, denoted with $[a]_n$, is the set of integers congruent to a modulo n:

$$[a]_n = \{b \in \mathbb{Z} \mid b \equiv_n a\} = \{a + kn \mid k \in \mathbb{Z}\}.$$

Example 14.3.4 For example, if we take $n = 4$ we have that:

$$[0]_4 = \{0, 4, -4, 8, -8 \dots\}$$

$$[1]_4 = \{1, 5, -3, 9, -7 \dots\}$$

$$[2]_4 = \{2, 6, -2, 10, -6 \dots\}$$

$$[3]_4 = \{3, 7, -1, 11, -5 \dots\}$$

Note that:

$$[4]_4 = [0]_4 = [-4]_4 = \dots$$

$$[5]_4 = [1]_4 = [-3]_4 = \dots$$

$$[6]_4 = [2]_4 = [-2]_4 = \dots$$

$$[7]_4 = [3]_4 = [-1]_4 = \dots$$

We note that in this example:

- there are no elements in common in two different congruence classes;

- the union of the congruence classes is the set of all integers;

- there is a finite number of congruence classes.

These facts, as we shall see, are valid in general.

Proposition 14.3.5 *Let $[a]_n$ and $[b]_n$ be two congruence classes modulo n. Then $[a]_n = [b]_n$ or $[a]_n$ and $[b]_n$ are disjoint, that is they do not have common elements.*

Proof. Assume there is a common element c between $[a]_n$ and $[b]_n$, that is $c \equiv_n a$ and $c \equiv_n$. So, by the transitive property of congruences $a \equiv_n b$, and then, again by the transitive property, $[a]_n = [b]_n$. ■

The proof of the following proposition is immediate.

Proposition 14.3.6   *1. Let r be the remainder of the division of a by n. Then $[a]_n = [r]_n$.*

2. $[0]_n, [1]_n, \ldots, [n-1]_n$ *are all the distinct congruence classes modulo* n.

3. $[0]_n \cup [1]_n \cup \ldots \cup [n-1]_n = \mathbb{Z}$.

We have come to the most important definition of this chapter: the set $\mathbb{Z}_n$.

Definition 14.3.7 The set of *integers modulo* n, denoted with $\mathbb{Z}_n$, is the set of congruence classes modulo n:

$$\mathbb{Z}_n = \{[0]_n, [1]_n, \ldots, [n-1]_n\}.$$

Definition 14.3.8 We define the following sum and product operations on the set $\mathbb{Z}_n$:

$$[a]_n + [b]_n = [a+b]_n, \qquad [a]_n\,[b]_n = [ab]_n.$$

Observation 14.3.9 The operations just defined do not depend on the numbers a and b chosen to represent the congruence classes which we add or multiply, but only from their congruency class. In this case, it is said that the operations are *well defined*. For example, in $\mathbb{Z}_4$ we have: $[1]_4 = [5]_4$ and $[2]_4 = [6]_4$. By definition, $[1]_4 + [2]_4 = [3]_4 = [11]_4 = [5]_4 + [6]_4$.

Example 14.3.10 We compute the tables of addition and multiplication for $\mathbb{Z}_3$ and $\mathbb{Z}_4$, inviting the student to practice in building the analogue tables for $\mathbb{Z}_5$ and $\mathbb{Z}_6$:

$\mathbb{Z}_3$

$+$	$[0]_3$	$[1]_3$	$[2]_3$
$[0]_3$	$[0]_3$	$[1]_3$	$[2]_3$
$[1]_3$	$[1]_3$	$[2]_3$	$[0]_3$
$[2]_3$	$[2]_3$	$[0]_3$	$[1]_3$

$\cdot$	$[0]_3$	$[1]_3$	$[2]_3$
$[0]_3$	$[0]_3$	$[0]_3$	$[0]_3$
$[1]_3$	$[0]_3$	$[1]_3$	$[2]_3$
$[2]_3$	$[0]_3$	$[2]_3$	$[1]_3$

$\mathbb{Z}_4$

$+$	$[0]_4$	$[1]_4$	$[2]_4$	$[3]_4$
$[0]_4$	$[0]_4$	$[1]_4$	$[2]_4$	$[3]_4$
$[1]_4$	$[1]_4$	$[2]_4$	$[3]_4$	$[0]_4$
$[2]_4$	$[2]_4$	$[3]_4$	$[0]_4$	$[1]_4$
$[3]_4$	$[3]_4$	$[0]_4$	$[1]_4$	$[2]_4$

$\cdot$	$[0]_4$	$[1]_4$	$[2]_4$	$[3]_4$
$[0]_4$	$[0]_4$	$[0]_4$	$[0]_4$	$[0]_4$
$[1]_4$	$[0]_4$	$[1]_4$	$[2]_4$	$[3]_4$
$[2]_4$	$[0]_4$	$[2]_4$	$[0]_4$	$[2]_4$
$[3]_4$	$[0]_4$	$[3]_4$	$[2]_4$	$[1]_4$

We note some very important facts: in $\mathbb{Z}_3$ each element other than $[0]_3$ admits an *inverse*, that is, for every $[a]_3 \neq [0]_3$ there is an element $[b]_3$ such that $[a]_3[b]_3 = [1]_3$. This inverse is denoted with $[a]_3^{-1}$. So we have: $[1]_3^{-1} = [1]_3$, $[2]_3^{-1} = [2]_3$. This property does not apply in the case of $\mathbb{Z}_4$. In fact, the multiplicative table shows that there is no inverse of the class $[2]_4$. As we will see in detail in the next paragraph, this diversity is linked to the fact that 3 is a prime number while 4 is not.

14.4 CONGRUENCES

In this section, we aim at solving linear equations in which the unknown belongs to the set $\mathbb{Z}_n$ introduced in the previous section.

Let us start by examining the structure of $\mathbb{Z}_p$, with p a *prime* number.

Proposition 14.4.1 *The following statements are equivalent:*

(1) p is a prime number.

(2) The equation $[a]_p x = [1]_p$, with $[a]_p \neq [0]_p$, has a solution in $\mathbb{Z}_p$, that is, every element $[a]_p \neq [0]_p$ in $\mathbb{Z}_p$ admits an inverse.

(3) If $[a]_p[b]_p = [0]_p$ in $\mathbb{Z}_p$ then $[a]_p = [0]_p$ or $[b]_p = [0]_p$.

Proof. (1) $\implies$ (2): since $[a]_p \neq [0]_p$, p does not divide a, so $\gcd(a, p) = 1$. Then by Theorem 14.2.8 there exists $u, v \in \mathbb{Z}$ such that $1 = au + pv$. Taking the classes congruence modulo p, we have: $[1]_p = [a]_p[u]_p + [p]_p[v]_p = [a]_p[u]_p$, therefore $x = [u]_p \in \mathbb{Z}_p$ is a solution of $[a]_p x = [1]_p$.

(2) $\implies$ (3): we have $[a]_p[b]_p = [0]_p$ with $[a]_p \neq [0]_p$. By hypothesis there is an inverse of $[a]_p$ that is, there is an element $[u]_p$ such that $[u]_p[a]_p = [1]_p$. Multiplying both members of the equality $[a]_p[b]_p = [0]_p$ by $[u]_p$ we get: $[u]_p[a]_p[b]_p = [u]_p[0]_p = [0]_p$, i.e. $[b]_p = [0]_p$.

(3) $\implies$ (1): we assume that $p = ab$ and we show that necessarily a and b are equal to ± 1 or $\pm p$, that is the only divisors of p are, up to changing the sign, p itself and 1. Considering the absolute values, we observe that $|a||b| = |ab| = |p| = p$ so $|a| \leq p$ and $|b| \leq p$. The equality $p = ab$, translated in $\mathbb{Z}_p$, becomes the equality $[p]_p = [a]_p[b]_p$ i.e. $[a]_p[b]_p = [0]_p$. By hypothesis we know that either $[a]_p = [0]_p$ or $[b]_p = [0]_p$. If $[a]_p = [0]_p$ then $a = \pm p$ and $b = \pm 1$. If $[b]_p = [0]_p$ then $b = \pm p$ and $a = \pm 1$. ■

Corollary 14.4.2 *If p is a prime number the equation $[a]_p x = [b]_p$, with $[a]_p \neq [0]_p$, has a single solution in $\mathbb{Z}_p$.*

Proof. By property (2) of the previous proposition we know that $[a]_p$ is invertible. Multiplying the given equation by $[a]_p^{-1}$ we get $x = [a]_p^{-1}[b]_p$, which shows at the same time that the solution exists and is unique. ■

The following result is very useful in the resolution of the exercises.

Proposition 14.4.3 *If $\gcd(a, n) = 1$ the equation $[a]_n x = [b]_n$ has a unique solution in $\mathbb{Z}_n$.*

Proof. Since $\gcd(a,n) = 1$, by Theorem 14.2.8 there exist $u, v \in \mathbb{Z}$ such that $au+nv = 1$, so $[a]_n$ is invertible in $\mathbb{Z}_n$, with inverse $[u]_n$. Arguing as in the proof of the previous corollary we get the result. ■

At this point, it is easy to characterize all elements in $\mathbb{Z}_n$ having an inverse.

Proposition 14.4.4 *The element $[a]_n$ has an inverse in $\mathbb{Z}_n$ if and only if $\gcd(a,n) = 1$.*

Proof. Assume that $\gcd(a,n) = 1$. Then by Proposition 14.4.3 the equation $[a]_n x = [1]_n$ has a unique solution $[c]_n$ in $\mathbb{Z}_n$. Thus $[c]_n$ is precisely the inverse of $[a]_n$. Viceversa, assume that there exists an element $[c]_n \in \mathbb{Z}_n$ such that $[a]_n [c]_n = [1]_n$, thus $n|1 - ac$, that is, $1 - ac = nr$ for some $r \in \mathbb{Z}$. Let $d = \gcd(a,n)$; we have that $d|a$, $d|n$, so $d|ac + nr = 1$, that is $d = 1$, as we wanted to prove. ■

Let us now examine the general case of Proposition 14.4.3.

Theorem 14.4.5 *Let $a, n \in \mathbb{Z}$, $n > 0$, and let $d = \gcd(a,n)$. Then:*

(1) the equation $[a]_n x = [b]_n$ has solution in $\mathbb{Z}_n$ if and only if $d|b$;

(2) if $d|b$ then equation $[a]_n x = [b]_n$ has exactly d distinct solutions in $\mathbb{Z}_n$.

Proof. Assume that $[c]_n$ is a solution of the given equation, then: $[a]_n[c]_n = [ac]_n = [b]_n$ i.e. $ac \equiv_n b$ or, equivalently, $n|ac - b$. Consequently $d|b$ since $d|n$ and $d|a$. Assume now that d divides b and let $a = a'd$, $n = n'd$, $b = b'd$. Observe that $\gcd(a', n') = 1$, otherwise d would not be the greatest common divisor between a and n. Then by Proposition 14.4.3, the equation $[a']_{n'} x = [b']_{n'}$ has a unique solution in $\mathbb{Z}_{n'}$, let it be $[c]_{n'}$. Thus we have: $[a']_{n'}[c]_{n'} = [a'c]_{n'} = [b']_{n'}$, i.e. $n'|a'c - b'$, so $n = n'd|a'dc - b'd = ac - b$, i.e. $[c]_n$ is a solution of the equation $[a]_n x = [b]_n$. This shows (1).

It remains to show that if $d|b$, there are d distinct solutions in $\mathbb{Z}_n$. Let $[c]_n$ be the solution found in point (1). If $[e]_n$ is another solution of the given equation then we have $[a]_n[c]_n = [b]_n = [a]_n[e]_n$, thus $n|ac - a$ and, that is $n'd|a'dc - a'de$ and therefore $n'|a'c - a'e$. It follows that $[a']_{n'}[e]_{n'} = [a']_{n'}[c]_{n'}$ and then $[e]_{n'}$ is the solution of the equation $[a']_{n'} x = [b']_{n'}$. By Proposition 14.4.3 we have $[e]_{n'} = [c]_{n'}$, i.e. $e = c + kn'$, with $k \in \mathbb{Z}$.

Then it is easy to verify that $[e]_n \in \{[c]_n, [c+n']_n, [c+2n']_n, \ldots, [c+(d-1)n']_n\} = X$ and that the elements of X are all distinct and are all solutions of the equation $[a]_n x = [b]_n$. This shows what we wanted. ■

Example 14.4.6 We want to determine all solutions in $\mathbb{Z}_{74}$ of the equation $[33]_{74}x = [5]_{74}$.

We use Euclid's algorithm:

$$74 = 2 \cdot 33 + 8$$

$$33 = 4 \cdot 8 + 1$$

$$8 = 8 \cdot 1$$

As $(33, 74) = 1$, we know that the solution exists and is unique, and the calculations just made allow us to compute the inverse of $[33]_{74}$. We have in fact that:

$$1 = 33 - 4 \cdot 8 = 33 - 4 \cdot (74 - 2 \cdot 33) = (-4) \cdot 74 + 9 \cdot 33.$$

We have therefore found: $1 = (-4) \cdot 74 + 9 \cdot 33$. We write this equation in $\mathbb{Z}_{74}$:

$$[1]_{74} = [9]_{74}[33]_{74}.$$

So the solution is $x = [9]_{74} \cdot [5]_{74} = [45]_{74}$.

Now let us see how Theorem 14.4.5 also allows us to solve linear congruences.

Definition 14.4.7 A *linear congruence* (modulo n) in the unknown x is a congruence of the type:

$$ax \equiv_n b$$

with $a, b, n \in \mathbb{Z}$, $n > 0$.

Observation 14.4.8 It is clear that an integer c is a solution of the linear congruence $ax \equiv_n b$ if and only if $[c]_n$ is a solution of the equation $[a]_n x = [b]_n$.
Let us go back to the proof of Theorem 14.4.5 for a moment, using the same notation. We have that the equation $[a]_n x = [b]_n$ has solution if and only if d divides n, where $d = \gcd(a, n)$. In this case, set $n = n'd$; if $[c]_n$ is a solution of the equation $[a]_n x = [b]_n$, found for example with the method used in point (1), the integers e such that $[e]_n$ is a solution of the equation $[a]_n x = [b]_n$ are all those of the type $e = c + kn'$, with $k \in \mathbb{Z}$. Thus the solutions of the linear congruence $ax \equiv_n b$ are precisely only those of the type $e = c + kn'$, with $k \in \mathbb{Z}$.

14.5   EXERCISES WITH SOLUTIONS

14.5.1 Prove by induction that a set with n elements has 2^n subsets.

Solution. Let us show $P(0)$. If a set is empty then it has zero elements and therefore only one subset (itself); therefore it has $1 = 2^0$ subsets. We now assume that the statement $P(k - 1)$ is true, and we want to show $P(k)$, with $k \geq 1$. The statement $P(k)$ says that a set S with k elements has 2^k subsets. As S contains at least an element x, we can think of S as the disjoint union of one of its subsets S' and the subset $\{x\}$. The subsets of S that do not contain x are also subsets of S', and there are 2^{k-1} of them by the inductive hypothesis $P(k-1)$, since S' contains $k - 1$ elements. On the other hand, a subset of S containing x is given by $T \cup \{x\}$, where T is a subset of S'. It is easy to convince ourselves that there are exactly 2^{k-1} such subsets. Therefore the subsets of S are $2^{k-1} + 2^{k-1} = 2^k$, and this concludes the proof by induction.

14.5.2 We want to determine all the solutions of the congruence $63x \equiv_{375} 24$.

Solution. Since $3 = \gcd(63, 375)$ divides 24, the congruence has solutions in $\mathbb{Z}$. We have that $63 = 3 \cdot 21$, $375 = 3 \cdot 125$, $24 = 3 \cdot 8$, thus we solve the equation $[21]_{125}\, x = [8]_{125}$. We want to find the inverse of $[21]_{125}$, and to do so we first use Euclid's algorithm to compute $\gcd(125, 21)$:

$$125 = 5 \cdot 21 + 20$$

$$21 = 1 \cdot 20 + 1$$

$$20 = 20 \cdot 1.$$

Proceeding backwards:

$$1 = 21 - 1 \cdot 20 = 21 - (125 - 5 \cdot 21) = -125 + 6 \cdot 21.$$

We therefore obtain that:

$$[1]_{125} = [6]_{125}[21]_{125}.$$

So the solution of the equation $[21]_{125}\, x = [8]_{125}$ is $x = [6]_{125} \cdot [8]_{125} = [48]_{125}$. At this point the solutions of the congruence are given by: $x = 48 + k125$, with $k \in \mathbb{Z}$.

14.6   SUGGESTED EXERCISES

14.6.1 Prove that $1 + 2^2 + 3^2 + \ldots n^2 = n(n+1)(2n+1)/6$.

14.6.2 Prove the Fundamental Theorem of Arithmetic 14.2.12 using the principle of complete induction.

14.6.3 Prove (by induction) that if n it is a non-negative integer then $2^n > n$.

14.6.4 Say if there are two classes $[a]_{37}$, $[b]_{37}$ in $\mathbb{Z}_{37}$, both nonzero such that $[a]_{37}[b]_{37} = [0]$. If they exist compute them, if they do not exist explain why. Answer the same question replacing $\mathbb{Z}_{37}$ with $\mathbb{Z}_{36}$.

14.6.5 Consider the two congruence classes: $[0]_6$ e $[3]_{12}$. Say if they are the same or if they are different from each other or if one is contained in the other.

14.6.6 Solve the following equations (if possible):

i) $[23]x = [7]$ in $\mathbb{Z}_{40}$
ii) $[3]x = [13]$ in $\mathbb{Z}_{17}$
iii) $[15]x = [9]$ in $\mathbb{Z}_{53}$
iv) $[4]x = [1]$ in $\mathbb{Z}_{400}$
v) $[18]x = [30]$ in $\mathbb{Z}_{42}$
vi) $[16]x = [36]$ in $\mathbb{Z}_{56}$
vii) $[20]x = [56]$ in $\mathbb{Z}_{178}$

14.7   APPENDIX: ELEMENTARY NOTIONS OF SET THEORY

In this appendix, we recall some elementary notions of set theory used in the text.

The concept of *set* and *membership* are *primitive*, so we do not give a rigorous definition of them. Informally, a set is a collection of objects, and to assign a set we can list its elements, for example:

$$X = \{3, 4, 5, 6, 7\},$$

or we can assign a property of its elements, for example, in the previous case, X is the set of natural numbers greater than 2 and smaller than 8, and can be referred to as:

$$X = \{x \mid x \text{ is a natural number and } 2 < x < 8\}.$$

To denote that an element belongs to a set, we use the symbol $\in$, whose negation is $\notin$. For example, in the previous case we have $5 \in X$, $9 \notin X$.

Some sets often used in the text are:

$$\mathbb{N} = \{0, 1, 2, \ldots, n, n + 1, \ldots\} \quad \text{set of natural numbers,}$$
$$\mathbb{Z} = \{0, \pm 1, \pm 2, \ldots, \pm n, \ldots\} \quad \text{set of integer numbers,}$$
$$\mathbb{R} \quad \text{set of real numbers.}$$

Two sets X and Y are the same if they have the same elements, and in this case we write $X = Y$.

Definition 14.7.1 If X and Y are sets, let us say X is a *subset* of Y if each element of X is also an element of Y, and we write $X \subseteq Y$.

Then there is a special set, the *empty set*, that is the set with no elements and is denoted with $\emptyset$. Note that the empty set it is a subset of any set X. We have for example:

$$\{x \mid x \in \mathbb{R}, \ x^2 = -1\} = \emptyset,$$

because no real number raised to the square gives -1 as a result. Care must be taken; for example $X = \{0\}$ is the subset of the real numbers that contains the single element zero, however, it is not the empty set because it contains an element.

Let us now recall the two fundamental operations that can be carried out between sets.

- The *union* of two sets X and Y is the set of all the elements that belong to X or to Y and is denoted with $X \cup Y$.

- The *intersection* of two sets X and Y is the set of all the elements that belong to both X and Y and is denoted with $X \cap Y$.

Complex Numbers

In this appendix, we introduce the set of complex numbers, necessary for a deeper understanding of the question of finding solutions of algebraic equations. All linear algebra results we describe in this book concerning real vector spaces, are also true replacing real numbers with complex numbers, without any modification to the theory. Since this topic involves an additional difficulty, we prefer to present our treatment of linear algebra limiting ourselves to the case of real scalars and leaving the complex number case in this appendix.

A.1   COMPLEX NUMBERS

Let $\mathbb{C}$ be the set of complex numbers: it represents an extension of the set of real numbers, coming mainly from the need to be able to find the square root of a negative real number. We therefore introduce a new symbol, denoted by i, called *imaginary unit*, with the property that $i^2 = -1$. We define the set of complex numbers as

$$\mathbb{C} = \{a + bi \mid a, b \in \mathbb{R}\}$$

with the sum and product operations defined as follows:

- $(a + bi) + (c + di) = (a + c) + (b + d)i$,

- $(a + ib)(c + id) = (ac - bd) + (ad + bc)i$,

where the real numbers $a+c, b+d, ac-bd, ad+bc$ are obtained with the usual operations of addition and product between real numbers. By definition, given $a + bi, c + di \in \mathbb{C}$, we have that $a + bi = c + di$ if and only if $a = c$ and $b = d$.

An easy way to remember the product operation between two complex numbers is to multiply them as if they were two polynomials in i and to remember that $i^2 = -1$, so:

$$(a + ib)(c + id) = ac + adi + bci + bdi^2 = (ac - bd) + (ad + bc)i.$$

Example A.1.1 Given the two complex numbers $1 + 2i$, $3 - i$ we want to compute the sum and the product.

$$(1 + 2i) + (3 - i) = (1 + 3) + (2 - 1)i = 4 + i$$
$$(1 + 2i)(3 - i) = 3 + 6i - i - 2i^2 = 3 + (6 - 1)i - 2(-1) = 5 + 5i$$

The set of real numbers is a subset of $\mathbb{C}$, because we can write any real number $a \in \mathbb{R}$ in the form $a = a + 0i$. We call complex numbers of the type $bi = 0 + bi$ *purely imaginary*. If $z = a + bi$ is a complex number, the real numbers a and b are called the *real part* and *imaginary part* of z, respectively.

We can represent complex numbers in the Cartesian plane as follows: we associate to $a + bi$ the pair of real numers (a, b). In this plane, the x-axis represents the real numbers and it is called *real axis*, while the y-axis represents pure imaginary complex numbers and we call it *imaginary axis*.

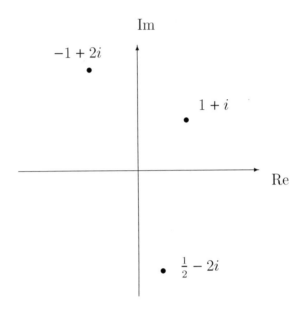

Given a complex number $\alpha = a + bi$, we define its *complex conjugate (or conjugate)* $\overline{\alpha}$ as $\overline{\alpha} = a - bi$. We also define the *modulus* of a complex number α as

$$|\alpha| = \sqrt{\alpha\overline{\alpha}} = \sqrt{a^2 + b^2}.$$

In the Cartesian representation, the modulus of α is the distance of the point $P = (a, b)$, representing α, from the origin.

The conjugation operation, which associates to a complex number its complex conjugate, satisfies some properties, which we leave as an easy exercise:

- $\overline{(\overline{\alpha})} = \alpha$ for each $\alpha \in \mathbb{C}$,

- $\overline{\alpha} = \alpha$ if and only if $\alpha \in \mathbb{R}$,

- $\overline{\alpha + \beta} = \overline{\alpha} + \overline{\beta}$ for each $\alpha, \beta \in \mathbb{C}$,

- $\overline{\alpha\beta} = \overline{\alpha}\overline{\beta}$ for each $\alpha, \beta \in \mathbb{C}$.

One of the most important properties of complex numbers is that the inverse α^{-1} of

$\alpha \in \mathbb{C}$, $\alpha \neq 0$ always exists, and it is a complex number. This inverse is obtained as $\alpha^{-1} = \alpha/|\alpha|^2$. More explicitly:

$$\alpha^{-1} = \frac{a}{a^2 + b^2} - \frac{b}{a^2 + b^2} i.$$

This allows us to immediately compute the quotient of two complex numbers. Instead of remembering the formula, we invite the student to understand the procedure described in the following example.

Example A.1.2 Let us consider the quotient of complex numbers: $\frac{3-2i}{1-i}$. We want to express this quotient as $a + bi$ for appropriate a and b. We proceed by multiplying the numerator and denominator by the complex conjugate of the denominator. The student will recognize the analogy with the procedure to rationalize the denominator of a fraction:

$$\frac{3 - 2i}{1 - i} = \frac{3 - 2i}{1 - i} \cdot \frac{1 + i}{1 + i} = \frac{(3 - 2i)(1 + i)}{|1 - i|^2} = \frac{3 - 2i + 3i - 2i^2}{2} = \frac{5}{2} + \frac{1}{2}i.$$

We conclude this section with a list of the properties of operations in complex numbers, the verification of which is left to the reader as an easy exercise.

- *commutativity of the sum*: $\alpha + \beta = \beta + \alpha$ for each $\alpha, \beta \in \mathbb{C}$;

- *associativity of the sum*:
  $(\alpha + \beta) + \gamma = \alpha + (\beta + \gamma)$ for each $\alpha, \beta, \gamma \in \mathbb{C}$;

- *existence of the neutral element 0 for the sum*:
  $\alpha + 0 = \alpha$ for each $\alpha \in \mathbb{C}$, with $0 = 0 + 0i$;

- *existence of the opposite* $-\alpha$ *of a complex number* α: $\alpha + (-\alpha) = 0$ for each $\alpha = a + bi \in \mathbb{C}$, with $-\alpha = -a - bi$;

- *product commutativity*: $\alpha\beta = \beta\alpha$ for each $\alpha, \beta \in \mathbb{C}$;

- *product associativity*:
  $(\alpha\beta)\gamma = \alpha(\beta\gamma)$ for each $\alpha, \beta, \gamma \in \mathbb{C}$;

- *existence of the neutral element 1 for the product*:
  $1\alpha = \alpha$ for each $\alpha \in \mathbb{C}$, with $1 = 1 + 0i$;

- *existence of the inverse* α^{-1} *of a complex number* α: $\alpha\alpha^{-1} = 1$ for each $\alpha \in \mathbb{C}$, $\alpha \neq 0$;

- *distribution of the sum with respect to the product*:
  $(\alpha + \beta)\gamma = \alpha\gamma + \beta\gamma$ for each $\alpha, \beta, \gamma \in \mathbb{C}$;

A.2 POLAR REPRESENTATION

We now want to represent a complex number in the Cartesian plane using the *polar* coordinate system. In this system, each point of the P plane is identified by two coordinates (ρ, θ), called respectively *radial coordinate* and *angular coordinate*. The coordinate ρ represents the distance of the point P from the origin O. The θ coordinate represents the angle between the x-axis and the ray OP:

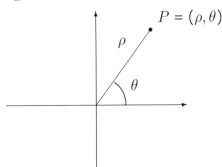

By definition of the trigonometric functions sine and cosine, we can immediately express the Cartesian coordinates (x, y) of a point P in terms of its polar coordinates (ρ, θ):

$$x = \rho \cos \theta \qquad y = \rho \sin \theta.$$

Conversely, given the Cartesian coordinates (x, y) of P, thanks to the Pythagorean theorem and the definition of arctangent, we can immediately write:

$$\rho = \sqrt{x^2 + y^2}, \qquad \theta = \arctan (y/x),$$

where $\rho \geq 0$ and θ is determined up to multiples of 2π (if $\rho \neq 0$).

If we now consider a complex number $\alpha = a + ib$ and its representation in the Cartesian plane, we immediately see that in polar coordinates we have $a = \rho \cos \theta$ and $b = \rho \sin \theta$. So we can also write

$$\alpha = \rho(\cos \theta + \sin \theta i).$$

This expression is called *trigonometric or polar representation* of the complex number $\alpha = a + bi$. As we have seen, ρ is the modulus of α, while the angle θ is called the *argument* of α.

If we use the trigonometric representation, the formula for the product of two complex numbers $\alpha = \rho_1(\cos \theta_1 + \sin \theta_1 i)$ e $\beta = \rho_2(\cos \theta_2 + \sin \theta_2 i)$ becomes particularly elegant. We have, in fact:

$$\begin{aligned} \alpha\beta &= \rho_1\rho_2(\cos \theta_1 + \sin \theta_1 i)(\cos \theta_2 + \sin \theta_2 i) \\[2mm] &= \rho_1\rho_2\big((\cos \theta_1 \cos \theta_2 - \sin \theta_1 \sin \theta_2) + (\cos \theta_1 \sin \theta_2 + \sin \theta_1 \cos \theta_2)i\big) \quad \text{(A.1)} \\[2mm] &= \rho_1\rho_2\big(\cos(\theta_1 + \theta_2) + \sin(\theta_1 + \theta_2)i\big). \end{aligned}$$

So, for two complex numbers we have that: the modulus of the product is the product of the moduli and the argument of the product is the sum of the arguments.

The trigonometric form of a complex number allows us to compute its n-th roots fairly quickly, through *De Moivre's formula*. Thanks to formula (A.1) we can compute the powers of a complex number:

$$
\begin{aligned}
\alpha &= \rho(\cos\theta + i\sin\theta) \\
\alpha^2 &= \rho^2(\cos 2\theta + \sin 2\theta\, i) \\
\alpha^3 &= \rho^3(\cos 3\theta + \sin 3\theta\, i) \\
&\vdots = \\
\alpha^n &= \rho^n(\cos n\theta + \sin n\theta\, i).
\end{aligned}
\tag{A.2}
$$

The last equality is called *De Moivre's formula*.

Therefore, we can immediately determine the n-th roots of a complex number $\alpha = \rho(\cos\theta + \sin\theta\, i)$:

$$
\alpha^{1/n} = \rho\Big(\cos[(\theta + 2k\pi)/n] + \sin[(\theta + 2k\pi)/n]\, i\Big)
\tag{A.3}
$$

with $k = 0, \ldots, n-1$, in other words we have n n-th complex roots of a given nonzero complex number.

Let us see an example.

Example A.2.1 We want to determine all the cube roots of $1 + i$. According to the formula (A.3) they are given by:

$$
\sqrt 2\Big\{ \cos[(\pi/4 + 2k\pi)/3] + \sin[(\pi/4 + 2k\pi)/3]\, i\Big\}
$$

with $k = 0, 1, 2$. More precisely:

$$
\alpha_1 = \sqrt 2\Big\{ \cos[(\pi/4)/3] + \sin[(\pi/4)/3]\, i\Big\} = \sqrt 2\big(\cos\tfrac{\pi}{12} + \sin\tfrac{\pi}{12}\, i\big)
$$

$$
\alpha_2 = \sqrt 2\Big\{ \cos[(\pi/4 + 2\pi)/3] + \sin[(\pi/4 + 2\pi)/3]\, i\Big\} =
$$

$$
= \sqrt 2\Big\{ \cos\tfrac{3\pi}{4} + \sin\tfrac{3\pi}{4}\, i\Big\}
$$

$$
\alpha_3 = \sqrt 2\Big\{ \cos[(\pi/4 + 4\pi)/3] + \sin[(\pi/4 + 4\pi)/3]\, i\Big\} =
$$

$$
= \sqrt 2\Big\{ \cos\tfrac{17\pi}{12} + \sin\tfrac{17\pi}{12}\, i\Big\}.
$$

We can represent such roots in the complex plane as follows:

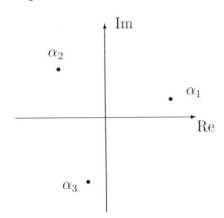

We conclude this section by stating a very important result: the *Fundamental Theorem of Algebra*, whose proof is particularly difficult. Since it is beyond the scope of this book, we refer the reader to one of several specific texts (see for instance S. Lang, *Algebra* [3]).

Theorem A.2.2 *Any polynomial of degree n with complex coefficients*

$$p(x) = a_n x^n + a_{n-1} x^{n-1} + \cdots + a_1 x + a_0, \quad a_n, \ldots, a_o \in \mathbb{C}$$

can be factored into a product of n linear factors (not necessarily distinct):

$$p(x) = a_n(x - \alpha_1)(x - \alpha_2)\cdots(x - \alpha_n),$$

with $\alpha_1, \ldots, \alpha_n \in \mathbb{C}$.

This specifically implies that a polynomial equation of degree n with coefficients in $\mathbb{C}$ always has n complex solutions, although not necessarily distinct.

Let us see an example.

Example A.2.3 We want to find all the solutions of the equation $x^4 - 16 = 0$. We can immediately factor the polynomial as:

$$x^4 - 16 = (x^2 - 4)(x^2 + 4) = (x - 2)(x + 2)(x + 2i)(x - 2i).$$

Therefore the zeros of the polynomial, corresponding to the solutions of the given equation, are: ± 2, $\pm 2i$. We can obtain the same result also by applying formula (A.3):

$$2 = 2[\cos(0) + \sin(0)\, i],$$

$$2i = 2[\cos(2\pi/4) + \sin(2\pi/4)\, i],$$

$$-2 = 2[\cos(4\pi/4) + \sin(4\pi/4)\, i],$$

$$-2i = 2[\cos(6\pi/4) + \sin(6\pi/4)\, i],$$

remembering that in polar coordinates: $-1 = \cos(\pi) + \sin(\pi)\, i$ and that $-i = \cos(6\pi/4) + \sin(6\pi/4)i$.

Solutions of some suggested exercises

Chapter 1: Introduction to linear systems

1.6.1 a) $x = y = 0$, $z = 1$. b) The system has no solutions.

1.6.3 The system has a unique solution for $k \neq 0, 1$. For $k = 0$, the system has infinitely many solutions depending on one parameter, for $k = 1$ the system does not have solutions.

Chapter 2: Vector Spaces

2.6.1 a), b), d), e), f), l) are subspacess. c), g), h), i), m) are not subspaces.

2.6.4 X is not a vector subspace.

Chapter 3: Linear Combinations and linear indipendence

3.4.1 a), d), e) are linearly independent sets. b), c) are linearly dependent sets.

3.4.3 Yes.

3.4.4 $k = \pm\sqrt{3}$.

3.4.5 $k \neq -1/2$.

3.4.6 $k = -2/5$.

3.4.9 a) $k \neq 2, -1$, b) $k \neq 2$.

3.4.10 a) The system has a unique solution for $k \neq 0, 1$, it has no solutions for $k = 0, 1$. b) $k \neq 0, 1$.

3.4.12 The given vectors are always linearly dependent.

3.4.13 a) $k \neq 0, 5/3$. For $k = 0$ we have $\mathbf{v}_2 = \mathbf{v}_3$.

3.4.14 a) $k \neq 0, -2$. b) $k \neq 0, -2$.

3.4.15 $k \neq 3$.

3.4.16 a) $\{(-2, 1, 0), (0, 0, 1)\}$.

Chapter 4: Basis and dimension

4.5.2 $k \neq 2$.

4.5.4 a) it is basis, b) it is a set of generators, c) it is not a set of generators.

4.5.6 $k \neq -1$.

4.5.7 The vectors are linearly independent for $k \neq 0$ and $k \neq 1$. For $k = 1$, $(1, -2, -6) \in \langle \mathbf{v}_1, \mathbf{v}_2, \mathbf{v}_3 \rangle$.

4.5.8 For all $k \in \mathbb{R}$.

4.5.10 $\mathbf{v}_1, \mathbf{v}_2, \mathbf{v}_3$ generate $\mathbb{R}^3$ for $k \neq \frac{3}{4}$; for $k = 0$ the vector $(4, -1, 6)$ belongs to $\langle \mathbf{v}_1, \mathbf{v}_2, \mathbf{v}_3 \rangle$.

4.5.11 $k = -5, k = 2$.

4.5.12 $k \neq \pm 6$.

4.5.13 $k \neq -\frac{3}{2}$.

4.5.17 (a) $k \neq 0, k \neq 1/10$.

Chapter 5: Linear maps

5.9.1 For all $k \in \mathbb{R}$.

5.9.2 For no value of $k \in \mathbb{R}$.

5.9.6 (a) F is isomorphism. (b), (c) the linear maps are not isomorphisms.

5.9.13 $k \neq -2$.

5.9.14 A basis of $\operatorname{Ker}(T)$ is $\{(-2, 0, 1)\}$ and a basis of $\operatorname{Im}(T)$ is $\{(1, 0, 2, 3), (0, 1, 3, -1)\}$, $\dim(\operatorname{Ker} T) = 1$, $\dim(\operatorname{Im} T) = 2$ and $\dim(\mathbb{R}^3) = 3 = 1 + 2 = \dim(\operatorname{Ker} T) + \dim(\operatorname{Im} T)$, hence T is not injective.

5.9.19 $k = \pm 2$.

5.9.20 $k \neq -3$.

5.9.21 F is injective for all $k \in \mathbb{R}$.

Chapter 6: Linear Systems

6.4.2 $k \neq -5$.

6.4.4 S is a subspace.

6.4.5 F exists and is unique.

Chapter 7: Determinant and Inverse

7.8.1 a) F is an isomorphism, because the determinant of the matrix associated to F with respect to the canonical basis is equal to 1.

7.8.2 $G(x, y, z) = (x + y - z, -y - z, -z)$.

7.8.3 The matrix is invertible for $a \neq 3/2$.

7.8.7 a) The inverse is:
$$\begin{pmatrix} -1 & 1 & -1 \\ -2 & 1 & -2 \\ 0 & 0 & 1 \end{pmatrix}$$

Chapter 8: Change of basis

8.5.1 $(4, 6, 9)$.

8.5.5 a) $k \neq \pm 1$, $\ker(T) = \langle (2/3, 0, 1) \rangle$.
b) $\begin{pmatrix} 3/2 & 1/2 & -1 \\ 6 & 1 & -4 \end{pmatrix}$

8.5.6 a) $k \neq 4$, $k \neq -1/2$.
b) With respect to the canonical bases in domain and codomain, G is associated to the matrix:
$$\begin{pmatrix} 0 & -1 & 0 \\ -1/4 & -1/4 & -1/2 \\ 0 & 0 & 1 \end{pmatrix}$$

Chapter 9: Eigenvalues and Eigenvectors

9.4.1 (a) Eigenvalues 2, 3, the matrix is not diagonalizable. (b) Eigenvalues 1, 4, L is diagonalizable. (c) Eigenvalues ± 1, 2, L is diagonalizable. (d) Eigenvalues 0, 7, L is diagonalizable. (e) L has not real eigenvalues.

9.4.2 a) Eigenvalues 0, 5, 7, A is diagonalizable.

9.4.4 Eigenvalues 2, 4, A is diagonalizable.

9.4.5 Eigenvalues -1, 2, A is not diagonalizable.

9.4.6 Eigenvalues 9, 3, A is diagonalizable.

9.4.7 Eigenvalues 0, 6, -1, A is diagonalizable.

9.4.8 Eigenvalues 2, 5, T is not diagonalizable.

9.4.9 Eigenvalues -2, 3, T is diagonalizable.

9.4.10 a) $k = -1$. b) For all $k \in \mathbb{R}$.

9.4.11 A is diagonalizable for $k = 0$.

9.4.12 F is diagonalizable for $k \neq 2$.

9.4.13 a) A is diagonalizable for $k \neq 5$. b) For $k = 0$.

Chapter 10: Scalar Products

10.8.2 a) Basis for $W^\perp$: $\{(1,2,2,-1)\}$. **b)** Basis for W: $\{(-2,1,0,0),(-2,0,1,0),$ $(5,0,0,1)\}$. Basis for $W^\perp$: $\{(-2/\sqrt{5},1/\sqrt{5},0,0),(-2/3\sqrt{5}),-4/3\sqrt{5},\sqrt{5}/3,0),(5/3\sqrt{34}),$ $5\sqrt{2/17}/3,5\sqrt{2/17}/3,3/\sqrt{34})\}$.

10.8.3 a) Basis for $W^\perp$: $\{(0,2,1,0),(-1,-1,0,1)\}$. **b)** Orthogonal basis for W: $\{(1/\sqrt{2},0,0,1/\sqrt{2}),(-1/\sqrt{22},\sqrt{2/11},-2\sqrt{2/11},1/\sqrt{22})\}$.

10.8.4 Basis for W: $\{(0,-1,1,0),(1/2,-1/2,0,1)\}$. Orthonormal basis for W: $\{(0,-1/\sqrt{2},1\sqrt{2},0),(2/\sqrt{22},-1/\sqrt{22}.-1/\sqrt{22},4/\sqrt{22})\}$. Orhonormal basis for $W^\perp$: $\{(0,1/\sqrt{3},-1/\sqrt{3},1/\sqrt{3})\}$. **c)** No.

10.8.7 a) Basis for $W^\perp$: $\{(0,1,-1,1)\}$. **b)** Orthonormal basis for W: $\{(1/\sqrt{3},-1/\sqrt{3},0,$ $1/\sqrt{3}),(\sqrt{2/3},1/\sqrt{6},0,-1/\sqrt{6}),(0,1/\sqrt{6},\sqrt{2/3},1/\sqrt{6})\}$.

Chapter 11: Spectral Theorem

11.6.1 a) no, **b)** no, **c)** yes, **d)** yes.

11.6.4 1) Eigenvalues: 5, 2. Basis of eigenvectors: $\mathbf{v}_1 = (1,1,1)$, $\mathbf{v}_2 = (-1,0,1)$, $\mathbf{v}_3 = (-1,1,0)$.

$$P^{-1} = \begin{pmatrix} 1/\sqrt{3} & 1/\sqrt{3} & 1/\sqrt{3} \\ -1/\sqrt{2} & 0 & 1/\sqrt{2} \\ 1/\sqrt{6} & \sqrt{2/3} & 1/\sqrt{6} \end{pmatrix}$$

2) Eigenvalues: $1/2(3 \pm \sqrt{5})$. Eigenvectors: $\mathbf{v}_1 = ((-1-\sqrt{5})/2,1)$, $\mathbf{v}_2 = ((-1+\sqrt{5}/2,1)$.
4) Eigenvalues: 5, 2, 0. Eigenvectors: $\mathbf{v}_1 = (0,-1,2)$ $\mathbf{v}_2 = (1,0,0)$ $\mathbf{v}_3 = (0,2,1)$.

$$P^{-1} = \begin{pmatrix} 0 & -1/\sqrt{5} & 2/\sqrt{5} \\ 1 & 0 & 0 \\ 0 & 1/\sqrt{5} & 1/\sqrt{5} \end{pmatrix}$$

Chapter 12: Applications of Spectral Theorem and Quadratic Forms

12.5.3 The eigenvalues of C are 8, ± 1, hence q and the associated scalar product is non-degenerate, but not positive definite. The signature of q is $(2,1)$. The basis in which the scalar product is associated to a diagonal matrix is given by eigenvectors of norm 1: $v_1 = (2/3,1/3,2/3)$, $v_2 = (1/\sqrt{18},-4/\sqrt{18},1/\sqrt{18})$, $v_3 = (-1/\sqrt{2},0,1/\sqrt{18})$.

12.5.4 1) $< (x,y),(x',y') >= xx' + (5/2)xy' + (5/2)yx'$. The associated matrix in the canonical basis is:

$$C = \begin{pmatrix} 1 & 5/2 \\ 5/2 & 0 \end{pmatrix}.$$

The eigenvalues of C are: $1/2(1 \pm \sqrt{26})$, hence the quadratic form is non-degenerate, but not positive definite. The signature of q is $(1,1)$. **2)** The curve is an hyperbole.

Chapter 13: Lines and Planes

13.6.1 r_1, r_2 are skew. r_1, r_3 are skew. r_1, r_4 are coincident. r_2, r_3 are parallel and their distance is $2\sqrt{6}/3$. r_2, r_4 are skew. r_3, r_4 are skew.

13.6.4 a) The line r is given by $(2 + 3t, 1 - t, -6t)$. b) The plane π' is given by $5x - 9y + 4z = 1$.

13.6.6 a) Parametric equations for r: $x = t$, $y = 1+t$, $z = -1/3-(2/3)t$. b) Cartesian equation for π: $x - 3y - 3z + 2 = 0$. c) Parametric equations for s: $x = 1+t$, $y = -3t$, $z = 1 - 3t$.

13.6.7 a) Parametric equations for r: $x = -12+2t$, $y = 4$, $z = t$. b) They are parallel. c) The distance is $(2/3)\sqrt{6}$.

Chapter 14: Discrete Mathematics

14.6.4 Such classes do not exist in $\mathbb{Z}_{37}$ because 37 is prime. In $\mathbb{Z}_{36}$ we have $[4]_{36}[9]_{36} = [0]_{36}$.

14.6.5 They are different, one is not contained into the other.

14.6.6 b) $[10]$, d) The congruence has no solutions.

Bibliography

[1] T. W. Hungerford. *Algebra*. Springer, 1974.

[2] T. W. Hungerford. *Abstract Algebra, an Introduction*. Cengage Learning, Inc, 2012.

[3] S. Lang. *Algebra*. Springer, 2002.

[4] S. Lang. *Undergraduate Algebra*. Springer Science & Business Media, 2005.

[5] S. Lang. *Introduction to Linear Algebra*. Springer Science & Business Media, 2012.

Index

$\mathbb{R}^2$, 26
$\mathbb{Z}_n$, 243

basis, 58
basis change for scalar products, 182
Bezout identity, 239
bilinear application, 177
bilinear form, 177

canonical basis, 61
Cartesian equations
    of a line, 228
    of a plane, 228
characteristic polynomial, 160
complete induction principle, 236
Completion Theorem, 61
complex numbers, 204
components, 62
congruence, 244
congruence classes, 241
congruence relationship, 241
coordinates, 62

determinant, 113, 128
diagonalization, scalar products, 209
division algorithm, 237

eigenvalue, 158
    calculation, 159
eigenvectors, 158
    calculation, 159
elementary row operations, 65
equivalence relation, 241
Euclid's algorithm, 238

finitely generated, 58
function, 78

Gaussian algorithm, 1, 64
gcd, 238
generators, 43

good ordering axiom, 235
Gram-Schmidt algorithm, 188
greatest common divisor, 238

hermitian product, 204
    non-degenerate, 205
    positive definite, 205
homogeneous equation, 1

image, 86
    calculation, 92
induction principle, 235
inverse, calculation of, 123
isomorphism, 91

kernel, 86
    calculation, 92

linear combination, 41
linear equation, 1
linear independence, 47
linear system, 1
linear transformation, 77, 78
    diagonalizable, 156
    orthogonal, 193
    symmetric, 198

matrix, 30
multiplicative additive tables of $\mathbb{Z}_3$ and
    $\mathbb{Z}_4$, 243

norm of a vector, 183

orthogonal basis, 187
orthogonal matrices, 194
orthogonal projection, 187
orthonormal basis, 187

parametric equations
    of a line, 226
    of a plane, 229